Measurement Techniques for Radio Frequency Nanoelectronics

Connect basic theory with real-world applications with this practical, cross-disciplinary guide to radio frequency measurement of nanoscale devices and materials.

- Learn the techniques needed for characterizing the performance of devices and their constituent building blocks, including semiconducting nanowires, graphene, and other two-dimensional materials such as transition metal dichalcogenides.
- Gain practical insights into instrumentation, including on-wafer measurement platforms and scanning microwave microscopy.
- Discover how measurement techniques can be applied to solve real-world problems, in areas such as passive and active nanoelectronic devices, semiconductor dopant profiling, subsurface nanoscale tomography, nanoscale magnetic device engineering, and broadband, spatially localized measurements of biological materials.

Featuring numerous practical examples, and written in a concise yet rigorous style, this is the ideal resource for researchers, practicing engineers, and graduate students new to the field of radio frequency nanoelectronics.

T. Mitch Wallis is a physicist in the Applied Physics Division at the National Institute of Standards and Technology, Boulder, Colorado. He is also the Chair of the IEEE Microwave Theory and Techniques Society's Technical Committee on Radio Frequency Nanotechnology.

Pavel Kabos is a physicist in the Applied Physics Division at the National Institute of Standards and Technology. He is the author of *Magnetostatic Waves and Their Applications* (Chapman and Hall, 1993) and a fellow of the IEEE.

The Cambridge RF and Microwave Engineering Series

Series Editor

Steve C. Cripps, Distinguished Research Professor, Cardiff University

Editorial Advisory Board

James F. Buckwalter, UCSB
Jenshan Lin, University of Florida
John Wood, Maxim Integrated Products

"This book represents a state-of-the-art look at measurement techniques of nanoelectronic devices in the RF and microwave frequency range. This field is of growing importance because of higher CMOS clock speeds approaching the GHz range as well as shrinking device dimensions, down to the 10 nm scale and below. The fundamental physical challenges of measuring and characterizing devices with these length scales, which approach atomic dimensions, are clearly laid out and presented in this book. The book begins with fundamental network analysis theory based on Maxwell's equations for radiation and transmission lines, progresses to on-wafer semiconductor device characterization in the RF and microwave to mm-wave frequency range, and progresses to apply these fundamentals to an increasingly challenging set of measurements. High impedance devices (up to and greater than the resistance quantum) are covered in detail, with the latest on-wafer calibration procedures laid out clearly. Scanning microwave microscopy as a complementary technique for high impedance devices is also covered. Materials characterization, including the beginnings of a new field of scanning microwave microscopy for tomography (a nanoscale version of synthetic aperture radar), is also covered. Applications to nanowires, nanotubes, and 2D materials such as graphene and WS2, in both passive and active modes, are clearly presented. I expect this book will be of great interest to beginning graduate students and senior undergraduates entering the field, as well as senior researchers with an interest in the latest techniques for measuring these tiny devices, which tend to have high impedances due to the quantum nature of electricity at this atomic length scale."

Peter Burke, University of California, Irvine

"This is a remarkable reference on high frequency nanoelectronics measurements and scanning microwave microscopy that includes applications for nano-devices and advanced materials. The basics of radio frequency (RF) measurements for extreme impedances and nanoscale-sized RF probes are laid out very well, as well as advanced concepts in modeling and RF calibration. This accessible book will be useful for a wide readership, including researchers and students in microwave engineering, semiconductor electronics, materials science, and microscopy."

Ferry Kienberger, Keysight Laboratories, Keysight Technologies Inc.

Measurement Techniques for Radio Frequency Nanoelectronics

T. MITCH WALLIS

National Institute of Standards and Technology, Boulder

PAVEL KABOS

National Institute of Standards and Technology, Boulder

CAMBRIDGE
UNIVERSITY PRESS

CAMBRIDGE
UNIVERSITY PRESS

University Printing House, Cambridge CB2 8BS, United Kingdom

One Liberty Plaza, 20th Floor, New York, NY 10006, USA

477 Williamstown Road, Port Melbourne, VIC 3207, Australia

4843/24, 2nd Floor, Ansari Road, Daryaganj, Delhi – 110002, India

79 Anson Road, #06-04/06, Singapore 079906

Cambridge University Press is part of the University of Cambridge.

It furthers the University's mission by disseminating knowledge in the pursuit of education, learning, and research at the highest international levels of excellence.

www.cambridge.org
Information on this title: www.cambridge.org/9781107120686
DOI: 10.1017/9781316343098

© Cambridge University Press 2017

First published 2017

Printed in the United Kingdom by TJ International Ltd. Padstow Cornwall

A catalogue record for this publication is available from the British Library.

ISBN 978-1-107-12068-6 Hardback

Contents

Acknowledgments

In many ways, this book represents the culmination of our collaborative work in the field of RF nanoelectronics over the past thirteen years. During that span of time, we have had the good fortune to work with many talented engineers and scientists. First and foremost, we are grateful to the students and postdoctoral associates that have contributed time and effort to our work in RF nanoelectronics at the National Institute of Standards and Technology (NIST): Sam Berweger, Joe Brown, Chien-Jen Chiang, Jonathan Chisum, Alex Curtin, Kristen Genter, Atif Imtiaz, Kichul Kim, Simone Lee, Sang-Hyun Lim, and Joel Weber. We also thank our NIST collaborators in the GaN nanowire growth project: Kris Bertness, Norman Sanford, Paul Blanchard, Matt Brubaker, Todd Harvey, Lorelle Mansfield, Alexana Roshko, and Bryan Spann. We are also grateful to many other past and present NIST colleagues, including Kevin Coakley, Dazhen Gu, Joe Kopanski, Paul Rice, Stephen Russek, and Karl Stupic. Finally, we wish to thank Professors Victor Bright, Dejan Filipovic, Zoya Popovic, and Y.-C. Lee at the University of Colorado, Boulder.

In preparing this manuscript, we have benefitted from the insights of many colleagues. We appreciate the helpful comments and suggestions of all our colleagues who read the manuscript: Joel Weber, Paul Blanchard, Ron Ginley, Bill Riddle, Sam Berweger, Arek Lewandowski, Haris Votsi, Peter Aaen, Jan Obrzut, Abhishek Sahu, Matt Brubaker, Dazhen Gu, Chris Long, Jason Killgore, Andrew Gregory, Yaw Obeng, Johannes Hoffmann, Claude Weil, Jim Randa, Matt Pufall, Mike Schneider, Georg Gramse, Marco Farina, Joe Dragavon, Kris Bertness, and Norman Sanford. We also appreciate the guidance of Julie Lancashire, Karyn Bailey, and Heather Brolly at Cambridge.

Abbreviations

2DEGs	Two-dimensional electron gases
AFM	Atomic force microscopes
AM-EFM	Amplitude-modulated, electrostatic force microscopy
CNT	Carbon nanotubes
CPW	Coplanar waveguide
DOS	Density of states
DUT	Device under test
EPR	Electron paramagnetic resonance
ESR	Electron spin resonance
FEM	Finite-element modeling
FET	Field effect transistors
FMR	Ferromagnetic resonance
GaN	Gallium nitride
GPR	Ground penetrating radars
GS	Ground-signal
GSG	Ground-signal-ground
LRM	Line-reflect-match
LRRM	Line-reflect-reflect-match
MEMS	Microelectromechanical systems
MESFET	Metal semiconductor field effect transistor
MFM	Magnetic force microscope
MIS	Metal-insulator-semiconductor
MOS	Metal-oxide-semiconductor
MRFM	Magnetic resonance force microscopy
NMR	Nuclear magnetic resonance
NSMM	Near-field scanning microwave microscope
NSOM	Near-field scanning optical microscopy
PALM	Photoactivated localization microscopy
RF	Radio Frequency
SCM	Scanning capacitance microscope
SEM	Scanning electron microscope
SiO_2	Silicon dioxide
SKPM	Scanning kelvin probe microscope

SOLT	Short-open-line-thru
SSRM	Scanning spreading-resistance microscope
STED	Stimulated emission depletion
STM	Scanning tunneling microscope
STS	Scanning tunneling spectroscopy
SUT	Sample under test
TE	Transverse electric
TEM wave	Transverse electromagnetic wave
TIRF	Total internal reflection fluorescent microscopy
TM	Transverse magnetic
TMD	Transition metal dichalcogenides
TRL	Thru-reflect-line
VED	Vertical electric dipole
VNA	Vector network analyzer
YIG	Yttrium iron garnet

1 An Introduction to Radio Frequency Nanoelectronics

1.1 Radio Frequency Nanoelectronics

The field of radio frequency (RF) nanoelectronics focuses on the fundamental study and engineering of devices that are enabled by nanotechnology and operate within a frequency range from about 100 MHz to about 100 GHz. This range includes frequencies traditionally identified as "radio frequencies," as well as microwaves and, at the high end of the frequency range, millimeter-waves. This emerging field sits at the intersection of two commercially vital trends in technology. The first trend is the ongoing shrinking of electronics to smaller length scales. Though this trend was initiated by the semiconductor and storage industries' pursuit of Moore's Law, the discoveries that followed have influenced a wide range of disciplines in addition to electronics, such as biological sensing, nanoelectromechanical systems, and low-dimensional materials science. The second overarching trend is the ever-increasing presence of devices that operate at radio frequencies, which has arisen in conjunction with the explosion of wireless connectivity. For the foreseeable future, communications technology will rely heavily on microwave and millimeter-wave transmission and in turn, devices that transmit, receive, and process signals at corresponding frequencies. In addition, current and foreseeable operating frequencies of integrated semiconductor electronics lie in the microwave frequency range. The ultimate goal of RF nanoelectronics is to leverage the new materials and new phenomena that have been revealed by scaling down to the nanoscale world in order to investigate new RF devices that will be of interest both for fundamental study and eventual commercial application.

In the past few decades, the emergence and growth of nanotechnology has proceeded hand in hand with the discovery and investigation of new forms of matter. Examples of nanoscale material systems, spanning from atomically thin two-dimensional materials to semiconducting nanowires to individual atoms, are shown in Fig. 1.1. From the outset, nanomaterials based on carbon have played a particularly important role. Indeed, one of the seminal moments in the brief history of nanotechnology was the synthesis of Buckminsterfullerenes in 1985 [1]. This achievement ignited the vigorous investigation of additional carbon-based nanomaterials, particularly graphene and carbon nanotubes (CNTs), which continues today [2]. Graphene is a stable, one-atom-thick sheet of carbon, while a

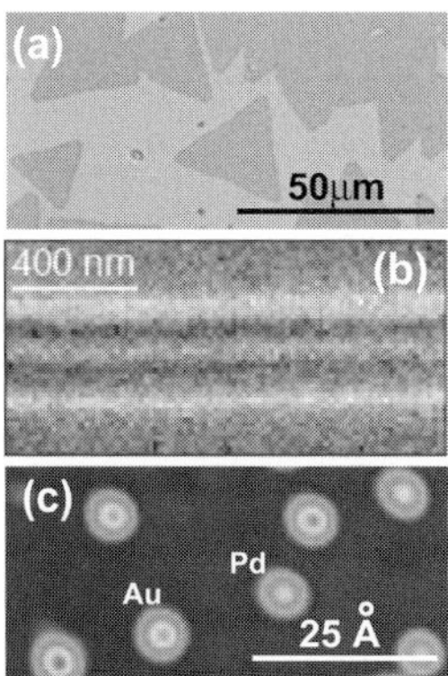

Figure 1.1.   Nanomaterial systems.
(a) Optical microscope image of triangular shaped MoS_2 flakes. (Optical image courtesy of Prof. Xiaobo Yin, University of Colorado, Boulder.) (b) Near-field scanning microwave microscope image of a GaN nanowire. (c) Scanning tunneling microscope image of individual Pd and Au atoms. (STM image courtesy of Prof. Wilson Ho, University of California, Irvine.)

single-walled CNT may be conceptually understood as a ribbon of graphene that has been rolled into a tube. These materials have remarkable mechanical, electrical, and thermal properties. Furthermore, by altering the geometry of the constituent carbon atoms in fullerenes and CNTs, one can tune their particular material properties. While these carbon-based materials were initially seen as exotic, it's important to notice that the chemistry of carbon bonds naturally leads to these morphologies. In the words of Nobel laureate Richard Smalley, "Carbon has this genius of making a chemically stable two-dimensional, one-atom-thick membrane in a three-dimensional world" [3]. It was only with the development of nanoscale fabrication and measurement techniques that we were able to recognize these previously unseen materials.

Among the remarkable properties of CNTs and graphene, their capacity for high current densities is particularly appealing for RF nanoelectronic applications. One illustrative example of the potential application of these materials is the use of CNTs as interconnects in RF electronics. While copper has historically been the material of choice for electronic interconnects due to its low resistivity, the resistivity of copper increases due to surface scattering effects at dimensions lower than 100 nm [4]. In fact, the resistivity of a copper nanowire with a diameter of 60 nm is about ten times higher than the resistivity of bulk copper [5]. In contrast, the properties of metallic, single-walled CNTs are superior to copper nanowires in many ways. Conduction of electrons in metallic CNTs is ballistic, leading to current densities greater than 10^9 A/cm^2 and electron mean free paths greater than 10^3 nm [6]. For copper, the corresponding values are three and two orders of magnitude smaller, respectively. As a result, CNTs are promising candidates for low-loss interconnects operating at gigahertz frequencies. Recently, integrated circuits have been

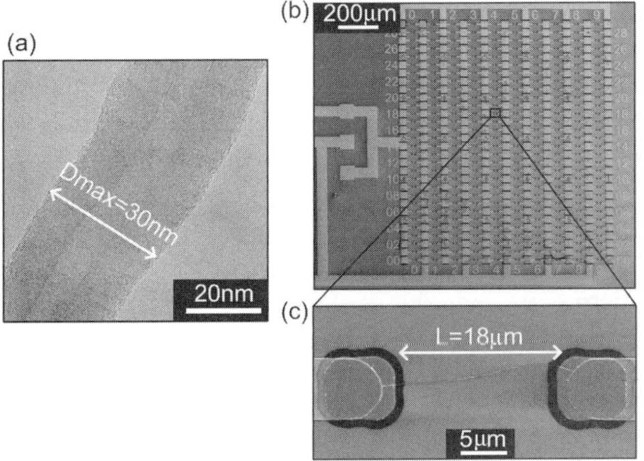

Figure 1.2. Multiwalled CNT interconnects.
(a) Transmission electron microscope image of a 30 nm multiwalled CNT. (b) Scanning electron microscope image of an array of multiwalled CNT interconnects. (c) Scanning electron microscope image of an individual interconnect. © 2009 IEEE. Reprinted, with permission from G. F. Close, S. Yasuda, B. Paul, S. Fujita, and H.-S. P. Wong, *IEEE Transactions on Electron Devices* 56 (2009) pp. 43–49.

demonstrated that combine silicon complementary-metal-oxide-semiconductor (CMOS) transistors with individual multiwalled CNT interconnects operating at one gigahertz [7], [8], as shown in Fig. 1.2.

Beyond the families of carbon-based materials, the palette of materials utilized in nanotechnology is wide, varied, and ever-expanding. Two-dimensional, graphene-like materials such as transition metal dichalcogenides (TMDs), though atomically thin, can extend to microscopic or even macroscopic lateral dimensions [9]. Nanowires have been grown for many types of materials, finding applications not only in RF nanoelectronics [10], but also in ultrasensitive, low-power sensing [11] and optoelectronics [12]. At the spatial limit, individual molecules and atoms represent the ultimate nanoscale material system. One of the central concepts of nanotechnology in general and nanoelectronics in particular is that these individual nanoscale material systems can serve as building blocks, from which complex devices with novel functionalities may be assembled. As examples, one can imagine multilayered stacks of two-dimensional materials or ensembles of molecules organized via self-assembly as customized material systems whose properties may be tuned and engineered.

While new material systems are one of the hallmarks of modern nanotechnology, they are not sufficient in and of themselves to distinguish nanoscience from other disciplines. Indeed, long before the term "nanotechnology" was coined, chemists were synthesizing new molecules and solid state physicists were engineering microelectronic devices by assembling semiconductor materials and components. One extraordinary feature of nanotechnology that distinguishes it from such earlier endeavors is access to and control of individual nanoscale building blocks. In order for RF nanoelectronic devices to be realized, individual nanoscale components must be placed in precise arrangements and individual elements must be addressable via high-quality electrical contacts. As a result of recent advances, many examples now exist of such capability and control. It is now possible to align individual nanowires with electrical contacts by use of dielectrophoresis [13] or fabricate graphene nanoribbons with ion beams [14]. At even smaller length scales, mechanical break junctions make it possible to isolate and measure individual molecules [15], while scanning probe microscopes are able to fabricate, manipulate, and characterize nanostructures one atom at a time [16]. Furthermore, industrially scalable processes for materials synthesis and processing have emerged that will make nanodevice fabrication compatible with bulk manufacturing [17].

1.2   Measurement Problems in RF Nanoelectronics

While the synthesis of new nanomaterials and the control of nanoscale building blocks at the spatial limit are critical to realizing RF nanotechnology, the promise of RF nanoelectronic devices will only be realized with accurate measurement science. The ultimate application and commercialization of RF nanoelectronic devices requires reproducible measurements for optimization of performance and informed selection between competing designs. Furthermore, reliable, quantitative determination of measurement uncertainties is desirable throughout all stages of RF device engineering. At a more fundamental level, measurements are vital to developing and testing quantitative models of underlying physics. As fundamental discoveries lead to the development of devices, quantitative measurements provide necessary insights and feedback for evaluating innovative device concepts. Finally, in an emerging field such as RF nanoelectronics, measurements can serve as a means for finding a common framework of terminology, calibration, and standard benchmarking among different research efforts.

Considered separately, both RF device measurements and nanoscale measurements are extremely challenging fields. In the development of measurement techniques for RF nanoelectronics, we must build upon both of these branches of metrology, adapting existing techniques while also forging new methods. Such blending of techniques naturally leads to cross-pollination of traditionally separate disciplines. This multidisciplinary nature of nanotechnology is part of what makes the field exciting. However, when trained specialists cross into unfamiliar disciplines, there is always some risk of misunderstanding. For many individuals who come to the field of RF nanoelectronics with minimal experience in RF and microwave measurements, there is much to learn about the art of microwave engineering in general and that of microwave measurements in particular. Conversely, those who have mastered microwave measurements, but have minimal experience in nanoscale measurements, have much to learn about material fabrication and workhorse characterization techniques such as scanning probe microscopy and spectroscopy.

Historically, progress in nanoscience and nanotechnology research has proceeded in step with progress in nanoscale measurement science. For example, the inventions of scanning tunneling microscopy [18] and atomic force microscopy [19] have enabled the visualization and characterization of surfaces with atomic-scale resolution. Electron microscopy techniques such as transmission electron microscopy and scanning electron microscopy also continue to reveal the beauty and complexity of matter at length scales from micrometers down to Ångstroms. Moreover, as such measurement tools have matured, they have become more versatile, providing chemical sensitivity, electronic and vibrational spectroscopic capabilities, as well as nanomanipulation and nanofabrication capabilities. With this versatility has come a breadth of application areas and specialized measurement modes.

The historical development of microwave measurements has also led to the emergence of a substantial number of subdisciplines, including measurements of noise, power, and impedance, as well as antenna characterization and other free space measurement techniques. The subdisciplines that are most relevant to RF nanoelectronics relate to the measurement of complex scattering parameters in guided-wave systems, such as waveguides and coaxial transmission lines [20]. Also, measurements of RF nanoelectronic devices rely heavily on the extension of techniques for on-wafer measurement of scattering parameters. For all guided-wave measurements, the development of the six-port reflectometer was an important milestone, paving the way for contemporary vector network analyzers, which are critical tools for nearly all of the measurement methods that are discussed in this book.

In the previous section, the application of CNTs as RF interconnects was presented as an illustrative example of a promising application of RF nanoelectronics. This example also illustrates how advances in measurement science are required in order for such applications to be realized. The recent development of microwave metrology for CNTs in general [21], [22], and CNT interconnects in particular [23], has revealed a number of substantial measurement challenges. How can microwave measurements be extended to systems with extremely high impedance? How can the intrinsic RF properties of nanoscale building blocks such as CNTs be de-embedded from contact impedance, parasitic capacitance, and other properties of the host device? Further, once de-embedding is possible, how can physical properties of the nanoscale components be estimated from the microwave measurements? Finally, what are the uncertainties in these measurements? In an effort to answer such questions, the research and development of RF nanoelectronic systems has necessarily led to the development of new measurement methods, which continues as the field extends to new materials, new length scales, new modeling approaches, and new frequencies.

1.3 Measurement Techniques for RF Nanoelectronics

The initial measurement techniques that were developed for RF nanoelectronics extended guided-wave microwave measurement techniques to devices and circuits that incorporated nanoscale building blocks [21], [24]–[27]. The measurands for these techniques are the frequency-dependent, complex scattering parameters. In general, these techniques require calibration approaches that allow the scattering parameters of the device under test (DUT) to be de-embedded from the effects of the test equipment. A further requirement for many RF nanoelectronic DUTs is that the technique must account for the extreme impedance mismatch between the nanoelectronic DUT and the test equipment. In many instances, a further objective is to use modeling and simulation to extract intrinsic properties of the nanoscale components in a measured device, as well as circuit parameters of interest such as contact impedance and other sources of parasitic reactance.

In this book, Chapters 2 through 6 describe several approaches to the extension of established guided-wave measurement techniques to RF nanoelectronic DUTs. First, Chapter 2 reviews the core concepts of guided-wave measurements techniques. Chapter 3 adapts these techniques to the general case of extreme impedance DUTs while Chapter 4 narrows the focus to on-wafer measurements of RF nanoelectronics. Chapter 5 covers modeling and simulation of RF nanoelectronics with emphasis on validation and circuit parameter extraction. In general, many aspects of RF nanoelectronic device development, including simulation, fabrication, and measurement, are difficult. As a result, a reliable framework for broadband characterization of nanoscale components necessarily incorporates multiple aspects such as specially fabricated test structures, calibration techniques, and validation through numerical simulation. To illustrate this multifaceted characterization strategy, Chapter 6 describes a case study that highlights strategies and challenges related to implementing a specific RF nanoelectronic device measurement, namely the broadband measurement of a two-port, on-wafer GaN nanowire device.

Beyond global device characterization, there is a need for approaches that provide microwave measurements that are spatially localized within a device, providing insight into the impacts of defects, interfaces, and other localized features upon device performance. In addition to intra-device measurement capabilities, it is also highly desirable to make nondestructive, RF measurements of the intrinsic properties of individual building blocks in a contact-free environment. One effective approach to both these measurement problems is to combine the nanometer-scale spatial resolution of scanning probe microscopy with broadband sensitivity in the frequency range from 100 MHz to 100 GHz. This combination can be realized by the integration of a one-port microwave network into the signal path of a scanning probe instrument, such as an atomic force microscope (AFM). In this book, we will refer to such an instrument as a "near-field scanning microwave microscope" (NSMM), but there are a number of closely related techniques that are described in the literature, including scanning capacitance microscopy and scanning impedance microscopy. An example of an NSMM [28] is shown in Fig. 1.3. As in other scanning probe microscope systems, NSMM requires that a probe be positioned on the order of a nanometer above a sample surface. A distance-following feedback mechanism is required to maintain a precise tip-sample separation as the probe tip is rastered across the sample surface.

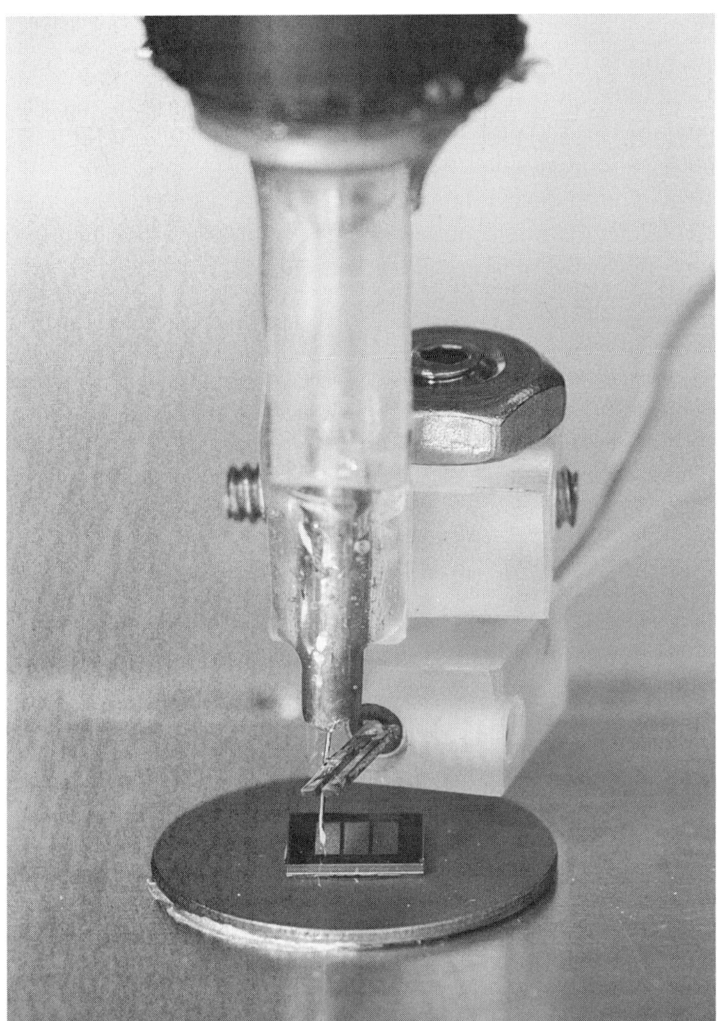

Figure 1.3. Near-field scanning microwave microscope (NSMM).
This photograph shows one of many different possible implementations of an NSMM. The system shown here has a needle-shaped probe extending from a truncated coaxial cable and uses a tuning-fork-based feedback system for distance following. Further details can be found in Reference [28]. Photograph by C. Suplee, NIST.

Ultimately, the outputs of NSMM and related techniques include high-resolution images and spatially localized spectroscopic measurements. Measurands include local impedance, capacitance, dopant concentration, sheet resistance, complex permittivity, and complex permeability. Further development of systems with multiple probes offers the opportunity to measure intra-device and intra-material transport as well as RF coupling between separated nanoscale components. Finally, the microwave skin depth effect enables an NSMM to measure subsurface electronic properties of materials and devices.

NSMM techniques are covered in Chapters 7 through 9. In Chapter 7, NSMM instrumentation is reviewed and a variety of different NSMM implementations are compared and contrasted. Chapter 8 presents a model of the tip-sample interaction in an NSMM. Building upon that model, the chapter also presents strategies for extracting calibrated, quantitative measurements, such as absolute capacitance measurements, from NSMM data. Chapter 9 introduces the fundamental concepts of electromagnetic materials measurements, then narrows its focus to a review of applications of NSMM to materials measurements.

Advances in nanoscience and nanotechnology have impacted many research and application areas across multiple disciplines, including nanoelectronics, optoelectronics, and biomedicine. Similarly, the measurement techniques developed for RF nanoelectronics have found applications in many fields. Thus, the final five chapters of the book cover specific measurement problems that are of ongoing interest. These areas serve as practical examples of how the measurement techniques for RF nanoelectronics are extended and customized for specific problems and applications. Chapter 10 discusses the broadband characterization of active nanotransistor devices in the RF range, including approaches for de-embedding intrinsic device properties from measurements. Chapter 11 presents approaches to spatially resolved dopant profiling of semiconductors by use of NSMM and related techniques. Subsurface measurements made by use of NSMM are then covered in Chapter 12. The subsurface imaging capability of NSMM is an emerging field of metrology that takes advantage of the microwave skin depth effect, thus providing a nondestructive approach to the characterization of subsurface interfaces and defects. However, quantitative subsurface measurements require complex mathematical approaches to inverse problems. Chapter 13 discusses measurements of nanoscale magnetic systems. The natural time scale for the dynamics of nanomagnetic systems falls in the microwave regime, providing an opportunity for the application and adaptation of measurements developed for RF nanoelectronics. Chapter 14 concludes with a discussion of nanoscale electromagnetic measurements for life science and medical applications.

References

[1]   H. W. Kroto, J. R. Heath, S. C. O'Brien, R. F. Curl, and R. E. Smalley, "C60: Buckminsterfullerene," *Nature* 318 (1985) pp. 162–163.

[2]   H.-S. P. Wong and D. Akinwande, *Carbon Nanotube and Graphene Device Physics* (Cambridge University Press, 2011).

[3]   R. E. Smalley, "Discovering the Fullerenes," Nobel Lecture, December 7, 1996.

[4]   W. Steinhogl, G. Schindler, G. Steinlesberger, M. Traving, and M. Engelhardt, "Comprehensive Study of the Resistivity of Copper Wires with Lateral Dimensions of 100 nm and Smaller," *Journal of Applied Physics* 97 (2005) art. no. 023706.

[5]   M. E. Toimil Molares, E. M. Hohberger, C. Schaeflein, R. H. Blick, R. Neumann, and C. Trautmann, "Electrical Characterization of Electrochemically Grown Single Copper Nanowires," *Applied Physics Letters* 82 (2003) pp. 2139–2141.

[6] A. Javey, P. Qi, Q. Wang, and H. Dai, "Ten- to 50-nm-long Quasi-Ballistic Carbon Nanotube Devices Obtained without Complex Lithography," *Proceedings of the National Academy of Sciences of the United States of America* 101 (2004) pp. 13408–13410.

[7] G. F. Close, S. Yasuda, B. Paul, S. Fujita, and H.-S. P. Wong, "A 1 GHz Integrated Circuit with Carbon Nanotube Interconnects and Silicon Transistors," *Nano Letters* 8 (2008) pp. 706–709.

[8] G. F. Close, S. Yasuda, B. Paul, S. Fujita, and H.-S. P. Wong, "Measurement of Subnanosecond Delay through Multiwall Carbon-Nanotube Local Interconnects on a CMOS Integrated Circuit," *IEEE Transactions on Electron Devices* 56 (2009) pp. 43–49.

[9] G. Fiori, F. Bonaccorso, G. Iannaccone, T. Palacios, D. Nuemaier, A. Seabaugh, S. K. Banerjee, and L. Colombo, "Electronics Based on Two-Dimensional Materials," *Nature Nanotechnology* 9 (2014) pp. 768–779.

[10] X. Miao, K. Chabak, C. Zhang, P. K. Mohseni, D. Walker, and X. Li, "High-Speed Planar GaAs Nanowire Arrays with f_{max} > 75 GHz by Wafer-Scale Bottom-Up Growth," *Nano Letters* 15 (2015) pp. 2780–2786.

[11] F. Gu, L. Zhang, X. Yin, and L. Tong, "Polymer Single-Nanowire Optical Sensors," *Nano Letters* 8 (2008) pp. 2757–2761.

[12] X. Duan, Y. Huang, Y. Cui, J. Wang, and C. M. Lieber, "Indium Phosphide Nanowires as Building Blocks for Nanoscale Electronic and Optoelectronic Devices," *Nature* 409 (2001) pp. 66–69.

[13] A. Motayed, M. He, A. V. Davydov, J. Melngailis, and S. N. Mohammad, "Realization of Reliable GaN Nanowire Transistors Utilizing Dielectrophoretic Alignment Technique," *Journal of Applied Physics* 100 (2006) art. no. 114310.

[14] J.-F. Dayen, A. Mahmood, D. S. Golubev, I. Roch-Jeune, P. Salles, and E. Dujardin, "Side-Gated Transport in Focused-Ion-Beam-Fabricated Multilayered Graphene Nanoribbons," *Small* 4 (2008) pp. 716–720.

[15] D. Natelson, "Mechanical Break Junctions: Enormous Information in a Nanoscale Package," *ACS Nano* 6 (2012) pp. 2871–2876.

[16] W. Ho, "Single-Molecule Chemistry," *Journal of Chemical Physics* 117 (2002) pp. 11033–11061.

[17] N. Behabtu, C. C. Young, D. E. Tsentalovich, O. Kleinerman, X. Wang, A. W. K. Ma, E. A. Bengio, R. F. ter Waarbeek, J. J. de Jong, R. E. Hoogerwerf, S. B. Fairchild, J. B. Ferguson, B. Maruyama, J. Kono, Y. Talmon, Y. Cohen, M. J. Otto, and M. Pasquali, "Strong, Light, Multifunctional Fibers of Carbon Nanotubes with Ultrahigh Conductivity," *Science* 339 (2013) pp. 182–186.

[18] G. Binnig, H. Rohrer, Ch. Gerber, and E. Weibel, "Surface Studies by Scanning Tunneling Microscopy," *Physical Review Letters* 49 (1982) pp. 57–61.

[19] G. Binnig, C. F. Quate, and Ch. Gerber, "Atomic Force Microscope," *Physical Review Letters* 56 (1986) pp. 930–933.

[20] V. Teppati, A. Ferrero, and M. Sayed (Eds.), *Modern RF and Microwave Measurement Techniques* (Cambridge University Press, 2013).

[21] J. J. Plombon, K. P. O'Brien, F. Gstrein, V. M. Dubin, and Y. Jiao, "High Frequency Electrical Properties of Individual and Bundled Carbon Nanotubes," *Applied Physics Letters* 90 (2007) art. no. 063106.

[22] P. Rice, T. M. Wallis, S. E. Russek, and P. Kabos, "Broadband Electrical Characterization of Multiwalled Carbon Nanotubes and Contacts," *Nano Letters* 7 (2007) pp. 1086–1090.

[23] L. Hao, D. Cox, K. Lees, J. C. Gallop, P. See, R. Clarke, T. J. B. M. Janssen, R. F. Zhang, and F. Wei, "Fabrication and Characterization of Carbon Nanotubes as r.f.

Interconnects," *2012 12th IEEE Conference on Nanotechnology (IEEE NANO)* (2012) pp. 1–5.

[24] S. Li, Z. Yu, S.-F. Yen, W. C. Tang, and P. J. Burke, "Carbon Nanotube Transistor Operation at 2.6 GHz," *Nano Letters* 4 (2004) pp. 753–756.

[25] J. M. Bethoux, H. Happy, G. Dambrine, V. Derycke, M. Goffman, and J. P. Burgoin, "An 8-GHz f_t Carbon Nanotube Field-Effect Transistor for Gigahertz Range Applications," *IEEE Electron Device Letters* 27 (2006) pp. 681–683.

[26] S. Vandenbrouck, K. Madjour, D. Theon, Y. Dong, Y. Li, C. M. Lieber, and C. Gaquiere, "12 GHz F_{MAX} GaN/AlN/AlGaN Nanowire MISFET," *IEEE Electron Device Letters* 30 (2009) pp. 322–324.

[27] T. Wang, K. Jeppson, N. Olofsson, E. E. B. Campbell, and J. Liu, "Through Silicon Vias Filled with Planarized Carbon Nanotube Bundles," *Nanotechnology* 20 (2009) art. no. 485203.

[28] J. C. Weber, J. B. Schlager, N. A. Sanford, A. Imtiaz, T. M. Wallis, L. M. Mansfield, K. J. Coakley, K. A. Bertness, P. Kabos, and V. M. Bright, "A Near-Field Scanning Microwave Microscope for Characterization of Inhomogeneous Photovoltaics," *Review of Scientific Instruments* 83 (2012) art. no. 083702.

2 Core Concepts of Microwave and RF Measurements

2.1 Introduction

In this chapter we review the core concepts of microwave and RF propagation in both guided-wave and on-wafer environments. Because most of these concepts are well known, we will introduce only the terms and definitions that are necessary for the development and description of the material used throughout this book. For many, this chapter will serve as a whirlwind tour of familiar concepts. Readers interested in further details will find them in the referenced literature.

Guided waves are often discussed exclusively in terms of transmission line theory. Here, our approach will begin with Maxwell's equations, from which we will then transition to the transmission line approach. Readers who do not require a review of the fundamental physics of guided electromagnetic waves may wish to skip directly to Section 2.3, which provides an overview of transmission line theory. Building upon transmission line theory, we define the impedance, admittance, and scattering parameter matrices. Then, after a brief discussion of signal flow graphs, we discuss calibration and de-embedding. From there, the calibration approach is extended to multimode propagation. Finally, we introduce one-port calibration of scanning microwave microscopes.

2.2 Maxwell's Equations

2.2.1 Macroscopic Equations

Without derivation, we define Maxwell's equations as follows:

$$\nabla \times \boldsymbol{E} = -\frac{\partial \boldsymbol{B}}{\partial t} \tag{2.1}$$

$$\nabla \times \boldsymbol{H} = \boldsymbol{J} + \frac{\partial \boldsymbol{D}}{\partial t} \tag{2.2}$$

$$\nabla \cdot \boldsymbol{D} = \rho \tag{2.3}$$

$$\nabla \cdot \boldsymbol{B} = 0 \tag{2.4}$$

where E and H are the electric and magnetic field vectors, respectively. B and D are the magnetic induction and electric displacement vectors, respectively. J is a vector that represents the induced and enforced current densities and ρ is the charge density. E, H, B, D, J, and ρ are functions of position $r = (x,y,z)$ and time t [1].

Maxwell's equations are complemented by the general electromagnetic materials equations:

$$B = [\mu] \cdot H \tag{2.5}$$

$$D = [\varepsilon] \cdot H \tag{2.6}$$

$$J = [\sigma] \cdot (E + E_{enf}) \tag{2.7}$$

where $[\mu]$, $[\varepsilon]$, $[\sigma]$ are the permeability, permittivity, and conductivity tensors, respectively. For isotropic media, these tensors are reduced to scalar quantities. Here, we will discuss propagation of electromagnetic waves only in media that are linear, isotropic, and passive, unless otherwise specified. In addition to Maxwell's equations, one must enforce the continuity equation

$$\frac{\partial \rho}{\partial t} + \nabla \cdot J = 0. \tag{2.8}$$

Finally, the particular solution of a given electromagnetic problem will depend on the boundary conditions at the interface between two different materials, denoted below by subscripts 1 and 2. For tangential components of electric and magnetic fields [1]

$$E_{t1} = E_{t2} \tag{2.9a}$$

$$H_{t1} = H_{t2} + K_s, \tag{2.9b}$$

where K_s is the surface current density in A/m. For the normal components of the displacement and magnetic induction

$$B_{n1} = B_{n2} \tag{2.10a}$$

$$D_{n1} = D_{n2} - \rho_s, \tag{2.10b}$$

where ρ_s is surface charge density in C/m^2.

2.2.2   Vector and Scalar Potentials

To gain physical insight into the meaning of Maxwell's equation within a material, it is necessary to introduce the polarization vector P and magnetization vector M as [2]

$$D = \varepsilon_0 E + P \tag{2.11}$$

and

$$B = \mu_0 (H + M). \tag{2.12}$$

Then the Equations (2.2) and (2.3) can be rewritten in the form:

$$\nabla \times B = \mu_0 J + \frac{1}{c^2}\frac{\partial E}{\partial t} + \mu_0 \nabla \times M + \mu_0 \frac{\partial P}{\partial t} \tag{2.13}$$

$$\nabla \cdot E = \frac{1}{\varepsilon_0}(\rho + \nabla \cdot P), \tag{2.14}$$

where c is speed of light in vacuum, ε_0 is the permittivity of free space, and μ_0 is the permeability of free space. Equations (2.1) and (2.4) retain their form. The material-related terms in Equation (2.13) represent the effective currents due to presence of the material and in Equation (2.14) the bound charge due to presence of the material. We will discuss polarization and magnetization vectors and their relation to microscopic material parameters in Chapter 9.

From Equation (2.4), it follows that the magnetic field can be written in the form

$$B = \nabla \times A. \tag{2.15}$$

The vector field A is known as the vector potential. Combining this definition with Equation (2.1), we can define a scalar quantity ϕ called the scalar potential such that the electric field can be expressed as

$$E = -\frac{\partial A}{\partial t} - \nabla \phi. \tag{2.16}$$

Note that A and ϕ are both functions of r and t. This form of Maxwell's equations is indispensable for describing nanoscale electromagnetic interactions with matter. The introduction of potentials in (2.15) and (2.16) does not uniquely determine A and ϕ. By introducing the so-called Lorentz gauge condition

$$\nabla \cdot A + \frac{1}{c^2}\frac{\partial \phi}{\partial t} = 0, \tag{2.17}$$

A and ϕ can be uniquely determined from

$$(\nabla^2 - \frac{\partial^2}{\partial t^2})A = -\mu_0 J \tag{2.18}$$

and

$$(\nabla^2 - \frac{\partial^2}{\partial t^2})\phi = -\frac{1}{\varepsilon_0}\rho. \tag{2.19}$$

The general solutions for the scalar and vector potentials in the Lorentz gauge have the following respective forms [3]–[5]:

$$\varphi(r,t) = \left[\frac{1}{4\pi\varepsilon_0}\int_{-\infty}^{\infty}\frac{\rho\left(r',t-\frac{|r-r'|}{c}\right)}{|r-r'|}d^3r'\right] \tag{2.20}$$

and

$$A(r,t) = \left[\frac{\mu_0}{4\pi}\int_{-\infty}^{\infty}\frac{J\left(r',t-\frac{|r-r'|}{c}\right)}{|r-r'|}d^3r'\right]. \tag{2.21}$$

These potentials are sometimes referred to as retarded potentials. Note the presence of the expression for the so-called retarded time, $t - |r - r'|/c$. Also note that the current densities and charge densities in these equations are assumed to contain all contributions including sources. We will return to this approach and its consequences when we discuss near-field interactions.

2.2.3   Hertz Vector Potentials

In order to solve Maxwell's equations in guided-waves systems, it is useful to introduce the Hertz potential π that is related to the vector and scalar potentials through [6]

$$A(r,t) = \varepsilon\mu\frac{\partial\pi(r,t)}{\partial t}. \tag{2.22}$$

Using the Lorentz gauge condition (2.17)

$$\varphi(r,t) = -\nabla\cdot\pi. \tag{2.23}$$

Note that if one can find an arbitrary solution for π, the electric and magnetic fields obtained from the Hertz vector fulfill all of Maxwell's equations and therefore describe the solution of the problem.

Inspection of Maxwell's equations for guided waves propagating in the z direction reveals two special solutions: one when $E_z = 0$ and the other when $H_z = 0$, where E_z and H_z are the components of the electric and magnetic fields along the direction of propagation. The first of these solutions has the electric field perpendicular to the direction of propagation and is called the transverse electric (TE) mode. The latter of these solutions has the magnetic field perpendicular to the direction of propagation and is called the transverse magnetic (TM) mode. The solution for these two cases simplifies if we introduce special forms of Hertz vector potentials: the electric Hertz vector potential $\boldsymbol{\pi}_e$ and magnetic Hertz vector potential $\boldsymbol{\pi}_m$. Formulas that correspond to propagation of guided waves in the positive z direction are introduced later in this chapter. For the propagation in the negative z direction they have to be modified appropriately [7]. The electric Hertz vector is defined such that:

$$E = \nabla \times \nabla \times \boldsymbol{\pi}_e, \tag{2.24a}$$

$$H = \varepsilon \frac{\partial}{\partial t}(\nabla \times \boldsymbol{\pi}_e). \tag{2.24b}$$

The magnetic Hertz vector is defined as:

$$H = \nabla \times \nabla \times \boldsymbol{\pi}_m, \tag{2.25a}$$

$$E = -\mu \frac{\partial}{\partial t}(\nabla \times \boldsymbol{\pi}_m). \tag{2.25b}$$

Both vectors satisfy the wave equation

$$\nabla^2 \, \boldsymbol{\pi}_{e,m} - \mu\varepsilon \frac{\partial^2}{\partial t^2}\left(\boldsymbol{\pi}_{e,m}\right) = 0. \tag{2.26}$$

When the time dependence of the electric and magnetic fields is harmonic, i.e., in the form $\exp(j\omega t)$, one can replace $\dfrac{\partial}{\partial t}$ by $j\omega$, where ω is the radial frequency of the harmonic signal and $j = \sqrt{-1}$.

The solution of (2.26), as mentioned previously, defines all components of guided-wave electromagnetic fields through (2.24) and (2.25). The utility of the Hertz vector is demonstrated by the fact that one can easily obtain the transverse component of the magnetic field, the so-called TM field, from the component of the electric Hertz vector in the direction of propagation. In a similar way, one can easily obtain the transverse component of the electric field, the so-called TE field, from the magnetic Hertz vector.

If the propagating electromagnetic field has both electric and magnetic field components in the plane perpendicular to the direction of propagation and these fields are a function of only one coordinate variable and time, then this field configuration is called a transverse electromagnetic or TEM wave. Transverse electromagnetic waves play an important role in microwave engineering because the form of propagating TEM wave equations is similar to that of transmission line equations, as will be shown in the following subsection.

2.2.4   Transition from Fields to Transmission Lines

Following the approach presented in Reference [7], we transition from the electromagnetic field representation to the quasi-equivalent transmission line approach. The transmission line model is widely used as it represents complex electromagnetic fields through conceptually simpler voltages and currents. Here, the case of TE waves is described in detail, but the approach is also valid for TM and TEM waves.

Assuming the propagation direction is in the z direction for TE waves, the $\boldsymbol{\pi}_m$ vector can be expressed as:

$$\boldsymbol{\pi}_m = T_e\left(x, y\right) L(z)\boldsymbol{u}_z, \tag{2.27}$$

where is $\boldsymbol{u}_z$ is a unit vector in the z direction and the functions T_e and L represent the transverse and longitudinal field components, respectively. Inserting (2.27) into (2.26) and separating the variables, we get two differential equations

$$\nabla^2 T_e(x, y) + K^2 T_e(x, y) = 0, \tag{2.28a}$$

$$\frac{d^2 L(z)}{dz^2} + \gamma^2 L(z) = 0, \tag{2.28b}$$

where $\gamma^2 = \varepsilon\mu - K^2$ and K is the separation constant. The solution of (2.28b) is in the form

$$L(z) = Ae^{-\gamma z} + Be^{\gamma z}, \tag{2.29}$$

which represents the wave propagation as a superposition of waves propagating in the positive and negative z directions. The propagation constant is a complex number, $\gamma = \alpha + j\beta$, where α is the damping parameter and β is a phase constant. One subsequently can introduce $\lambda_g = \dfrac{2\pi}{\beta}$, the wavelength of the guided-wave mode.

Both Equations (2.28a) and (2.28b) have to be solved as eigenvalue problems with corresponding boundary conditions. We are not going to address the mathematical solution of such eigenvalue problems here. In general, the solution of the boundary value problem for TE and TM modes leads to an infinite number of solutions for each of the modes. The existence of this set of solutions and the guided-wave mode structure it represents are critical concepts for the understanding of guided waves.

Now we define the circuit variables for voltage v and current i in terms of the guided-wave electromagnetic fields. The TM and electric components of the fields are defined as

$$\boldsymbol{H}_t = i\boldsymbol{h}^0, \tag{2.30a}$$

$$\boldsymbol{E}_t = v\boldsymbol{e}^0. \tag{2.30b}$$

For the TE mode:

$$h^0 = C_2 \nabla T_e,$$

(2.31a)

$$e^0 = C_1 (u_z \times \nabla T_e),$$

(2.31b)

and for the TM mode:

$$h^0 = j\omega\varepsilon C_2 (\nabla T_m \times u_z),$$

(2.32a)

$$e^0 = C_1 \nabla T_m.$$

(2.32b)

C_1 and C_2 are constants. Combining Equation (2.32) with Equations (2.26) through (2.29) and assuming harmonic time dependence one gets

$$v = \frac{j\omega\mu L(z)}{C_1},$$

(2.33a)

$$i = \frac{dL(z)/dz}{C_2}.$$

(2.33b)

Note that v and i represent the maximum amplitudes of voltages and currents of the particular propagating mode under specified boundary conditions. It is assumed that the product $(v \cdot i^*)$ is proportional to power flow of the mode and the ratio $\frac{v}{i} = Z$ is the impedance of the mode. We can define the power in the usual way for electric circuits as $P = \frac{1}{2}\Re\left(v \cdot i^*\right)$. This definition of the power imposes the condition that only one of the constants C_1 and C_2 is arbitrary. The arbitrary constant is obtained from additional requirements representing normalization conditions, which are usually chosen such that the fields do not contradict basic physics. In the case of lines such as coaxial cables that have a principal mode, the TE field obeys the Laplace equation. Therefore, one can integrate along the path between the electrodes to obtain the voltage between them. This defines uniquely the outstanding, arbitrary constant. The interested reader can find further details in References [2], [5], and [7]–[11].

With i and v defined, we can now define the characteristic impedance of the mode. Begin with the following relation between the transverse field components

$$u_z \times H_t = u_z \times iC_2 \nabla T_e = \frac{iC_2}{vC_1} E_t = \frac{C_2}{C_1 Z} E_t.$$

(2.34)

For a TE mode propagating in air, it can be shown that

$$\frac{C_1}{C_2} Z = Z_{0TE} = \left(\frac{\mu_0}{\varepsilon_0}\right)^{1/2} \frac{\lambda_g}{\lambda_0}, \tag{2.35}$$

where Z_{0TE} is the characteristic impedance of the TE mode, and λ_0 is the free space wavelength. The characteristic impedance of the TM mode can be found in a similar way [12]

$$Z_{0TM} = \left(\frac{\mu_0}{\varepsilon_0}\right)^{1/2} \frac{\lambda_0}{\lambda_g}. \tag{2.36}$$

Note that this definition of the characteristic impedance is not unique. It is not possible to uniquely define the characteristic impedance of the guided wave in general, but this definition is a reasonable one.

Finally, for completeness, it is necessary to describe power flow in a guided-wave configuration. For electromagnetic fields the power flow is represented by Poynting's vector

$$\mathbf{S} = \mathbf{E} \times \mathbf{H}. \tag{2.37}$$

In a guided-wave structure, the power flow is obtained by integrating the normal component of the Poynting vector over the waveguide cross section. Up to now, we used a general approach that depends on the solution of a boundary value problem of arbitrary configuration. It has allowed us to introduce the concepts of currents and voltages in a general sense for the transverse components of an arbitrary guided-wave field configuration, paving the way for the introduction of transmission line theory.

2.3   Transmission Line Theory

Although the theory of guided electromagnetic waves can be fully developed from Maxwell's equations, the concepts of electrical circuits, including both lumped-element and distributed circuits, are widely used in microwave engineering. If the electromagnetic problem can be reduced to the propagation of a TEM wave, the circuit representation provides utility and fundamental insight. TEM-like waves are the principal modes of widely used waveguides such as coaxial cables, microstrip lines, and coplanar waveguides (CPWs). Here we will use the framework of circuit theory to introduce transmission line theory along with many key terms and concepts used throughout this book.

We will follow an approach introduced in the early stages of the development of microwave electronics [12]. Transmission line theory was originally developed by Heaviside [13] and the interested reader can find further details in a number of texts [14]–[17]. Having introduced current i and voltage v, propagation of a TEM

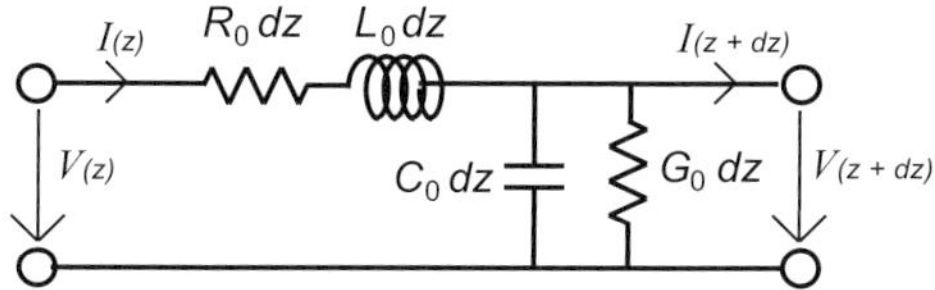

Figure 2.1. Transmission line model.
A schematic of a segment of a transmission line. The line of infinitesimal length dz is characterized by per-unit-length electrical circuit elements: resistance R_0, inductance L_0, capacitance C_0, and conductance G_0.

mode in one direction is reduced to the propagation of a voltage or current wave on a transmission line of a finite length l. Note that in the context of microwave metrology, the transmission line concept is advantageous over pure circuit theory as it allows for easier definition of measurement reference planes.

A transmission line is characterized by per-unit-length electrical circuit elements: resistance R_0, inductance L_0, capacitance C_0, and conductance G_0 as shown in Fig. 2.1. These parameters depend on transmission line dimensions and the materials used to construct the line. Applying Kirchoff's laws for this infinitesimally long element yields the Telegrapher's equations:

$$\frac{dV(z)}{dz} = -\left(R_0 + j\omega L_0\right) I(z), \tag{2.38a}$$

$$\frac{dI(z)}{dz} = -\left(G_0 + j\omega C_0\right) V(z). \tag{2.38b}$$

We use the capital italicized letters I and V to represent the current i and voltage v with assumed harmonic time dependence. Taking the expression for $I(z)$ from (2.38b) and inserting into (2.38a) gives the wave equation for voltage. The wave equation for current can be obtained in a similar way. The solution of this wave equation is in the form (2.29), but with $L(z)$ replaced by voltage $V(z)$:

$$V(z) = V_+ e^{-\gamma z} + V_- e^{\gamma z}. \tag{2.39}$$

As in Equation (2.29), the solution is a superposition of forward propagating wave (from the source) and backward propagating wave (from the load). Unique values of V_+ and V_- are obtained from the voltage and current and the load impedance at the end of the transmission line. If the transmission line of length l is terminated by a load impedance Z_L and we move the origin of the coordinate system there, then the expression $V_+ e^{-\gamma l}$ represents the voltage wave incident on the load and $V_- e^{\gamma l}$ represents the voltage wave propagating away from this load.

The propagation constant and characteristic impedance of the transmission line are functions of the per-unit-length circuit elements:

$$\gamma = \left[\left(R_0 + j\omega L_0\right)\cdot\left(G_0 + j\omega C_0\right)\right]^{1/2} \tag{2.40}$$

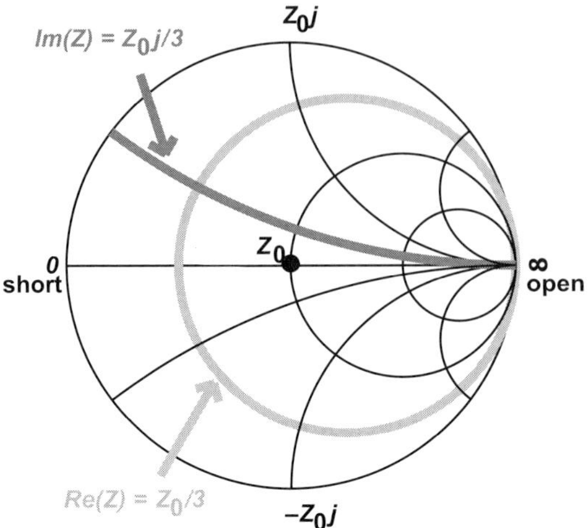

Figure 2.2.   The Smith chart.
The positions of several possible loads are shown: an open circuit, a short circuit, and a matched load Z_0. Examples of a constant resistance curve and a constant reactance curve are shown in gray.

$$P_W = \frac{1}{2}\Re\left(V(z_1)I^*(z_1)\right),\tag{2.52}$$

where the asterisk denotes the complex conjugate. Using the previously introduced forward and backward waves this can be rewritten as

$$P_W = \frac{1}{2}\Re\{V_{FW}(z_1)I^*_{FW}(z_1) - V_{BW}(z_1)I^*_{BW}(z_1) + (V_{BW}(z_1)I^*_{FW}(z_1) - V_{FW}(z_1)I^*_{BW}(z_1))\}.\tag{2.53}$$

It follows that the power transmitted through a transmission line is not simply the difference between the power transmitted by the forward and backward waves, but also includes the interaction of these waves on the transmission line. This result is not surprising since the superposition principle applies only to voltages and currents (in linear circuits), but not to power.

2.4   Impedance, Admittance, and Scattering Matrixes

From basic circuit theory one can write equations that describe the relationships between the voltages and currents at each port of a multiport device. In order to calibrate and analyze the microwave measurements it is useful to describe these relationships in terms of impedance, admittance, and scattering parameters. Here, we will develop this approach for a two-port configuration, but the approach can be generalized to any number of ports. Following convention, we introduce the currents at port

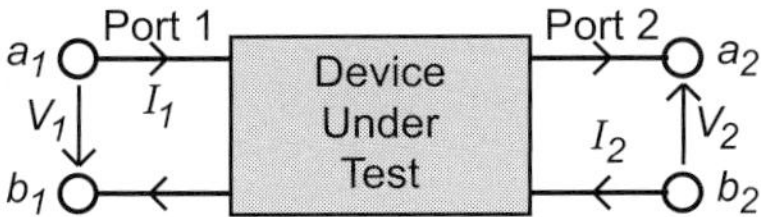

Figure 2.3. A two-port device.
A schematic of a two-port device defines the currents (I_1 and I_2), voltages (V_1 and V_2), and power waves (a_1, a_2, b_1, b_2).

1 (I_1) and port 2 (I_2) and the corresponding voltages (V_1 and V_2) as shown in Fig. 2.3. If the network is linear, then the voltages will be linear functions of currents:

$$V_1 = Z_{11}I_1 + Z_{12}I_2, \tag{2.54a}$$

$$V_2 = Z_{21}I_1 + Z_{22}I_2, \tag{2.54b}$$

where the variables Z_{ij} have the units of impedance and collectively form an impedance matrix. From the reciprocity principle for passive linear circuits, it follows that $Z_{12} = Z_{21}$. Alternative representations may be developed by choosing variables other than V_1 and V_2 to be the dependent variables. For example, equations could be written for I_1 and I_2 as functions of V_1 and V_2 with the matrix of corresponding coefficients representing admittances. As yet another alternative, the so-called h-matrix representation, equations could be written with the input voltage and the output current as the dependent variables.

At microwave frequencies, it is extremely difficult to directly measure voltages and currents. Therefore, a different approach had to be introduced that is more suitable for metrology at these frequencies. Assume that a two-port device is inserted (embedded) into a transmission line. We will call this device the DUT. Recall from Equations (2.46) that the voltages and currents at a port can be expressed as a superposition of waves propagating toward and away from the port. Here, the amplitudes of the waves propagating toward and away from a given port n are V_{n+} and V_{n-}. We can express the amplitudes of the waves propagating away from the port as functions of the amplitudes of the waves propagating toward the port n. Specifically, for a linear, two-port DUT:

$$V_{1-} = S_{11}V_{1+} + S_{12}V_{2+}, \tag{2.55a}$$

$$V_{2-} = S_{21}V_{1+} + S_{22}V_{2+}. \tag{2.55b}$$

The coefficients S_{ij} are called scattering parameters and collectively form a scattering matrix. The scattering matrix relates the waves reflected or scattered from the network to those incident upon the network. The scattering matrix parameters are sometimes referred to as "S-parameters." In a two-port device, the physical meaning of S_{11} is the input reflection coefficient when the output is matched ($V_{2+} = 0$), S_{21} is the forward transmission from port 1 to port 2, S_{12} is the reverse transmission from port 2 to port 1, and S_{22} is the reflection coefficient at port 2. An $n \times n$ matrix for an n-port device is considered reciprocal when $S_{ij} = S_{ji}$ and symmetric if it is reciprocal and $S_{ii} = S_{jj}$ for all values of i and j.

In commercial test equipment such as vector network analyzers, the scattering matrix parameters are usually normalized following the procedure introduced in Reference [18], in which the waves are defined in terms of the complex amplitudes of the incident and reflected power waves. This was done to make their definition consistent with the conservation of energy. Voltage amplitudes, on the other hand, have to be normalized to an arbitrary reference impedance. Usually, the characteristic impedance of the line is used as the normalization constant. Note that the characteristic impedance for each port, Z_{0n}, can differ from port to port. By convention, the reference impedance is 50 Ω for most commercial test equipment. The power waves have amplitudes a_n and b_n, which are related to the voltage amplitudes introduced in (2.46) as follows:

$$a_n = \frac{V_n + Z_{0n} I_n}{2\sqrt{|\Re(Z_{0n})|}} = \frac{V_{n+}}{\sqrt{|\Re(Z_{0n})|}};$$ (2.56a)

$$b_n = \frac{V_n - Z_{0n} I_n}{2\sqrt{|\Re(Z_{on})|}} = \frac{V_{n-}}{\sqrt{|\Re(Z_{on})|}}.$$ (2.56b)

Note that the amplitudes have the dimension of square root of power. From these definitions, the relation between the port voltages and currents for port n and the power waves is:

$$V_n = \sqrt{|\Re(Z_{on})|}\,(a_n + b_n),$$ (2.57a)

$$I_n = \frac{1}{\sqrt{|\Re(Z_{on})|}}(a_n - b_n).$$ (2.57b)

A set of equations analogous to Equation (2.55) expressed in terms of the two-port power waves a and b can be obtained:

$$b_1 = S_{11}a_1 + S_{12}a_2,$$ (2.58a)

$$b_2 = S_{21}a_1 + S_{22}a_2.$$ (2.58b)

In terms of the power waves, the incident power into port n is

$$P_n = \frac{1}{2}(a_n a_n^* - b_n b_n^*).$$ (2.59)

It is important to remember that the scattering matrix is well-defined only if all ports are matched, though Z_{0n} and thus the matching condition may generally vary from port to port. In practice, the scattering matrix formulation is convenient for measurements as well as simulations. Therefore, microwave network analyzers are designed to measure scattering parameters.

In order to perform meaningful, quantitative measurements, it is necessary to define reference planes. When a two-port DUT is embedded into a transmission line, one may define specific reference planes at the ports of the device. Sometimes it is not possible to measure the response of the DUT at these reference planes. In that case, one has to do measurements at different, accessible planes and then translate them to the ports of the DUT. Fortunately, the scattering matrix formulation is amenable to the translation of the reference planes within a DUT. If the distance between the new reference plane to port 1 of the device is l_1 and the distance to port 2 is l_2 then the relation between the scattering matrix measured at the reference plane S and the translated scattering matrix S' at the reference plane of the DUT are expressed as:

$$[S'] = \begin{bmatrix} e^{-\gamma_1 l_1} & 0 \\ 0 & e^{-\gamma_2 l_2} \end{bmatrix} [S] \begin{bmatrix} e^{-\gamma_1 l_1} & 0 \\ 0 & e^{-\gamma_2 l_2} \end{bmatrix}. \tag{2.60}$$

Despite these advantages, there are applications where the S-parameter representation is not optimal. For example, the scattering matrix representation is inconvenient for cascading multiple devices. Cascading of matrices is more easily accomplished by converting the scattering matrix parameters to transfer matrix parameters T_{ij}:

$$\begin{bmatrix} b_1 \\ a_1 \end{bmatrix} = \begin{bmatrix} T_{11} & T_{12} \\ T_{21} & T_{22} \end{bmatrix} \cdot \begin{bmatrix} a_2 \\ b_2 \end{bmatrix} = \begin{bmatrix} S_{12} - \dfrac{S_{22}S_{11}}{S_{21}} & \dfrac{S_{11}}{S_{21}} \\ -\dfrac{S_{22}}{S_{21}} & \dfrac{1}{S_{21}} \end{bmatrix} \cdot \begin{bmatrix} a_2 \\ b_2 \end{bmatrix}. \tag{2.61}$$

The transfer matrix parameters are sometimes referred to as "T-parameters." The interested reader can find other useful matrix transformations in Reference [8].

2.5   Signal Flow Graphs

Sometimes, it is useful to represent a system of linear equations in a graphical form. In the context of RF and microwave calibration procedures, this approach is particularly useful for the development of error models and error corrections, as we will see later in this chapter. Here, we briefly review the basic principles. The system of linear equations to be represented by this graphical approach has the general form [19], [20]

$$y = [M]x + [M']y \tag{2.62}$$

where $[M]$ and $[M']$ are square matrices with n columns and rows, the vector x represents the n independent variables, and the vector y the n dependent variables. This system of equations is quite general and can be applied to many systems, including

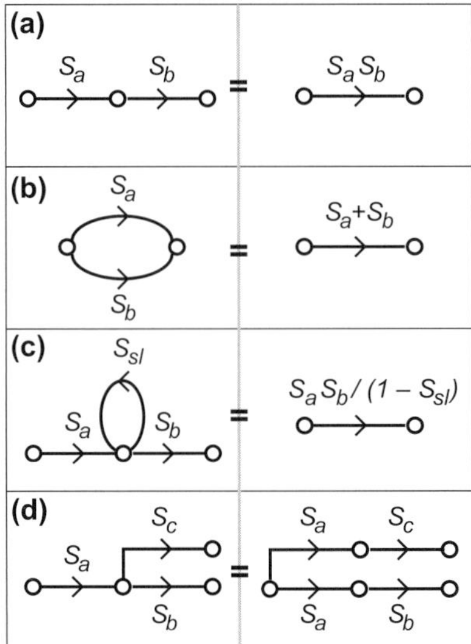

Figure 2.4.   Rules for simplifying signal flow graphs.
(a) Series rule; (b) parallel rule; (c) loop/self-loop rule; and (d) splitting rule.

circuits with closed signal loops. If there are no direct signal loops, Equation (2.62) simplifies to the standard scattering matrix in Equation (2.55). A signal flow graph consists of a set of nodes that are connected by branches. Each pair of nodes represents the amplitudes of an incident and an exiting wave: an "a" and a "b," as defined in Equation (2.56). The branches represent the complex S-parameters that relate the wave amplitudes. In other words, they represent the gains or losses along the path between two nodes. Note that the branches have a specified direction, denoted by an arrow, and that signals propagate only in the direction of arrows. For example, if a port is terminated by a load, then the corresponding pair of nodes is connected via an additional branch, with the load branch corresponding to the reflection coefficient of the load.

The transfer function of a signal flow graph may generally be determined by application of the so-called Mason's rules. It is often helpful to simplify the graph by use of four simple rules that govern the algebra of signal flow graphs:

(1) *Series rule*: Two sections in series can be reduced to one with the resulting gain given by multiplication of the two S-parameters. (See Fig. 2.4(a).)
(2) *Parallel rule*: Two branches in parallel pointing into the same node can be replaced by one with the gain equal to the sum of the two S-parameters, or, more generally, the S-parameters of all branches entering a node may be summed. (See Fig. 2.4(b).)
(3) *Loop/self-loop rule*: Branches that begin and end at the same node are "self-loops." A self-loop can be eliminated by multiplying all branches feeding the self-loop node by $1/(1 - S_{sl})$ where S_{sl} is the gain of the self-loop (Fig. 2.4(c)).

(4) *Splitting rule*: If a node has exactly one incident branch and one or more exiting branches, the incoming branch can be "split" and directly combined with each of the exiting branches. This rule can be used to treat the loops (arrows in parallel branches point to different directions) (Fig. 2.4(d)).

With experience, one can learn when it is most advantageous to use the signal flow graph approach and when it is more advantageous to use a matrix formulation.

2.6 Device De-embedding and Calibration

2.6.1 De-embedding

A central topic of this book is the measurement of nanoscale devices at radio frequencies. Many nanoscale device measurements are implemented as one- or two-port scattering parameter measurements with a vector network analyzer (VNA). As a result, we will focus on de-embedding of devices, calibration techniques and simple error models for such measurements. Nanoscale devices are generally integrated with a larger test structure that includes host structures, probes, connectors, and contacts. We will refer to such blocks of elements external to the nanoscale DUT as test fixtures. Measurements with VNAs are usually done at reference planes that include both the fixture and the DUT, which we will refer to as the "coaxial reference plane," as this reference plane often coincides with a coaxial connector. To be able to accurately characterize the DUT, one needs to remove the test fixture characteristics from the measurements.

Broadly speaking, there are many different approaches for removing the effects of the fixtures. Fundamentally, each of these approaches may be classified either as a "direct measurement" or as a "de-embedding." In the first case there are two stages. First, a series of measurements are made with physical reference standards inserted into the fixture in place of the DUT. Subsequently direct measurement of the DUT is performed. The reference planes are positioned at the boundary of the fixture and DUT. This approach requires development of specialized physical reference standards. Thus, the precision of direct measurement results depends on the quality of these physical standards.

By contrast, a de-embedding procedure uses models of test fixtures. These models are either mathematical or obtained experimentally, especially in the case of on-wafer measurements, which are discussed later in this chapter. Using these models, we can analytically remove the fixtures from the measurement. The precision of this approach depends once again on the accuracy of the model used. We will use both the direct measurement and de-embedding approaches throughout this book.

The concept of scattering parameters together with the flow graph approach is especially useful for the development of the theory of de-embedding fixtures from the measurements. Figure 2.5 shows the signal flow graph for a two-port fixtured measurement with both the fixtures and the DUT represented by S-parameters. The outer and inner pairs of dashed lines represent the coaxial and DUT reference

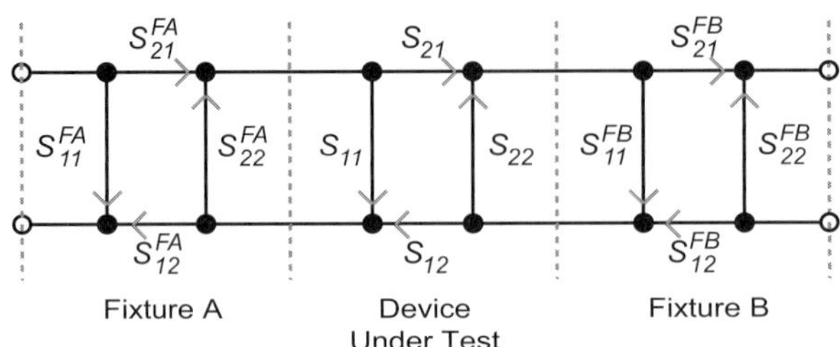

Figure 2.5.   A fixtured, two-port measurement.
The properties of the DUT are represented by S-parameter matrix elements S_{ij}. The properties of the two fixtures are represented by S-parameter matrix elements S^{FA}_{ij} and S^{FB}_{ij}. The four gray, dashed lines represent reference planes. The outer pair represents the coaxial reference planes while the inner pair represents DUT reference planes.

planes, respectively. The properties of the two test fixtures are described by the scattering parameter matrix elements S^{FA}_{ij} and S^{FB}_{ij}.

Matrix algebra provides the simplest approach to de-embed the scattering parameters of the DUT from the measurements. First, one needs to convert the S-parameter matrices of the measurement, both test fixtures and the DUT, to T-parameter matrices using Equation (2.46). Then

$$T^m = T^{FA}\, T^{DUT}\, T^{FB}, \tag{2.63}$$

where T^m is the T matrix from the VNA measurements, and T^{FA}, T^{FB}, and T^{DUT} are the T matrices for fixture A, fixture B, and the DUT, respectively. Multiplying by the inverse T matrices of fixtures yields

$$T^{DUT} = \left[T^{FA}\right]^{-1} T^m \left[T^{FB}\right]^{-1}. \tag{2.64}$$

The S-parameters of the device are then obtained by converting T^{DUT} back to an S-parameter matrix.

2.6.2   Multiline TRL and Other Calibration Techniques

In general, there is no "ideal" measurement test equipment. Therefore, the measurement strategy is to evaluate deviations from ideal behavior and to remove these systematic deviations from the measurements through calibration, thus significantly improving the accuracy of network analyzer measurements. This in turn provides the most accurate picture of device performance.

Many approaches have been developed for calibrated scattering parameter measurements. Detailed descriptions of coaxial-plane calibration techniques can be found in manufacturers' applications notes [21]–[23] as well as published papers [24]–[26]. Many of these approaches were initially developed for a coaxial environment, but have since been adapted to the on-wafer environment. In this book, considerable use is made of the multiline thru-reflect-line (TRL) calibration procedure and therefore we will focus on that approach. Multiline TRL offers a high degree of precision and utilizes an easily implemented set of calibration standards, however many alternative calibration techniques exist. For example, the short-open-line-thru (SOLT) calibration utilizes symmetric open lines, symmetric shorted lines, a line of known length, and a thru line that is short enough that one can assume that the transmission is unity. Another possibility is the line-reflect-match (LRM) calibration, which supplements the SOLT calibration with a symmetric 50 Ω load. There is one important limitation all these standard calibration procedures have in common: they are valid only under the assumption that the waves represent a single propagating mode within the calibration standard and the DUT. If this condition is not satisfied, a multimode calibration procedure must be introduced.

Returning to the TRL calibration procedure, it is useful to describe the general principles and implementation of the multiline TRL calibration before discussing the specific case of on-wafer multiline TRL. The TRL calibration procedure was originally introduced in Reference [24], for calibration of dual six-port network analyzers, but now is commonly used in conventional VNA calibrations. For calibration of two-port VNA measurements, it is convenient to introduce the concept of error boxes that describe deviations of the VNA from ideal behavior. In this model, incident (outgoing) waves going in to (out of) the ideal VNA are entering (exiting) the error boxes at some fictitious reference planes. This approach is justified because the four-port reflectometers of the network analyzer can be reduced to a cascade of two-port equations [25]. For a two-port measurement, the error boxes take the form of a two by two matrix. In turn, the calibration problem takes a form similar to Equations (2.48) and (2.64), where the fixture matrices are replaced by error boxes. In contrast to the fixture matrices, the error boxes do not in general satisfy the reciprocity requirement. The error boxes may be determined by solving a set of equations for set calibration standards, each of which is in the form of Equation (2.55).

As originally conceived, three calibration standards were used in the TRL calibration. First, a "thru" standard was established by directly connecting ports 1 and 2 at the coaxial reference planes (the outer pair of reference planes in Fig. 2.5). Second, each port is terminated at the coaxial reference planes by a short circuit, open circuit, or any impedance that has a load out of center of the Smith chart and is the same for both ports. The third standard is established by connecting a line of a known length between the coaxial reference planes. Note that this line has a different length than the thru line.

Early implementations of TRL faced a number of difficulties that were ultimately overcome by extending the technique by use of multiple, redundant lines. For instance, early implementations of TRL were band limited due to the fact that the line length had to differ from $\lambda/2$, where λ is the wavelength of the source signal. For broadband measurements, additional lines had to be introduced, but this approach introduced continuity problems at the boundaries of the frequency bands. Further, the approach did not take advantage of the fact that multiple lines provided redundant measurements that could potentially reduce measurement errors.

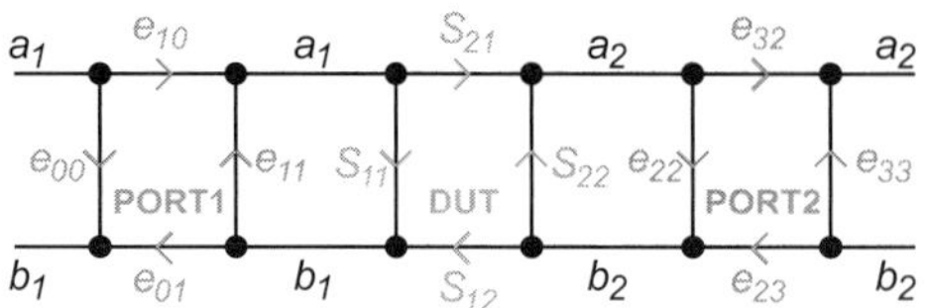

Figure 2.6.   Two-port, eight-term error model.
The S-parameters of the DUT are S_{ij} and the eight error terms are e_{ij}.

These problems were ultimately solved by Bianco et al. [27] and Marks [26], who utilized multiple redundant line standards in the TRL calibration procedure, culminating in the technique we now know as multiline TRL. In multiline TRL, the calibration standards consist of a set of transmission lines that differ only in length, the shortest of which serves as the thru standard, as well as reflection standards, which are assumed to be the same for both port connections. The procedure is based on estimation of the propagation constant of the transmission line standards at each measured frequency. Then the S-parameter correction coefficients are calculated using the accurate estimate of the propagation constant. The multiline TRL approach may be represented by an error box formulation as shown in Fig. 2.6.

Importantly, the so-called switch terms must be measured in addition to the calibration standards. The switch terms are specific to a given network analyzer, and account for differences between the forward and reverse match conditions. The switch terms can be measured only if all four wave parameters are accessible via the measurement instrument. Typically, the switch terms are measured simultaneously with the thru standard. The forward (Γ_{FW}) and reverse (Γ_R) switch terms are defined as

$$\Gamma_{FW} = \frac{a_{2m}}{b_{2m}}\bigg|_{Source=Port1} \quad ; \Gamma_R = \frac{a_{1m}}{b_{1m}}\bigg|_{Source=Port2} . \tag{2.65}$$

Multiline TRL is in general done in two steps. In the first step, the propagation constant and line corrections are obtained. In the second step, the corrected lines are used to get the error box parameters. This is accomplished by analytically solving the eigenvalue problem following from the cascaded ports in Fig. 2.6. Several software packages are available for this calibration, including the NISTcal and STATISTIcal among other programs.

Note that while the S-parameters are defined relative to the characteristic impedance of the transmission line standards, Z_0, multiline TRL obtains the propagation constant and the correction parameters without knowledge of the characteristic impedance. Once Z_0 is ultimately known, the correction coefficients can be transformed to any reference impedance. In addition, it is useful to know the length of the calibration lines or at minimum their relative length differences. This enables the selection of line pairs at each frequency that in turn avoids singularities in the calculation and improves the accuracy of the propagation constant and error box parameters by averaging the results from the line pairs.

2.6.3 On-Wafer Calibration

Now we turn to on-wafer calibration procedures. Many nanoscale devices are incorporated into "on-chip" or "on-wafer" devices. Thus, we will introduce the basic concepts of on-wafer calibration and discuss some of the simpler error correction schemes. Our discussion in this chapter will be limited to the extension of multiline TRL to an on-wafer environment. In Chapter 4, on-wafer measurement instrumentation and other practical considerations are discussed and in Chapter 6, an example of a calibrated, on-wafer measurement of a nanowire device is given.

On-wafer waveguide structures are usually in the form of a microstrip line or CPW. These structures are contacted with a set of specially designed probes. During calibration, the probes are treated as part of the test fixture. The tips of the probe now define the position of the reference plane that we have up until now referred to as the "coaxial" reference plane. For the on-wafer multiline TRL calibration, the standards must have the same contact layout and geometry as the "fixture lines" of the DUTs. As in the multiline calibration approach described earlier, we once again obtain the propagation constant, the error boxes, and the corrected lines and we use the error boxes to correct the S-parameters of the DUT. The error boxes include the properties of both the probes and the test platform, including the VNA. As long as the connecting transmission lines within the DUT have the same configuration as the calibration standards, one can translate the reference planes as described in Equation (2.60). Reference plane translation is often utilized in RF nanoelectronic devices to move the reference plane position as close as possible to the nanoscale building block(s) within the device.

2.7 Multimode Calibration

A significant limiting factor of standard calibration procedures is that they require single-mode propagation at the reference planes. This may not be always the case: multimode propagation can easily occur when coupled or multiline waveguides are investigated and may also occur in some nanoelectronic devices. Therefore, it is important to address the more complicated case of calibration when there are multiple modes propagating at the reference planes. The philosophy of multimode calibration follows basic multiport calibration techniques. Multimode calibration assumes that each mode propagates from its own effective port. Under this assumption the problem of multimode calibration is recast as a multiport calibration procedure where the number of physical ports is multiplied by the number of propagating modes. Instead of solving for propagation constants and error boxes at each port, one solves for unique propagation constants and error boxes corresponding to each propagating mode. The multimode TRL calibration technique was first introduced in Reference [28], but we will follow the approach introduced in Reference [29]. The multimode TRL calibration can be divided into three main steps. As in the multiline TRL calibration, the first step focuses on determination of the propagation constants of all of the propagating modes based on measurements of the thru and line standards. In the second step the error box matrices T^A, T^B are partially determined. In the third step, a reflect measurement is used to reduce the number of unknowns in these matrices.

Here, we will describe the two-port device case, which can be extended to an arbitrary number of ports. We introduce the generalized reverse cascade matrix approach, which may simplify some problems in which symmetry is present. Such cases are not that common generally, but are present in some de-embedding cases that are discussed in following chapters. In a generalized scattering matrix of a multimode two-port device, all incident waves at port 1, incorporating power waves for N modes, are represented by an incident wave vector A_1. Similarly, all reflected waves at port 1 are represented by the vector B_1. The vectors A_2 and B_2 represent the incident and reflected waves at port 2 of the generalized two-port device. The vectors are defined:

$$A_1 = \begin{bmatrix} a_1 \\ \vdots \\ a_N \end{bmatrix}; \qquad A_2 = \begin{bmatrix} a_{N+1} \\ \vdots \\ a_{2N} \end{bmatrix}, \tag{2.66a}$$

and

$$B_1 = \begin{bmatrix} b_1 \\ \vdots \\ b_N \end{bmatrix}; \qquad B_2 = \begin{bmatrix} b_{N+1} \\ \vdots \\ b_{2N} \end{bmatrix}. \tag{2.66b}$$

This definition enables a scattering matrix definition that includes transmission between N modes at port 1 and N modes at port 2. Thus, this definition also includes the mixing of the modes from different ports. Further, it can also incorporate evanescent modes.

The generalized scattering matrices are related to the generalized incident and reflected wave vectors by

$$\begin{bmatrix} B_1 \\ B_2 \end{bmatrix} = \begin{bmatrix} S_{11} & S_{12} \\ S_{21} & S_{22} \end{bmatrix} \begin{bmatrix} A_1 \\ A_2 \end{bmatrix}, \tag{2.67}$$

where S_{ij} are $N \times N$ submatrices. In a similar way one can introduce the generalized left to right cascade (transmission) matrix

$$\begin{bmatrix} B_1 \\ A_2 \end{bmatrix} = \begin{bmatrix} T_{11} & T_{12} \\ T_{21} & T_{22} \end{bmatrix} \begin{bmatrix} A_2 \\ B_2 \end{bmatrix}. \tag{2.68}$$

The conversion from the generalized scattering matrix to the generalized cascade matrix is similar to Equation (2.61) with scalar S-parameters replaced by S submatrices. This conversion is possible only if S_{21} is nonsingular. One can also introduce the left-to-right reverse cascade matrix (or connected right to left). Simple arithmetic manipulation shows that

$$\begin{bmatrix} B_2 \\ A_2 \end{bmatrix} = \begin{bmatrix} \bar{T} \end{bmatrix} \begin{bmatrix} A_1 \\ B_1 \end{bmatrix}, \tag{2.69}$$

with the reverse cascade matrix

$$\bar{T} = PT^{-1}PP = \begin{bmatrix} 0 & I \\ I & 0 \end{bmatrix} \tag{2.70}$$

This matrix represents the mirrored (connected right-to-left) version of the original left-to right matrix. P is a $2N \times 2N$ permutation matrix with I a $N \times N$ unity matrix.

In this notation the calibrated two-port network analyzer measures the scattering parameter matrix M:

$$M = T^A T^{DUT} \bar{T}^B, \tag{2.71}$$

where T^{DUT} is the cascade matrix of the DUT and the matrices T^A and T^B represent the error boxes. With these definitions in place, the multimode calibration procedure is analogous to the single-mode case. Specifically, similar calibration standards are measured. It is assumed though that these standards can support multiple quasi-TEM propagating modes corresponding to the modes supported by the DUT. The complex propagation constants of the multiple modes, as well as the matrices T^A and T^B are determined in steps similar to those in the single-mode case. Additional details of the procedure can be found in References [28] and [29].

2.8 Calibration of a Scanning Microwave Microscope and Other One-Port Systems

We conclude this chapter with a specific calibration approach that is important for local, near-field probes such as NSMMs. The calibration of the near-field probing measurements is particularly challenging because the condition of single-mode propagation is not satisfied for near fields. Fortunately, this problem can be avoided for most NSMM measurements if they are done in a single-port configuration, i.e., if the microwave signal path in the NSMM can be represented as a one-port network. This is true for the vast majority of existing NSMMs that are based on atomic force and scanning tunneling microscope systems.

For single-port NSMMs the measurand is the complex reflection coefficient S_{11m}. The near-field interaction is localized at the tip and thus will be included in the

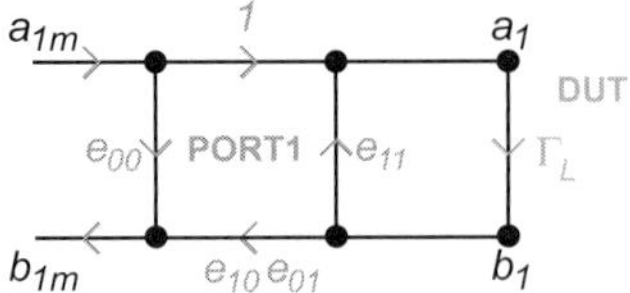

Figure 2.7.   Error model for an NSMM and other one-port measurements.
This model includes three error terms: e_{00} is directivity, the product $e_{10}e_{01}$ is tracking, and e_{11} is the port match.

de-embedded properties of the measured DUT. The effects of near-field interaction cannot be removed by calibration. Quantifying and understanding this interaction must be done by analyzing and interpreting the calibrated measurement results.

In the single-port case the signal flow graph shown in Fig. 2.6 is simplified to the graph shown in Fig. 2.7. The terms in Fig. 2.7 are defined as follows: e_{00} is directivity, the product $e_{10}e_{01}$ is tracking, and e_{11} is the port match [30]. Using the signal flow graph rules the relation between the measured and actual reflection coefficients is

$$S_{11m} = e_{00} + \frac{e_{01}e_{10}\Gamma_L}{1 - e_{11}\Gamma_L}, \tag{2.72}$$

where Γ_L is the de-embedded reflection coefficient of the DUT, as shown in Fig. 2.7. From this equation, it follows that to calibrate any one-port measurement, including one obtained with an NSMM, it is necessary to solve a system of equations for three complex variables which requires no fewer than three measurements of calibration standards. Specific approaches to NSMM calibration will be discussed in Chapter 7.

References

[1] P. A. Rizzi, *Microwave Engineering Passive Circuits* (Prentice Hall, 1988).

[2] R. F. Harrington, *Time-harmonic Electromagnetic Fields* (Wiley, 2001).

[3] O. Keller, *Quantum Theory of Near Field Electrodynamics* (Springer, 2011).

[4] J. D. Jackson, *Classical Electrodynamics* (John Wiley 1975).

[5] J. A. Stratton, *Electromagnetic Theory* (McGraw Hill, 1941).

[6] H. Hertz, "Die Kräfte electrischer Schwingungen, behandelt nach Maxwellschen Theorie," *Annalen der Physik* 36 (1888) pp. 1–22.

[7] D. M. Kerns, "Plane-wave Scattering Matrix Theory of Antennas and Antenna-Antenna Interactions," *National Bureau of Standards Monograph* 162 (1981).

[8] D. M. Pozar, *Microwave Engineering* (Wiley, 2004).

[9] K. C. Gupta, R. Garg, I. Bahl, and P. Bhartia, *Microstrip Lines and Slotlines* (Artech House, 1996).

[10] R. E. Collin, *Foundations of Microwave Engineering* (McGraw Hill, 1992).

[11] L. D. Landau and E. M. Lifshitz, *Classical Theory of Fields* (Pergamon Press, 1962).

[12] J. C. Slater, "Microwave Electronics," *Reviews of Modern Physics* 18 (1946) pp. 441–512.

[13] O. Heaviside, *Electromagnetic Theory,* Complete and Unabridged Edition v.1 no. 2 and v.3 (Dover, 1950).

[14] A. Sommerfeld, *Electrodynamics* (Academic Press, 1952). R. J. Collier, *Transmission Lines: Equivalent Circuits, Electromagnetic Theory, and Photons* (Cambridge University Press, 2013).

[15] F. Olyslager, *Electromagnetic Waveguides and Transmission Lines* (Oxford University Press, 1999).

[16] R. A. Chipman, *Theory and Problems of Transmission Lines* (McGraw-Hill, 1968).

[17] G. Mathaei, L. Young, and E. M. T. Jones, *Microwave Filters, Impedance-Matching Networks, and Coupling Structures* (Artech House, 1980).

[18] K. Kurokawa, "Power Waves and the Scattering Matrix", *IEEE Transactions on Microwave Theory and Techniques* 13 (1965) pp. 194–202.

[19] S. J. Mason, "Feedback Theory – Some Properties of Signal Flow Graphs," *Proceedings of the IRE* 41 (1963) pp. 1144–1156. F. Caspers, "RF Engineering basic concepts: S-parameters," arXiv:1201.2346 [physics.acc-ph].

[20] S. J. Mason, "Feedback Theory – Further Properties of Signal Flow Graphs," *Proceedings of the IRE* 44 (1956) pp. 920–926.

[21] "S-Parameter Design," *Agilent Application Note 154* (Agilent Technologies, 2006).

[22] "Vector Network Analyzer Primer," *Anritsu Application Note 11410-00387* (Anritsu Company, 2009).

[23] "Measuring Balanced Components with Vector Network Analyzer ZVB," *Rhode and Schwartz Application Note 1EZ53_0E* (Rhode and Schwartz, 2004).

[24] G. F. Engen and C. A. Hoer, "Thru-Reflect-Line: An Improved Technique for Calibrating a Dual Six-Port Automatic Network Analyzer," *IEEE Transactions on Microwave Theory and Techniques* 27 (1979) pp. 987–993.

[25] H.-J. Eul and B. Schiek, "A General Theory and New Calibration Procedures for Network Analyzer Self-calibration," *IEEE Transactions on Microwave Theory and Techniques* 39 (1991) pp. 724–731.

[26] R. Marks, "A Multiline Method of Network Analyzer Calibration," *IEEE Transactions on Microwave Theory and Techniques* 39 (1991) pp. 1205–1215.

[27] B. Bianco, M. Parodi, S. Ridella, and F. Selvaggi, "Launcher and Microstrip Characterization," *IEEE Transactions on Instrumentation and Measurement* 25 (1976) pp. 320–323.

[28] C. Sequinot, P. Kennis, J-F. Legier, F. Huret, E. Paleczny, and L. Hayden, "Multimode TRL – New Concept in Microwave Measurements: Theory and Experimental Verification," *IEEE Transactions on Microwave Theory and Techniques* 46 (1998) p. 536.

[29] M. Wojnowski, V. Issakov, G. Sommer, and R. Weige, "Multimode TRL Calibration Technique for Characterization of Differential Devices," *IEEE Transactions on Microwave Theory and Techniques* 60 (2012) p. 2220.

[30] "De-embedding and Embedding S-parameter Networks Using a Vector Network Analyzer," *Agilent Application Note 1364-1* (Agilent/Keysight Technologies, 2004).

3 Extreme Impedance Measurements

3.1 The Impedance Matching Challenge in RF Nanoelectronics

Microwave measurements of RF nanoelectronic devices present numerous challenges. Among these, perhaps the most difficult measurement challenge arises from the inherent, often extreme impedance mismatch between nanoelectronic systems and conventional commercial test equipment. The physical origin of this mismatch may be understood by comparing two physical constants: the free space impedance and the quantum resistance [1]. The impedance of free space Z_{free} is given by the ratio of the magnitude of the electric field component to the magnitude of the magnetic field component in a TEM (or far-field) electromagnetic wave:

$$Z_{free} = \frac{E}{H} = \sqrt{\frac{\mu_0}{\varepsilon_0}} = 377\,\Omega, \tag{3.1}$$

where μ_0 is the permeability of free space, ε_0 is the permittivity of free space, and E and H are the magnitudes of the electric and magnetic field components, respectively. Recall from Section 2.2 that the impedance of the TE mode in a waveguide is given by the product of Z_{free} times a ratio of the wavelengths of a given mode in free space and in a waveguide and that the impedance of the TM mode is given by Z_{free} divided by this same ratio. Thus, the value of Z_{free} sets the natural impedance range for both free space and guided waves. As a result, the relevant commercial test equipment has been developed to match this impedance scale. By contrast, the quantum resistance is given by:

$$R_Q = \frac{h}{2e^2} = 12.9\,\text{k}\Omega \tag{3.2}$$

where h is Planck's constant and e is the charge of an electron. Experimentally, the conductance of a two-dimensional electron gas, such as those found in high electron mobility transistors or single-layer graphene, takes on quantized values proportional to $1/R_Q$ [2]. This effect is known as the quantum Hall effect and occurs only at cryogenic temperatures and in the presence of a high magnetic field. More broadly, the resistance of nanoelectronic devices with dimensions of the order of the de Broglie wavelength of an electron is on the order of the quantum resistance. In the particular case of a single-walled CNT, the resistance is $R_Q/2$ [1] (the factor of two arises from band structure degeneracy).

The vastly different scales of Z_{free} and R_Q illustrate the inherent impedance mismatch between the nanoelectronic world and conventional microwave measurements. As a consequence of impedance mismatch, directly connecting a calibrated network analyzer to an extreme impedance device will not usually yield meaningful measurements. Most of the signal incident on the extreme impedance device will be reflected back to the analyzer. Alternative approaches to extreme impedance measurements must be developed. Before discussing specific applications of such approaches to RF nanoelectronics in Chapter 4, we will discuss extreme impedance measurements more generally in the remainder of this chapter. Because the impedance of nanoelectronic devices is generally much greater than that of commercial test equipment, the methods we discuss have primarily been developed and verified for extremely high impedance measurements. The methods should be applicable to extremely low impedance measurements, but to date extension of these methods to extremely low impedances has been limited.

3.2   An Introduction to Extreme Impedance Measurements

Consider a one-port, microwave network terminated by a load impedance of $Z_L = a_L + b_L j$. The complex reflection coefficient from the load, Γ_L, is given by

$$\Gamma_L = \frac{Z_L - Z_0}{Z_L + Z_0},\qquad(3.3)$$

where Z_0 is the characteristic impedance of the guided-wave network that is connected to the load. For most guided-wave networks, including coaxial- and waveguide-based transmission, the characteristic impedance is typically $Z_0 = 50\ \Omega$ (or in a few cases $75\ \Omega$). For free space measurements, the characteristic impedance is $377\ \Omega$. Equation (3.3) is plotted in Fig. 3.1(a) for the case $Z_L = a_L$, with $b_L = 0$. In order to maximize the signal transmitted to the load, impedances of microwave devices are designed to match the characteristic impedance. In general, the matching condition is that the load impedance should be equal to the characteristic impedance:

$$Z_L = Z_0.\qquad(3.4)$$

Note that this condition differs from the condition for maximum power transfer to the load, which occurs when the load impedance is equal to the complex conjugate of the characteristic impedance. Historically, approaches to calibration and measurement of microwave devices have been developed, tested, and optimized for devices-under-test that are reasonably well-matched, meeting or nearly meeting the condition in Equation (3.4). By contrast, if the load impedance is much larger or much smaller than the characteristic impedance, the magnitude of Γ_L will approach one. In such cases, most of the incident microwave signal will be reflected from the device and little if any signal will be transmitted to the load.

For many RF nanoelectronic devices, the DUT does indeed possess an extreme impedance that is far from the reference impedance, giving rise to new measurement challenges. First, the measured values of Γ_L will differ only slightly from one. As a specific example, for a resistive device with impedance on the order of one resistance quantum (Z_L = 12.9 kΩ), $|\Gamma_L|$ = 0.9923. Over ninety-nine percent of the incident power is reflected from the device! In other words, $|\Gamma_L|$ will differ from an open circuit, for which $|\Gamma_L|$ is unity, by only a few parts in one thousand. Furthermore, measurements of Γ_L for high impedance devices are insensitive to changes in Z_L. Taking the derivative of Equation (3.3) with respect to the normalized impedance Z_L/Z_0 yields:

$$\frac{\delta \Gamma_L}{\delta \left(\dfrac{Z_L}{Z_0} \right)} = \frac{2Z_0}{(Z_L + Z_0)^2}. \tag{3.5}$$

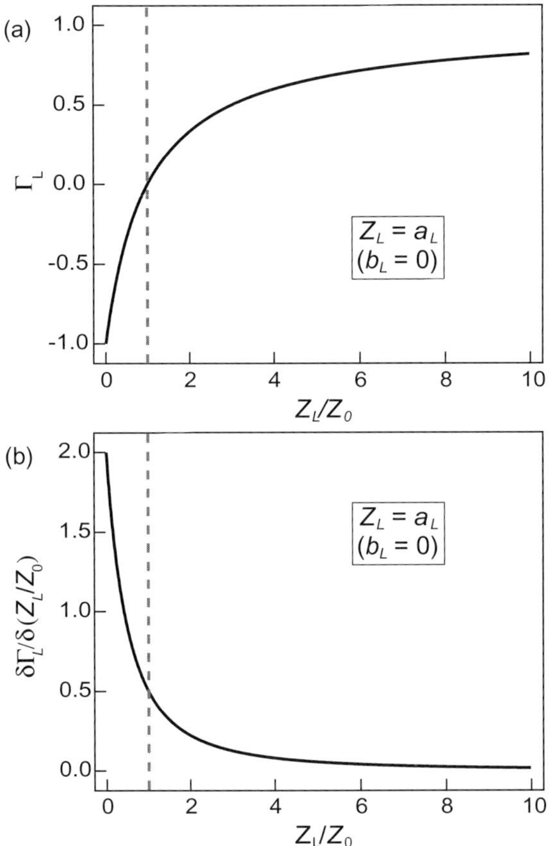

Figure 3.1.   Reflection coefficient Γ_L for a one-port network.
(a) The one-port reflection coefficient Γ_L of a one-port network connected to a load impedance Z_L as a function of normalized impedance Z_L/Z_0 for the case $Z_L = a_L$ ($b_L = 0$). (b) The derivative of Γ_L with respect to the normalized impedance Z_L/Z_0 for the case $Z_L = a_L$ ($b_L = 0$). The vertical dashed line corresponds to a matched impedance ($Z_L = Z_0$) in both (a) and (b).

Equation (3.5) is plotted in Fig. 3.1(b) for the case $Z_L = a_L$ with $b_L = 0$. Note that $d\Gamma_L/d(Z_L/Z_0)$ falls off as Z_L is increased from Z_0 to higher values, rapidly approaching zero. This indicates that a one-port measurement of Γ_L will be insensitive to changes in Z_L if Z_L is greater than or on the order of 10 Z_0, even if those changes are quite substantial.

One way to visualize the extreme impedance problem is through the Smith chart, as illustrated in Fig. 3.2. Recall from Chapter 2 that for a one-port network, the center of the Smith chart corresponds to the point where the load is perfectly impedance-matched and the reflection coefficient is zero. By contrast, extreme impedance measurements represented on a Smith chart will lie near the edge of the chart, as shown by the shaded region in Fig. 3.2. In the particular case of RF nanoelectronics measurements, the impedance will lie near the right-hand edge of the chart, corresponding to extremely high impedance. In this region of the Smith chart, the magnitude of the reflection coefficient approaches one. In addition, there is a high density of constant resistance circles and constant reactance curves in this region of the Smith chart. This visually illustrates that substantial changes in extremely high Z_L will result in only a small translation on the Smith chart.

In summary, the closeness of the one-port reflection coefficient Γ_L to unity and the relative insensitivity of Γ_L to changes in Z_L, present significant measurement challenges. Note that these challenges arise from the attempt to directly measure Γ_L. One alternative strategy is to reconfigure the measurement platform in order to measure an alternate quantity that is better matched and more sensitive to Z_L. Graphically, this reconfiguration can be envisioned as a translation from the

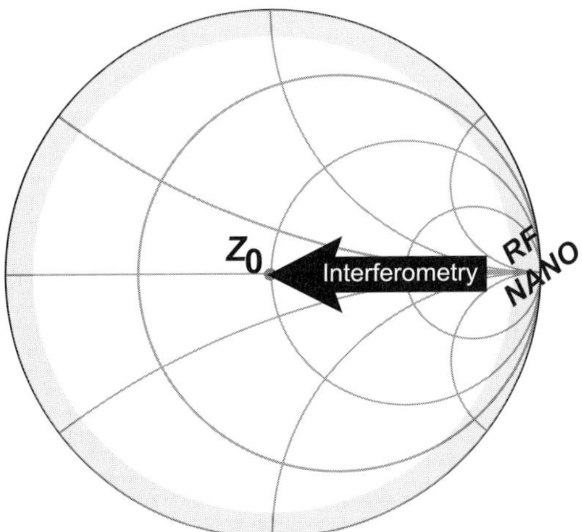

Figure 3.2. Smith chart representation of the extreme impedance measurement problem. Extreme impedance loads are represented by the shaded gray region near the circumference of the Smith chart, with most RF nanoelectronics devices (*RF NANO*) falling near the right-hand edge of the chart. Interferometry is one strategy for moving the measurand from the edge of the chart closer to the characteristic impedance Z_0.

edge of the Smith chart to the center, as shown by the black arrow in Fig. 3.2. Interferometry is one effective approach to moving the measurand from the edge to the center of the Smith chart. A drawback of this strategy is that additional calibration and analysis steps may be required to obtain the desired measurement from the alternate, well-matched measurand. Four specific approaches will be discussed here. The first is a simple impedance matching network, which may be useful particularly when the nanoelectronic device has a narrow operating bandwidth. The other three approaches are interferometric: a reflectometer based on a power splitter, a reflectometer based on a hybrid coupler, and an interferometer with active signal injection. In this chapter, the discussion of these four approaches will be restricted to guided-wave structures, but the approaches can be generalized to an on-wafer environment.

3.3   Impedance Matching Networks

Perhaps the most common, traditional approach to impedance mismatch problems is the introduction of an impedance matching network. Consider the measurement of a DUT with impedance Z_{DUT}, as shown in Fig. 3.3(a). An impedance matching network has been inserted between the test equipment and the DUT. In general, there are a wide range of possible circuit configurations for matching networks and a great deal of engineering has historically been devoted to the impedance matching problem. Here, we will consider only a simple case consisting of a series impedance Z_S and a parallel impedance Z_P, as shown in Fig. 3.3(b). This simple configuration is useful for cases where Z_{DUT} is much larger than Z_0, which is nearly always the case for RF nanoelectronic devices. With this matching network in place, the total load impedance is now

$$Z_L = Z_S + \frac{Z_P Z_{DUT}}{Z_P + Z_{DUT}}. \tag{3.6}$$

In order for the inserted elements to function as an impedance matching network, the values of Z_S and Z_P are chosen so that Equation (3.4) is satisfied.

When the extreme impedance of the DUT is well known, the implementation of a fixed impedance matching network may be an effective tool for facilitating the measurement over a finite bandwidth. Combining Equations (3.4) and (3.6), we find

$$Z_0^* = Z_S + \frac{Z_P Z_{DUT}}{Z_P + Z_{DUT}}. \tag{3.7}$$

Note that the asterisk denotes complex conjugation. Consider a simple implementation of the matching network shown in Fig. 3.3(b) might consist of an inductance L as the series element ($Z_S = \omega L$) and a capacitance C as the parallel element ($Z_P = 1/\omega C$). In this case, the complex-valued Equation (3.7) may be solved for the parameters L and C. For example, for a purely resistive load impedance of $Z_L = 12.9$ kΩ

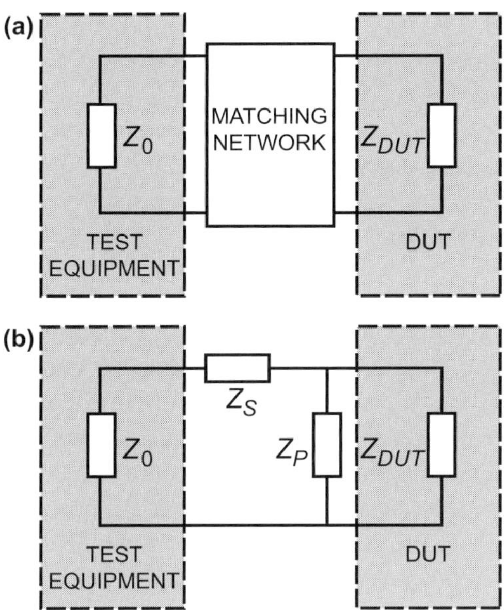

Figure 3.3. Schematic of an impedance matching network.
(a) A matching network is inserted between the test equipment (reference impedance Z_0)
and the device under test (impedance Z_{DUT}). (b) One example of an implementation of a
matching network.

at an operating frequency of ten gigahertz, we find $L = 12.8$ nH and $C = 19.8$ aF.
Historically, graphical approaches based on the Smith chart have provided a practical approach to determining the necessary circuit elements for an impedance
matching network [3]. More recently, commercial software packages have enabled
modeling and design of matching circuits by automating the determination of the
circuit elements as well as the optimization of the frequency response. In addition,
the advent of automated tuners opens the possibility for real-time, adaptive matching networks.

While the implementation of impedance matching networks remains a workhorse
technique for microwave and RF circuit design, traditional impedance matching networks may have limited value for broadband measurements in RF nanoelectronic
environments for several reasons. First, while matching networks function effectively only over a limited bandwidth, the targeted characterization of RF nanoelectronic devices and their constituent materials often spans tens of gigahertz. Second,
the introduction of the matching network introduces additional circuit elements to
the measurement platform, further complicating the de-embedding of the intrinsic
response of the nanostructure under test. Keep in mind that de-embedding of the
broadband nanostructure properties is at times the sole objective of the measurements, particularly in a research and development environment. By contrast, the
purpose of a matching network is to maximize the amount of power transferred
from the source to the DUT. Finally, where production of novel nanomaterials is

not yet uniform or optimized, variations in material components may necessitate tuning or customizing the matching network for each prototype DUT.

3.4   Reflectometer Methods for One-Port Devices

3.4.1   Implementation with a Power Splitter

Interferometric techniques provide an effective approach to extreme impedance measurements in the microwave regime. One interferometric approach is to integrate the extreme impedance DUT into a reflectometer [4], [5]. This is achieved by use of a multiport device, such as a power splitter or a hybrid coupler, with one output port of the device terminated by the high impedance DUT and another output port terminated by a high impedance reference device (or series of reference devices). In this configuration, the reflection coefficient of the entire reflectometer will be proportional to a simple algebraic combination of the reflection coefficients of the two high impedance structures. With a proper choice of the reference devices, the reflection coefficient of the entire reflectometer will be close to zero, and thus will present a measurand that is suitable for commercial, 50 Ω RF and microwave test equipment.

Consider the case of a one-port reflectometer based on a three-port power splitter shown in Fig. 3.4. The reference device and the high impedance DUT terminate port 2 and port 3 of the splitter, respectively. The reference device has impedance Z_{ref} and reflection coefficient Γ_{ref} while the DUT has impedance Z_L and reflection coefficient Γ_L. The reflection coefficient of the entire reflectometer structure Γ_m is measured by connecting port 1 to a VNA.

If the splitter is an ideal, broadband, two-resistor power splitter, the three-port scattering parameter matrix of the splitter will be given by

$$S^{splitter} = \begin{bmatrix} 0 & 1/2 & 1/2 \\ 1/2 & 1/4 & 1/4 \\ 1/2 & 1/4 & 1/4 \end{bmatrix}. \tag{3.8}$$

It follows that Γ_m is given by

$$\Gamma_m = \frac{\dfrac{1}{4}(\Gamma_{ref} + \Gamma_L)}{1 - \dfrac{1}{4}(\Gamma_{ref} + \Gamma_L)}. \tag{3.9}$$

Note that for the case where $\Gamma_{ref} = -\Gamma_L$, Equation (3.9) simplifies to $\Gamma_m = 0$. In other words, if the reference device has equal magnitude and opposite phase shift to the high impedance device, the reflectometer will present a matched load to the VNA. By introducing the reflectometer with an appropriately chosen reference device, we have changed the measurand from Γ_L to Γ_m and effectively moved nearer to the center of the Smith chart. Γ_L can be determined from the measurement of Γ_m by

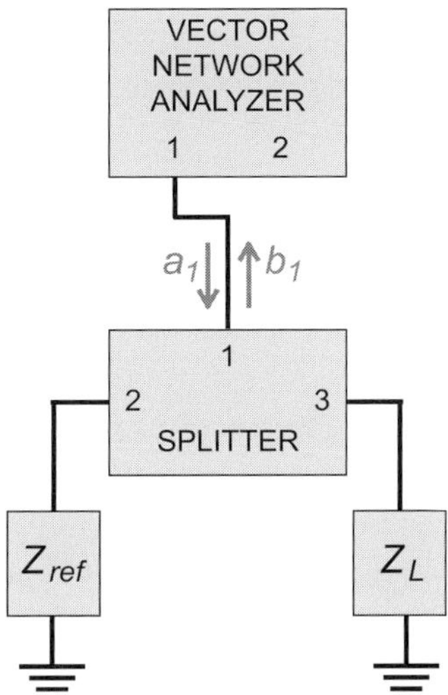

Figure 3.4. Schematic of a one-port reflectometer based on a power splitter. The reflectometer consists of a power splitter with a reference device (impedance Z_{ref}) and a device under test (impedance Z_L) terminating ports 2 and 3, respectively. The reflection coefficient of the reflectometer, $\Gamma_m = b_1/a_1$, is measured by connecting port 1 of the splitter to one port of a VNA. © 2008 IEEE. Adapted, with permission from A. Lewandowski, D. LeGolvan, T. M. Wallis, A. Imtiaz, and P. Kabos, 2008 72nd ARFTG Microwave Measurement Symposium (2008) pp. 45–49.

$$\Gamma_L = \frac{4\Gamma_m - (\Gamma_m + 1)\Gamma_{ref}}{\Gamma_m + 1}.$$

(3.10)

There are several important, underlying assumptions to note about this method. First, the reflection coefficient of the reference device, Γ_{ref}, must be known either through accurate modeling or, preferably, measurement. Calibration standards such as open circuits, short circuits, and offset short circuits are suitable reference devices, as they are often readily available and frequently measured in a calibration laboratory. Second, we assume that the incorporation of the extreme impedance DUT into the reflectometer does not alter Z_{ref} or Z_L. Third, we have assumed that the power splitter is ideal in Equations (3.8), (3.9), and (3.10). The approach can be improved by measurement of the scattering parameters of the power splitter, which can be used in place of the ideal scattering parameters in Equation (3.8) in the subsequent analysis.

3.4.2 Implementation with a Hybrid Coupler

An alternative implementation of a reflectometer for measuring a one-port device is shown in Fig. 3.5. This implementation incorporates a 180-degree, 3 dB hybrid

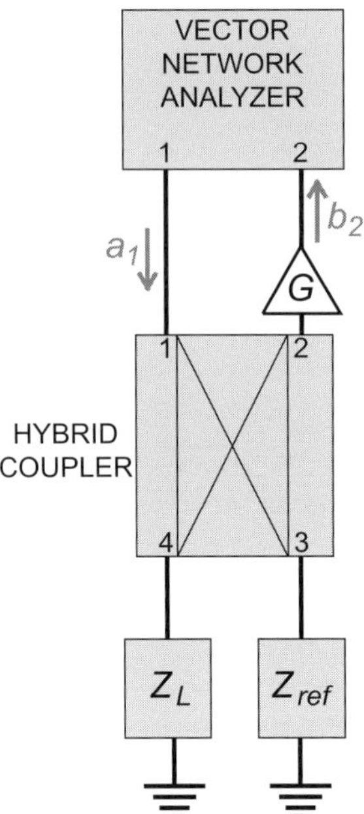

Figure 3.5.   Schematic of a one-port reflectometer based on a hybrid coupler. The measurement setup consists of a 180° 3 dB hybrid coupler with a reference device (impedance Z_{ref}) and a device under test (impedance Z_L) terminating ports 3 and 4 of the coupler, respectively. The transmitted scattering parameter, $S_{21} = b_2/a_1$, is measured by connecting ports 1 and 2 of the hybrid coupler to ports 1 and 2 of a VNA. The amplifier has gain G. © 2008 IEEE. Adapted, with permission from M. Randus and K. Hoffmann, *2008 72nd ARFTG Microwave Measurement Symposium* (2008) pp. 40–44.

coupler in place of a power splitter [5], [6]. Note that other passive, four-port elements, such as a 90-degree, 3 dB hybrid coupler, may be used in place of the 180-degree, 3 dB hybrid coupler, albeit with corresponding minor changes to the following analysis. The reference device and the high impedance DUT terminate port 3 and port 4 of the splitter, respectively. Port 1 of a VNA serves as the microwave signal source and is connected to port 1 of the hybrid coupler. The output at port 2 of the coupler is connected to an amplifier of gain G and then to port 2 of the VNA. Assuming that the hybrid coupler is ideal, the four-port scattering parameter matrix for the 180-degree hybrid coupler is given by

$$S^{hybrid} = \begin{bmatrix} 0 & 1/\sqrt{2} & 1/\sqrt{2} & 0 \\ 1/\sqrt{2} & 0 & 0 & -1/\sqrt{2} \\ 1/\sqrt{2} & 0 & 0 & 1/\sqrt{2} \\ 0 & -1/\sqrt{2} & 1/\sqrt{2} & 0 \end{bmatrix}. \tag{3.11}$$

It follows that the transmitted scattering parameter S_{21} measured by the VNA in Fig. 3.5 is

$$S_{21} = \frac{G}{2}\left(\Gamma_L - \Gamma_{ref}\right). \tag{3.12}$$

Here, as Γ_{ref} approaches Γ_L, the destructive interference within the reflectometer will be maximized and S_{21} will approach zero. The reflection coefficient of the device can be determined algebraically:

$$\Gamma_L = \frac{2S_{21}}{G} + \Gamma_{ref} \tag{3.13}$$

Comparing the two reflectometer methods, Equation (3.13) for the hybrid-coupler-based reflectometer provides a simpler form than Equation (3.10) and, in turn, more straightforward analysis. Further, the power-splitter-based reflectometer is based on a one-port reflection measurement while the hybrid-coupler-based reflectometer is based on a two-port transmission measurement. Note that although the hybrid-coupler-based measurement uses two ports of a VNA, the DUT is still a one-port device.

The underlying assumptions of the power-splitter-based reflectometer also apply to the hybrid-coupler-based reflectometer. It is also worthwhile to note that both Equation (3.9) and (3.12) involve simple linear combinations of the reflection coefficients of the high impedance DUT and a high impedance reference device. Physically, this is the result of direct interference of the signal reflected from the DUT with the signal reflected from the reference device. By satisfying the condition $\Gamma_{ref} = -\Gamma_L$, one insures that this interference is destructive. In general, such interometric approaches to extreme impedance measurements have been found to be effective and the implementation of such approaches is recommended when it can be implemented in the measurement platform. When interferometric approaches are unavailable or impractical, one must resort to analytical comparison of separate measurements, as we will discuss in Chapter 4.

3.5   Statistical Measurements

3.5.1   Use of Redundant Measurements in the Reflectometer Method

Before discussing an additional interferometric technique with active signal injection, it is useful to discuss some practical considerations, namely the use of statistical measurement techniques. In general, microwave measurement and calibration techniques require the measurement of a minimum number of reference devices in order to determine all of the unknown variables in the measurement process. For example, in order to calibrate a one-port, guided-wave system three known calibration standards must be measured in order to determine the three unknown calibration coefficients. When redundant measurements of reference devices are included in addition to the minimum number of required measurements, an overdetermined system results, which must be solved by one of many possible fitting or optimization techniques. Statistical techniques that make use of redundant measurements have proven to be an effective strategy for improving the statistical uncertainties associated with a given measurement process. Here, two relevant applications of statistical measurements are discussed: application to the reflectometer methods and application to the characterization of a three-port power splitter.

Consider the hybrid-coupler-based reflectometer method described in Section 3.4. From Equation (3.13), only one measurement is needed to find the reflection coefficient of the extreme impedance DUT, namely a measurement (S_{21}) made while a reference standard with a known reflection coefficient (Γ_{ref}) is connected to the reflectometer. Suppose instead of a single reference standard, we have a series of N reference standards indexed by $k = 1, 2... N$. Each reference standard has a known reflection coefficient, Γ^k_{ref}. The reference standards must all be chosen so that Γ^k_{ref} approaches Γ_L, ensuring complete (or nearly complete) destructive interference within the hybrid-coupler-based reflectometer. There will now be N measurements, one with each of the reference standards connected to the coupler, which will be designated S^k_{21}.

There are many strategies for obtaining Γ_L from the redundant measurements. One straightforward approach is to solve the set of de-coupled equations:

$$\Gamma^k_L = \frac{2S^k_{21}}{G} + \Gamma^k_{ref} \tag{3.14}$$

for N different values of Γ^k_L. Then, a value of Γ_L can be obtained from the average of the Γ^k_L values. If the uncertainties in Γ^k_{ref} vary significantly from standard to standard, a weighted average of the Γ^k_L values may be more appropriate. Alternatively, a cost function K may be defined. One possible implementation of K is

$$K = \sum_{k=0}^{N} \left| S^k_{21} - \frac{G}{2} \left(\Gamma_L - \Gamma^k_{ref} \right) \right|. \tag{3.15}$$

An optimization algorithm may then be used to find a value of Γ_L that minimizes K. A variety of automated optimization approaches exist, many of which can be easily implemented by use of commercial software. Once again, weighting coefficients may be added to the cost function in proportion to the uncertainties in the values of Γ^k_{ref}.

3.5.2   Use of Redundant Measurements to Characterize a Power Splitter

Another relevant application of statistical measurements is the characterization of a three-port power splitter. This approach is presented here as an additional example of a statistical measurement approach and as a method to improve the reflectometer method. Recall that the accuracy of the power-splitter-based reflectometer method may be improved by using the measured scattering parameters of the power splitter in place of Equation (3.8). Because of the wide availability of conventional two-port VNAs, several approaches have been developed that use a two-port VNA to characterize a three-port device [7], [8]. The technique in reference [7] is extendable to a statistical measurement technique that makes use of redundant measurements [9].

In order to perform the measurement, the two ports of a calibrated VNA are connected to the output terminals of the power splitter (ports 2 and 3), as shown in Fig. 3.6. Note that an adapter with known scattering parameters may be required in order to provide compatibility with an insertable two-port calibration. Known one-port reference standards Γ^k_{ref} are then connected to the input port of the power splitter (port 1). For each standard, a set of four scattering parameters S^M_{ij} are measured (i = 2, 3 and j = 2, 3). The measured S^M_{ij} are related to the scattering parameters of the power splitter, S_{ij}, by:

$$S^M_{ij} = S_{ij} + \frac{S_{1j}S_{i1}\Gamma^k_{ref}}{1 - S_{11}\Gamma^k_{ref}}. \tag{3.16}$$

This system of equations has eight unknowns: S_{11}, S_{22}, S_{33}, S_{23}, S_{32}, $S_{13}S_{31}$, $S_{12}S_{21}$, and $S_{12}S_{31}$ (note that $S_{13}S_{21}$ can be found from the other unknowns). Since there are four equations for each set of measurements made with a given one-port reference standard, the system will be overdetermined if more than two reference standards are used.

One approach to solving this system of overdetermined equations begins with multiplying Equation (3.16) by a factor of $1 - S_{11}\Gamma^k_{ref}$ in order to linearize the equations. The system of linear equations may then be solved by the method of least squares or another form of regression analysis. In turn, these solutions to the linearized problem may serve as initial guesses, or "seeds," for more sophisticated optimization routines.

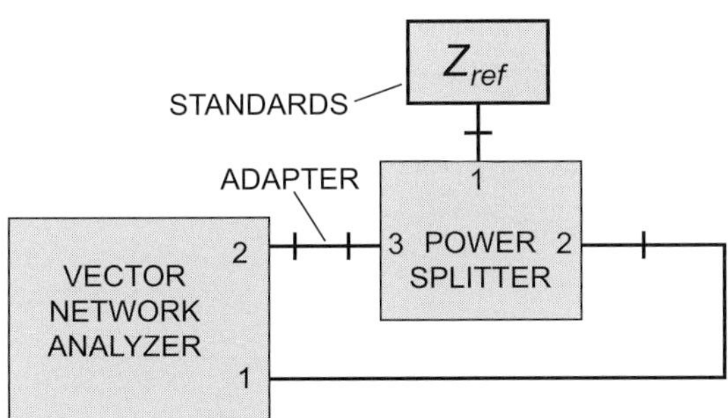

Figure 3.6. Schematic of power splitter measurement.
The output terminals of the power splitter, ports 2 and 3, are connected to a calibrated VNA. A series of reference standards are connected to the input port of the splitter, port 1, during measurement. An adapter is inserted to provide compatibility with an insertable two-port calibration. © 2008 IEEE. Adapted, with permission from T. M. Wallis and A. Lewandowski, *2008 72nd ARFTG Microwave Measurement Symposium* (2008) pp. 50–53.

3.6   Interferometer with Active Signal Injection

So far, we have discussed reflectometers based on passive components, in which the signal reflected from the extreme impedance DUT interfered destructively with the signal reflected from a known reference impedance. Alternatively, an interferometer can be implemented with an actively injected signal in place of the signal reflected from a known impedance [10]. This all-electronic approach eliminates the requirement for a mechanical reference impedance (or set of mechanical reference impedances).

A schematic of the interferometer with active signal injection is shown in Fig. 3.7. A comparison of this measurement platform with the coupler-based reflectometer shown in Fig. 3.5 reveals that both techniques use a hybrid coupler to generate interference between signals and a network analyzer to measure scattering parameters.

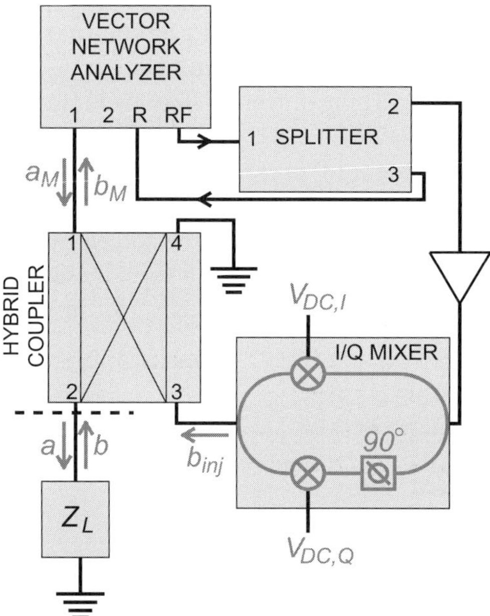

Figure 3.7.   Schematic of an interferometer with active signal injection.
The interferometer comprises a VNA, 180° 3 dB hybrid coupler, a power splitter, and an I/Q mixer. The signal is taken from the VNA source (labeled RF) and, after emerging from the power splitter, fed back to the VNA reference channel (labeled R). The measurement is made at port 1 of the VNA. A device under test (extreme impedance Z_L) terminates port 2 of the coupler. The injected signal (complex amplitude b_{inj}) is output from the I/Q mixer and interferes destructively with the signal reflected from the device under test (amplitude b). DC bias voltages $V_{DC,I}$ and $V_{DC,Q}$ control the amplitude and phase components of b_{inj}. The reflection coefficient of the reflectometer, $\Gamma_M = b_M/a_M$, is measured. © 2015 IEEE. Adapted, with permission from G. Vlachogiannakis, H. T. Shivamurthy, M. A. Del Pino, and M. Spirito, *2015 IEEE MTT-S International Microwave Symposium (IMS)* (2015) pp. 1–4.

The VNA reference channel (sometimes referred to as the "R" channel) provides the input signal for a power splitter. An amplified signal from port 2 of the power splitter provides the local oscillator (LO) drive signal for the I/Q mixer. Two DC voltages, $V_{DC,I}$ and $V_{DC,Q}$, are used to adjust the signal phase and amplitude of the output of the I/Q mixer, which is subsequently injected into port 3 of the hybrid coupler. For an ideal coupler, the complex amplitude of the injected signal, b_{inj}, is given by:

$$b_M = \frac{1}{2}(b - b_{inj})$$ (3.17)

where b is the complex amplitude of the signal reflected from the extreme imped-ance and b_M is the complex amplitude of the reflected wave measured by the VNA. Since b_{inj} can be tuned by the DC voltages $V_{DC,I}$ and $V_{DC,Q}$, a value of b_{inj} can be selected for which b_M approaches zero and complete (or nearly complete) destruc-tive interference occurs. Thus, the measured value $\Gamma_M = b_M/a_M$ will also approach zero, corresponding to a measurement near the center of the Smith chart. For a nonideal hybrid coupler, Equation (3.17) must be modified to account for nonideal coupling and loss [10], but the general conclusion remains: the injected signal may be tuned so the destructive interference occurs between it and the signal reflected from the extreme impedance DUT.

In order to extract the impedance of the extreme impedance device from the measured Γ_M, the system must be calibrated [11]. The first step is to carry out a one-port calibration at the DUT reference plane, which is shown as a dashed black line in Fig. 3.7. This one-port calibration is carried out with the injection signal turned off. A one-port calibration can be carried out by use of three stand-ards, such as a short, an open, and a matched load, for example. As described in Chapter 2, the one-port calibration determines three error terms: the directivity (e_{00}), the source match (e_{11}), and the reflection tracking ($e_{10} e_{01}$). Once these error terms are known from the calibration, the reflection coefficient at the reference plane Γ can be determined:

$$\Gamma = \frac{\Gamma_M - e_{00}}{e_{10}e_{01} + e_{11}(\Gamma_M - e_{00})}.$$ (3.18)

The second step is to optimize the amplitude and phase of the injected signal. The objective is to tune the injected signal such that the calibrated reflection coefficient of a reference extreme impedance device in the presence of the injected signal, Γ_{ref}, is close to the calibrated reflection coefficient of a matched load in the absence of the injected signal, $\Gamma_{matched}$. Keep in mind that Γ_{ref} and $\Gamma_{matched}$ are both complex quanti-ties. In order to obtain $\Gamma_{matched}$, a matched load is connected at the reference plane and Γ_M is measured with the injection signal turned off. $\Gamma_{matched}$ is then found via Equation (3.18). Subsequently, Γ_{ref} is obtained in a similar fashion, but now with a known extreme impedance connected at the reference plane and the injection signal turned on. The amplitude and phase of the injection signal is then tuned by adjust-ing $V_{DC,I}$ and $V_{DC,Q}$ until Γ_{ref} is nominally close to $\Gamma_{matched}$.

With the one-port calibration complete and the injection signal optimized, the extreme impedance DUT can now be measured. The extreme impedance DUT is connected at the reference plane and Γ_M is measured with the injection signal turned on and optimized. The calibrated reflection coefficient of the DUT Γ is once again found via Equation (3.18). Since the injection signal was optimized with the reference impedance Z_{ref} connected at the reference plane, the extreme impedance, the DUT impedance Z_L can be found via [11]:

$$Z_L = Z_{ref}\left(\frac{1-|\Gamma|}{1+|\Gamma|}\right). \tag{3.19}$$

As in the case of the reflectometer, the reference impedance Z_{ref} must once again be known, either through modeling or measurement. Once again, calibration standards such as open circuits, short circuits, and offset short circuits may serve as suitable reference devices. The design and fabrication of practical extreme impedance verification devices and standards is an area of ongoing investigation. For example, devices that integrate waveguides operating below cutoff with high-resistance shunts have been proposed as extremely high impedance standards [12].

References

[1] C. Rutherglen and P. Burke, "Nanoelectromagnetics: Circuit and Electromagnetic Properties of Carbon Nanotubes," *Small* 5 (2009) pp. 884–906.

[2] K. v. Klitzing, G. Dorda, and M. Pepper, "New Method for High-Accuracy Determination of the Fine-Structure Constant Based on Quantized Hall Resistance," *Physical Review Letters* 45 (1980) pp. 494–497.

[3] H. Jasik, *Antenna Engineering Handbook,* 1st edn (McGraw-Hill, 1961). Section 31.7.

[4] A. Lewandowski, D. LeGolvan, T. M. Wallis, A. Imtiaz, and P. Kabos, "Wideband Measurement of Extreme Impedances with a Multistate Reflectometer," *2008 72nd ARFTG Microwave Measurement Symposium* (2008) pp. 45–49.

[5] M. Randus and K. Hoffmann, "A Simple Method for Extreme Impedances Measurement – Experimental Testing," *2008 72nd ARFTG Microwave Measurement Symposium* (2008) pp. 40–44.

[6] M. Randus and K. Hoffmann, "A Method for Direct Impedance Measurement in Microwave and Millimeter-Wave Bands," *IEEE Transactions on Microwave Theory and Techniques* 59 (2011) pp. 2123–2130.

[7] M. Davidovitz, "Reconstruction of the S-matrix for a 3-port Using Measurements at Only Two Ports," *IEEE Microwave and Guided Wave Letters* 5 (1995) pp. 349–350.

[8] J. C. Tippet and R. A. Speciale, "A Rigorous Technique for Measuring the Scattering Matrix of a Multiport Device with a 2-Port Network Analyzer," *IEEE Transactions on Microwave Theory and Techniques* 30 (1982) pp. 661–666.

[9] T. M. Wallis and A. Lewandowski, "Statistical Measurement Techniques for Equivalent Source Mismatch of 1.85 mm Power Splitter," *2008 72nd ARFTG Microwave Measurement Symposium* (2008) pp. 50–53.

[10] G. Vlachogiannakis, H. T. Shivamurthy, M. A. Del Pino, and M. Spirito, "An I/Q-Mixer

Steering Interferometric Technique for High-Sensitivity Measurement of Extreme Impedances," *2015 IEEE MTT-S International Microwave Symposium (IMS)* (2015) pp. 1–4.

[11] F. Mubarak, R. Romano, and M. Spirito, "Evaluation and Modeling of Measurement Resolution of a Vector Network Analyzer for Extreme Impedance Measurements," *2015 86th ARFTG Microwave Measurement Symposium* (2015) pp. 1–4.

[12] M. Haase and K. Hoffmann, "Calibration/Verification Standards for Measurement of Extremely High Impedances," *2015 86th ARFTG Microwave Measurement Symposium* (2015) pp. 1–4.

4 On-Wafer Measurements of RF Nanoelectronic Devices

4.1 Broadband Characterization of RF Nanoelectronic Devices

The preceding chapters have introduced the core concepts and techniques of microwave measurements, in general, and techniques for microwave measurements of extreme impedance devices, in particular. Here, we narrow the focus further to on-wafer, microwave measurements of RF nanoelectronic devices. In this chapter, the term "nanoelectronic devices" refers to electronic, charge-based devices that incorporate nanoscale elements or nanomaterials, such as CNTs, semiconducting nanowires, or graphene. For now, discussion will be further limited to characterization of passive devices (characterization of active devices will be discussed in Chapter 10). In a device development environment, a priori knowledge of the electronic properties of such nanoscale building blocks may be limited. Further, physical properties of nanoscale material systems may vary strongly from building block to building block and, consequently, performance may vary strongly from device to device. As a result, fundamental device properties such as device impedance or cutoff frequency may be unknown and broadband measurements will be required to determine them.

The development of such broadband metrology is but one piece of a comprehensive measurement framework of RF nanoelectronic devices. The first element of such a framework is nanofabrication. It is only through considerable advances in the fabrication of nanomaterials and devices over recent years that RF nanoelectronic devices have become realizable. In the context of characterizing such devices at RF, the integration of nanoscale building blocks into RF-compatible structures such as CPWs is a necessary step. While a comprehensive discussion of nanofabrication is outside the scope of this book, clear understanding of nanofabrication techniques is an asset in the design and execution of nanoelectronic device metrology. The second aspect of the measurement framework is the suite of specific measurement techniques for RF nanoelectronic devices, as described in this chapter and the literature referenced herein. Third, appropriate circuit models need to be developed in order to extract physical, material, and electrical parameters from broadband measurements. Finally, the measurement framework must be validated by comparison to modeling and simulation. In Chapter 5, comprehensive modeling, parameter extraction, and finite-element simulation will be discussed, completing the measurement framework for broadband characterization of RF nanoelectronic devices. Within this measurement framework, the following objectives are realized: (a) calibrated measurements of the frequency-dependent scattering parameters and impedance of the nanoelectronic device are obtained; (b) the intrinsic properties of the nanoelectronic device are de-embedded from contact properties and other parasitic effects, including the stray capacitance; (c) circuit models that describe the nanoscale element and its contacts are developed and fully validated; (d) estimates of quantitative values of circuit parameters are extracted from the measurements. By realizing these objectives, the measurement framework may inform development and design of emerging RF nanoelectronics applications.

The focus of the present chapter is the measurement of de-embedded, complex scattering parameters of a two-port, passive nanoelectronic device over a broad bandwidth, typically from tens of megahertz to tens of gigahertz. Presently, most nanoelectronic devices are implemented on-wafer. The on-wafer measurement environment presents new challenges that are not present in the guided-wave measurement environment, including radiative loss and parasitic coupling between measurement probes. For RF nanoelectronic devices, these challenges are further augmented by the inherent impedance mismatch with commercial test equipment as well as design and fabrication challenges associated with integration of nanoscale building blocks into on-wafer, RF host structures.

In order to address these challenges, several strategies have been developed. Because of the inherent challenges of the on-wafer measurement environment, the user must observe best practices in order to obtain meaningful, on-wafer measurements. A number of such practical considerations are reviewed in the following section. To address the specific measurement problems presented by RF nanoelectronic devices, several different approaches have been developed, including the on-wafer application of the techniques described in the previous chapter on extreme impedance measurements. Below, three additional approaches to broadband, on-wafer measurements of nanoelectronic devices are discussed. The first approach is based on the integration of the nanoscale element into a Wheatstone bridge structure. The second approach augments traditional on-wafer calibration with the measurement of an additional, "empty" reference device. The third approach is based on fabrication of many nanostructures in parallel in order to produce an impedance-matched on-wafer device.

4.2   Practical Considerations for On-Wafer Measurements

In order for RF nanoelectronic applications to become widely deployed in the near future, they must be compatible with existing electronics technology. More specifically, they must be compatible with the engineering and fabrication of semiconductor-based CMOS devices. As a result, many prototype RF nanoelectronic devices are presently implemented on-wafer. A variety of fabrication techniques have been developed to integrate nanostructures such as nanowires, nanotubes, and two-dimensional materials into processes that are compatible with standard lithographic patterning of planar electronics. Though there are examples of nanostructures implemented in nonplanar geometries, such as CNT composites for shielding applications, our discussion will focus upon on-wafer measurements of the complex scattering parameters of RF nanoelectronic devices. In Chapter 2, we reviewed the core concepts and underlying theory of calibration and de-embedding, in general, and the on-wafer, multiline thru-reflect-line (TRL) technique, in particular. Here, we review practical aspects of a typical on-wafer test platform, from the network analyzer and cables, through the probes, and onto the DUT itself. For any application, the on-wafer measurement environment has historically been a challenging one [1], requiring appropriate methodology, sound theory [2], as well as a skilled, practiced user. In the case of nanoelectronic devices, these challenges are further compounded, particularly with respect to measurement sensitivity and repeatability.

A schematic of a typical on-wafer test platform is shown in Fig. 4.1. A stable measurement environment, free from both significant mechanical noise as well as fluctuations in temperature and humidity, is a necessity. The VNA is the heart of any test platform for scattering parameter measurements of coaxial, waveguide, or on-wafer DUTs. Broadband cables provide a signal path from the VNA to the probes, which in turn connect to the on-wafer DUT. As the probes will need to be repositioned throughout the measurement, flexible cables are usually required. Cables should be chosen to minimize temperature-, humidity-, and flexure-related changes in phase. In order to verify that the VNA and cables are in good working order, it is best practice to maintain a coaxial verification kit that consists of multiple, well-characterized DUTs that benchmark the measurement platform in different ways. A typical two-port verification kit might include a two-port attenuator, a two-port mismatch standard (also known as a Beatty standard), a one-port matched load, and a one-port flat short circuit. Before connecting the on-wafer contact probes, measurement of each of these verification standards at the coaxial reference planes with a calibrated system will establish that there are no systematic problems with the VNA or cabling.

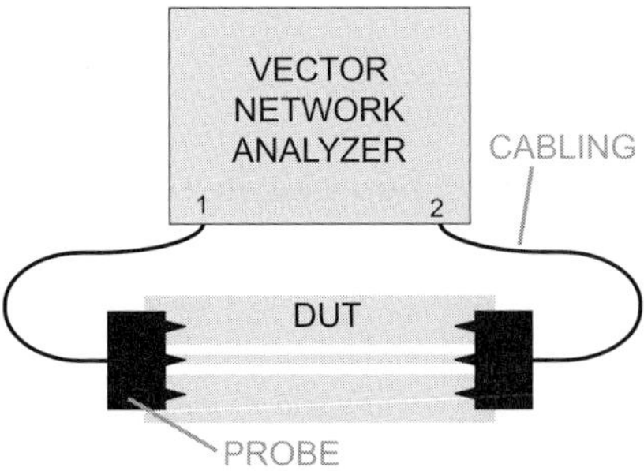

Figure 4.1.   Schematic of an on-wafer test platform.
Key components include a VNA, broadband cabling, and contact probes. The device under test (DUT) is illustrated here as a coplanar waveguide.

Before discussing on-wafer microwave probes, which effectively transform a guided-wave measurement platform into an on-wafer platform, it is useful to briefly review the properties of microwave and RF coaxial connectors, as most of the off-wafer interfaces in the test platform will be coaxial. In general, as the dimensions of a coaxial connector decrease, its operational frequency range will increase. Specific examples of precision connectors (and their nominal frequency cutoff) include 3.5 mm (33 GHz), 2.4 mm (50 GHz), and 1.0 mm (110 GHz). The increased frequency range of smaller connectors comes with an important trade-off: smaller connectors are mechanically more fragile and require delicate handling by skilled, experienced users to avoid rapid degradation or damage. Use of an appropriate torque wrench is an absolute requirement for repeatable measurements that do not damage connectors. Additionally, consistent use of the same connector type throughout the off-wafer measurement platform will improve the quality of measurements and the ease of de-embedding. Excessive use of intertype adapters introduces extra interfaces and unwanted impedance mismatches along the signal path. Finally, the connector life and measurement repeatability will both be maximized by regular inspection and cleaning of connectors.

Microwave probes provide a signal path from the cabling to the on-wafer DUT. On one side, a probe connects to the connectorized environment via a standard coaxial or waveguide interface. On the other side, the probe has sharp metallic points that are electrically connected to the ground and signal lines. Signal transmission to the on-wafer environment is achieved by bringing these points into direct mechanical contact with the signal and ground lines on a planar guided-wave structure, such as a CPW. Note that the probe points are generally designed to skate laterally along the DUT as they make mechanical contact. Typical lateral skate distances are on the order of ten to fifty micrometers. For CPW measurements, a ground-signal-ground (GSG) configuration of the probe points is required. Other configurations, such as ground-signal (GS), are available to accommodate alternative device geometries. The probe pitch is determined by the distance between the ground and signal connections, with typical values of the pitch in the range from tens of micrometers to millimeters. Clearly, the on-wafer device geometry must be designed to match available probes.

Commercial on-wafer probe stations provide mechanical support for the test platform and motion control in order to control the position and orientation of components. The DUT wafer is generally held in place on the smooth surface of a vacuum chuck. Probes are generally mounted on a three-axis (XYZ) translation stage as well as a goniometer. Further, the DUT wafer is usually also on a translation stage, so that once the probe orientation is optimized, access to different DUTs can be achieved by moving the wafer, with minimal repositioning of the probes. Given the mechanical dimensions, an optical microscope is necessary for relative positioning of both DUTs and probes.

Microwave measurements are particularly sensitive to the quality and repeatability of the mechanical interface between a microwave probe and the DUT. Two issues are critical: planarization and repeatable lateral positioning. If the points of the probe are not aligned in the plane of the device, the quality of contacts can differ between points, and in the worst case, one or more of the points may fail to contact the DUT surface, leading to stray capacitance and measurement errors. In practice, probe planarization can be checked by contacting a metal surface on the DUT wafer and examining the mechanical scratches in the metal surface that result from the probe points skating on the surface. Scratches of equal length and depth indicate proper planarization of the probe. In order to achieve repeatable measurements, it is also necessary to position the probes at the same position on all comparable DUTs for every measurement. In particular, the distance that the probes skate after contact must be consistent. One strategy to achieve this is to use an automated, programmable probe positioner. When automated positioning is not available, appropriate fiduciary marks can be incorporated into the DUT design, as shown in Fig. 4.2.

With all of the elements of the test platform in place, including the VNA, cabling, and microwave probes, informed choices can be made with respect to the design of the DUT wafer. Ideally, any necessary calibration structures will be fabricated on the same substrate as the DUTs. While calibration methods exist for situations where the calibration structures are on separate substrates [3], the introduction of additional uncertainty by use of such methods presents an unwanted complication

Figure 4.2.   Photograph of a CPW.
Note the alignment marks at the side of the structure, which are used to improve the repeatability of the probe contact position and the distance that the probes skate during repeated measurements.
Photograph by N. Orloff, NIST.

to sensitive measurements of RF nanoelectronics. Repeated measurements of both calibration standards and DUTs present an additional strategy for reducing statistical uncertainties. If possible, the implementation of a verification standard, such as an on-wafer mismatch standard, on the same wafer is also highly desirable.

4.3 Wheatstone Bridge Approach

4.3.1 The Wheatstone Bridge

In Chapter 3, interferometric methods were introduced to measure the scattering parameters of extreme impedance DUTs. While implementation of these methods in an on-wafer environment is possible in principle, experimental demonstration of such an on-wafer implementation is challenging to achieve. Note that a fully on-wafer implementation of such a method would require integration of the DUT as well as a well-known reference impedance Z_{ref} with an on-wafer power splitter or on-wafer hybrid coupler. Alternately, a connectorized splitter or hybrid coupler could be used, with the DUT and the reference impedance device (or multiple reference impedance devices) remaining on the wafer. A set of calibration structures would also be needed, ideally on the same wafer substrate.

A host device architecture, based on a Wheatstone bridge, provides an alternative approach to on-wafer measurements of high-impedance nanoelectronic systems, such as CNTs [4], [5]. Like the interferometric methods, the bridge-based method effectively reduces the impedance mismatch between the nanoelectronic device and the test equipment. This method reduces measurement error, though it requires that the nanoelectronic device be integrated into a specific test structure and enough wafer area must be available for the fabrication of the bridge structure as well as several bridge-based calibration structures. Before describing this method in detail, we will review the basic concepts of a Wheatstone bridge.

A schematic of a Wheatstone bridge is shown in Fig. 4.3. The bridge consists of a detector and four impedances, one in each of the four branches of the bridge: Z_1, Z_2, Z_3, and Z_L. The circuit is driven by an RF oscillator. One useful feature of this

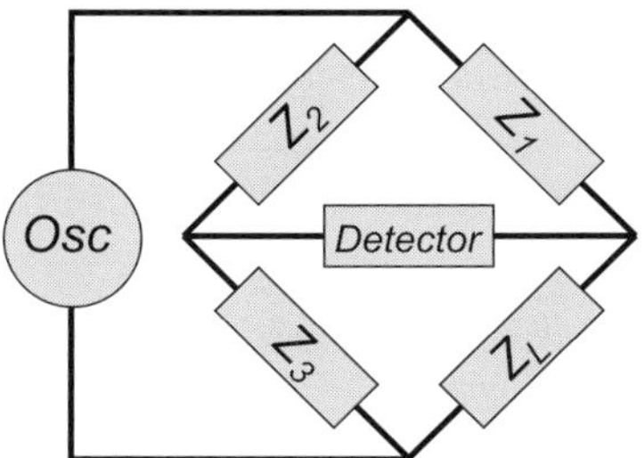

Figure 4.3.   Schematic of a Wheatstone bridge.
The bridge consists of a detector and four impedances: Z_1, Z_2, Z_3, and Z_L. The circuit is driven by an RF oscillator (*Osc*).

structure is that it can be used to measure an unknown impedance in one arm of the bridge, provided that the other three impedances are known. In general, the ratio of the detector signal, V_{det}, to the input signal V_{osc}, is given by

$$\frac{V_{det}}{V_{osc}} = \left(\frac{Z_3}{Z_2 + Z_3} - \frac{Z_L}{Z_1 + Z_L} \right). \tag{4.1}$$

If the signal at the detector is zero, the bridge is said to be balanced and the unknown impedance is given by

$$Z_L = \frac{Z_1}{Z_2} Z_3. \tag{4.2}$$

Ideally, the general strategy for impedance measurement with a Wheatstone bridge is implemented with a variable impedance in one of the arms of the bridge. Then, the variable impedance is tuned so that the signal at the detector is zero and the unknown impedance can be determined from Equation (4.2). In an on-wafer environment, it is much simpler to implement a fixed impedance than a variable impedance. As a result, an alternative measurement strategy will be developed in the following subsection.

4.3.2   Bridge-Based Measurements of a Nanoelectronic Device

The Wheatstone bridge is the basis for an elegant approach to measurement of an individual nanofiber such as a nanowire or CNT [4], [5]. In order to extend the Wheatstone bridge-based approach to an individual nanowire, nanotube, or other nanoscale building block, the nanoscale element must be integrated into an on-wafer Wheatstone bridge. Schematics and a signal flow diagram of such a measurement structure are shown in Fig. 4.4. As shown in Fig. 4.4(a), the nanoelectronic DUT is connected across one branch of the bridge, while resistors are connected across each of the other branches of the bridge. For the measurement of a nanoscale element with an impedance on the order of the resistance quantum (12.9 kΩ) such as an individual, single-wall CNT, the bridge resistance R_{br} should be on the order of 1 kΩ. This reduces the impedance mismatch between the bridge structure and commercial test equipment. On-wafer resistors can be fabricated by use of photolithographic patterning of thin metal films such as NiCr or PdAu.

The Wheatstone bridge structure acts as an impedance transformer. To see this, it is useful to insert the values for the bridge impedances shown in Fig. 4.4(a) into Equation (4.1):

$$\frac{V_{det}}{V_{osc}} = \frac{1}{2} - \frac{Z_L}{R_{br} + Z_L}. \tag{4.3}$$

If we define the measured reflection coefficient as $\Gamma_M = V_{det}/V_{osc}$ and define the reflection coefficient of the unknown load with respect to a reference impedance R_{br} as Γ_L, then Equation (4.3) can be rewritten as

$$\Gamma_M = \frac{1}{2}\left(\frac{R_{br} - Z_L}{R_{br} + Z_L}\right) = -\frac{1}{2}\Gamma_L \tag{4.4}$$

From Equation (4.4), we see that an ideal Wheatstone bridge acts as an impedance transformer. In practice, the actual bridge structure will be nonideal, but it can be represented with a general, bilinear transform,

$$\Gamma_M = e_{00} + \frac{e_{01}e_{10}\Gamma_L}{1 - e_{11}\Gamma_L}, \tag{4.5}$$

where e_{00} is directivity, the product $e_{10}e_{01}$ is tracking, and e_{11} is the port match, as originally introduced in Chapter 2.

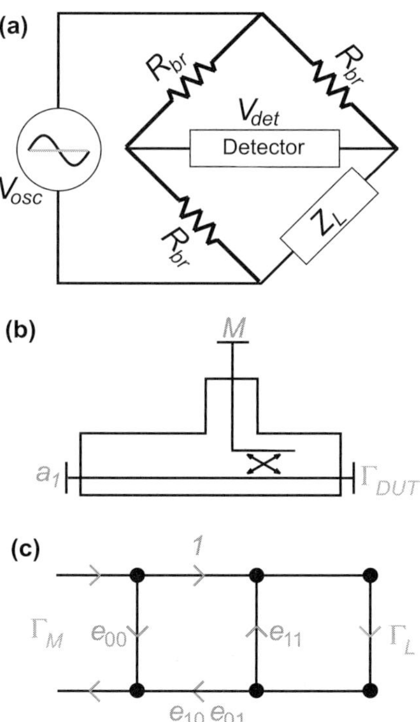

Figure 4.4. Wheatstone bridge for measuring RF nanoelectronic devices. (a) The bridge has resistors R_{br} in three of the branches and the nanoelectronic device with impedance Z_L in the fourth branch. The voltage supplied by the oscillator is V_{osc} and the voltage measured across the detector is V_{det}. (b) A schematic of a directional coupler with coupling factor α. (c) A signal flow graph representing a three-term error model for a one-port network. The measured reflection coefficient is Γ_M, the device reflection coefficient is Γ_L, and the error terms are e_{00}, $e_{10}e_{01}$, and e_{11}. Adapted from L. Nougaret, G. Dambrine, S. Lepillett, H. Happy, N. Chimot, and J.-P. Bourgoin, *Applied Physics Letters* 96 (2010) art. no. 042109, with permission from AIP Publishing.

Equation (4.5) can alternatively be introduced by considering Equation (4.3) in several special cases [4]. When the bridge is balanced, $Z_L = R_{br}$ and the detector signal is zero. When a short circuit is present at the unknown arm of the bridge, $Z_L = 0$ and $V_{det}/V_{osc} = 1/2$. Finally, when an open circuit is present at the unknown arm of the bridge, $1/Z_L = 0$ and $V_{det}/V_{osc} = -1/2$. These special cases suggest that the bridge structure acts as a directional coupler. An ideal direction coupler is shown in Fig. 4.4(b). Suppose that the unknown load of impedance Z_L and corresponding reflection coefficient Γ_L is connected to an ideal directional coupler. The measured output signal M is related to the incident signal a_1 and the coupling factor α by:

$$M = a_1 \Gamma_L \alpha. \tag{4.6}$$

Again, the actual bridge structure will be nonideal, but it can be represented as an equivalent one-port network. The ideal coupler shown in Fig. 4.4(b) can be represented by the signal flow graph shown in Fig. 4.4(c), which is a three-term error model for a one-port network. Following the methods summarized in Chapter 2, analysis of the signal flow graph leads directly to Equation (4.5).

As with other one-port calibrations, three standards must be measured in order to determine the three error terms. For the Wheatstone bridge-based technique, the three standards are a balanced bridge ($Z_L = Z_{br}$), an open circuit ($1/Z_L = 0$), and a short circuit ($Z_L = 0$), as shown in Fig. 4.5(a). Note that we have used the bridge-balancing impedance Z_{br} in place of the simple resistance R_{br}, as the nonideal fabricated structure may have some nonzero reactance. Values of the error terms e_{00}, $e_{10}e_{01}$ and e_{11} can be determined from measurements of these standards by use of Equation (4.5). This technique utilizes a differential measurement to measure each standard as well as any DUT. With a ground-signal-ground probe connected to the device, two sets of scattering parameters are measured. Two scattering parameters, S_{p1} and S_{p2}, are measured with a high impedance probe connected to the bridge, first at point $p1$ and then at point $p2$, as labeled on the balanced bridge structure in Fig. 4.5(a). A high impedance probe is used here to minimize its perturbation of the bridge circuit, but the signal from the high impedance probe must be amplified to compensate for the signal attenuation in the probe. The differential measurement is calculated by:

$$\Gamma_M = S_{p1} - S_{p2}. \tag{4.7}$$

An example of a single-wall CNT device that has been measured by this approach [4] is shown in Fig. 4.5(b). In this device, an individual, single-wall CNT supported by a silicon oxide layer serves as the nanoscale element in the Wheatstone bridge. Once the CNT device has been measured by use of the differential measurement described previously, the impedance of the CNT, Z_{CNT}, may be determined from:

$$Z_{CNT} = Z_{br} \frac{\Gamma_M + 1}{\Gamma_M - 1}, \tag{4.8}$$

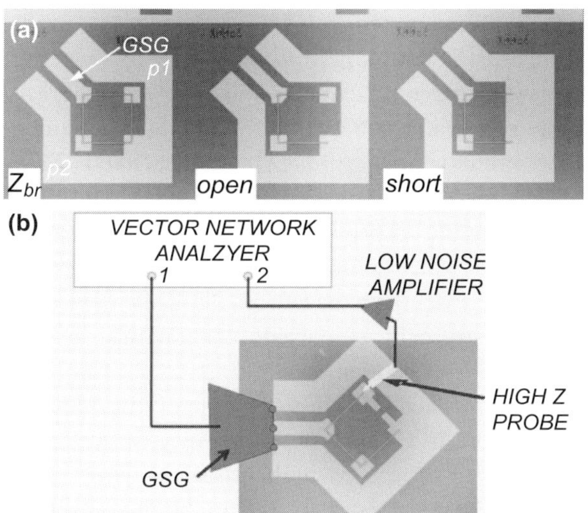

Figure 4.5. Wheatstone bridge standards and CNT device.
(a) Three standards used in the Wheatstone bridge approach. From left to right: a balanced bridge ($Z_L = Z_{br}$), an open circuit ($1/Z_L = 0$), and a short circuit ($Z_L = 0$). The contact points for the high impedance probe during the differential measurement are indicated in the image of the balanced bridge standard as *p1* and *p2*. The center conductor contact point for the ground-signal-ground probe is indicated by *GSG*. (b) Image of the single-wall CNT device with a schematic of the differential measurement setup.
Adapted from L. Nougaret, G. Dambrine, S. Lepillett, H. Happy, N. Chimot, and J.-P. Bourgoin, *Applied Physics Letters* 96 (2010) art. no. 042109, with permission from AIP Publishing.

provided that the bridge-balancing impedance Z_{br} is known. The impedance Z_{CNT} represents the total impedance of the entire branch of the bridge, including the CNT itself, the contacts to the CNT, and the electrical leads. In order to separate the contributions of the CNT from those of the contacts and the leads, further measurements as well as modeling and simulation are required.

4.4 Empty Device Approach

Many on-wafer measurement approaches, including the bridge-based method discussed earlier, require the integration of the nanoscale DUT into a specialized structure that enables the approaches to both measurement and calibration. As a matter of practice, it may not be feasible or efficient to integrate the nanoscale DUT into such a structure. Moreover, if the objective is to develop practical devices such as transistors and amplifiers in a device package, the design, fabrication, and optimization of a separate measurement host structure may be inconvenient. In this case, an alternative calibration approach must be developed to de-embed nano-scale devices from the measurement platform while simultaneously accounting for parasitic reactance in the extreme impedance device. Particularly in the early stages of device development, parasitic coupling in the device such as stray capacitances, may dominate the measured response. Here, we develop a calibration approach that augments established on-wafer calibration algorithms with the measurement of empty devices to account for and estimate parasitic effects. Measurement of an empty device, which is identical to the nanoscale DUT, except for the exclusion of the nanoscale building block, provides a reasonable approach to estimating the magnitude of stray capacitive coupling. The measurement of empty devices in order to estimate stray capacitive effects in nanometer-scale devices is reminiscent of similar approaches that historically were used to deal with stray capacitive coupling in microelectronic devices on Si substrates [6].

Here, we will base our approach on the extension of the multiline TRL calibration method [7], introduced in Chapter 2. The empty-device approach is more broadly applicable and may also be used with other on-wafer calibration methods. Note that emerging calibration approaches may be more suitable for a given nanoscale DUT or other extreme impedance device, but established methods such as TRL represent a practical starting point. The robustness of established on-wafer calibration methods has been confirmed through interlaboratory comparisons as well as the development of reference CPW calibration artifacts by measurement standards laboratories. By use of such reference samples, one may compare calibrated on-wafer measurements to those performed at the measurement standards laboratories and thus gain confidence in calibration methods and measurements. Further validation of this particular approach through simulation and modeling, as well as in-depth discussion of the limitations of this approach are included in the next chapter.

Consider an individual nanowire that has been integrated into an on-wafer, RF-compatible device. (We present a nanowire as the nanoscale device element in this discussion, but it is straightforward to extend it to any nanoscale building block.) One relatively straightforward option is to integrate the nanowire into a CPW [8]. The intrinsic properties of the contacted nanowire are de-embedded from the parasitic, stray capacitive effects as follows. The calibrated scattering parameters of the nanowire device S^{dev}_{ij} ($i = 1,2$; $j = 1,2$ where 1 and 2 correspond to ports 1 and 2, respectively) are obtained by use of the multiline TRL technique. Then, the calibrated scattering parameters of an empty, nanowire-free device S^{gap}_{ij} are also measured with the multiline TRL technique. As part of the calibration procedure, the reference planes may be translated as close to the nanowire as needed in order to remove the response of the host structure from the calibrated measurements. The objective is to isolate the scattering parameters of the nanowire and contacts, S^{nw}_{ij}. The response of the nanowire device is modeled as the parasitic coupling across the gap in parallel with the response of the contacted nanowire while the transmission through the empty device is modeled as due purely to parasitic coupling, as illustrated in Fig. 4.6. If the scattering parameters are transformed to an admittance parameter representation, this can be expressed in a simple algebraic form:

$$Y_{ij}^{nw} = Y_{ij}^{dev} - Y_{ij}^{gap},\tag{4.9}$$

where Y^{dev}_{ij} and Y^{gap}_{ij} are the calibrated measurements of the nanowire and empty device, respectively, transformed to an admittance representation. The intrinsic admittance of the nanowire Y^{nw}_{ij} can subsequently be found by simple algebra. It is important to remember that all of the terms in Equation (4.9) are complex valued. The chosen admittance representation not only leads to the simple algebraic form of Equation (4.9), but also is a natural representation given that nanowires are often represented by equivalent circuits with a pi structure, as we will discuss in Chapters 5 and 6.

Historically, this "empty device approach" was developed by a number of groups pursuing broadband characterization of CNT and nanowire devices [10]. Early work by Li et al. described the fabrication and one-port broadband measurement of single-walled CNT transistors [11]. Later, Bethoux et al. introduced a calibration procedure analogous to that described above in order to determine the cutoff frequency of a transistor consisting of a large number of single-wall CNTs [12]. Zhang et al. utilized a similar strategy to perform broadband, two-port transmission measurements of multiple CNTs [13]. Researchers at Intel extended two-port measurements to individual and bundled CNTs, placing particular emphasis on the potential application of CNTs as high-frequency interconnects [14]. More recently,

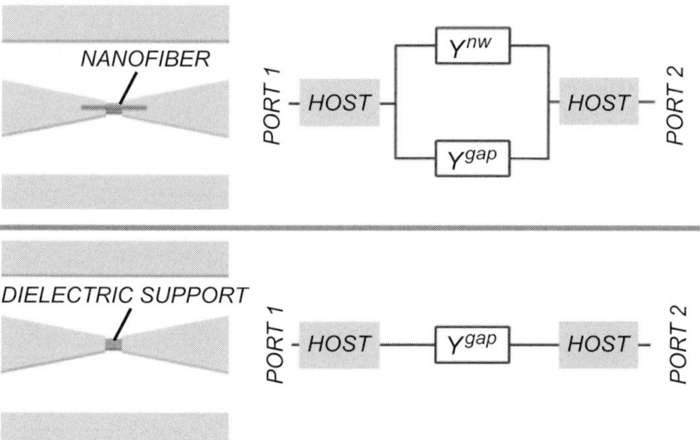

Figure 4.6. Comparison of a two-port nanowire device to an empty reference device. (a) Conceptual illustration of a nanowire device integrated with a CPW host structure. (b) Conceptual illustration of an "empty," nanowire-free device. The circuit model of the nanowire device includes the properties of the host structure, the intrinsic admittance of the nanowire Y^{nw}_{ij}, and the parasitic capacitance Y^{gap}_{ij}. By comparison, the circuit model for the empty device excludes Y^{nw}_{ij}. Any supporting structures that are present in the nanowire device, such as the dielectric labeled in (b), must also be present in the empty device. © 2011 IEEE. Reprinted, with permission from T. M. Wallis, K. Kim, D. S. Filipovic, and P. Kabos, *IEEE Microwave Magazine* 12 (2011) pp. 51–61.

Vandenbrouck et al. employed a similar strategy to perform broadband electrical characterization of a GaN/AlN/AlGaN transistor device [15].

The empty device approach has several limitations. For example, Equation (4.9) implies that the introduction of a nanoscale building block does not substantially alter the parasitic reactance beyond the introduction of a contact impedance. This is not always the case. For example, the welding of a multiwalled CNT into a host structure may alter the parasitic reactance of the host device by damaging or otherwise altering metallization layers in the host structure [16]. Furthermore, care must be taken to ensure that S^{dev}_{ij} and S^{gap}_{ij} are measured under controlled, identical conditions, as the parasitic coupling may depend sensitively upon a number of experimental variables, including temperature, optical illumination, and exposure to different gas environments. Finally, given that the uncertainties in on-wafer scattering parameter measurements are larger than those for connectorized on-wafer measurements, Y^{dev}_{ij} and Y^{gap}_{ij} may be equivalent within the experimental uncertainty, particularly if the resistance of the nanoscale element is extremely high. An alternative approach is to use modeling to extract an estimate of the effective parasitic capacitance from the measurements of the empty device, which can subsequently be used as an input into a model of the nanoelectronic device, as described in the next chapter.

4.5   Fabrication of Impedance-Matched On-Wafer Devices

One strategy that has emerged in the development of broadband, nanofiber-based devices is the use of massively parallel arrays of nanofibers. For instance, a massively parallel array of hundreds of single-wall CNTs results in an impedance close to 50 Ω without compromising the highly desirable qualities of CNTs that make them well-suited to RF interconnect and transistor applications [17], [18]. Figure 4.7 illustrates such an array of single-wall CNTs integrated into a one-port CPW. The 1.2 nm to 1.4 nm diameter CNTs were deposited in solution on the 1 μm gap in a lithographically defined host structure and then aligned by use of dielectrophoresis. After alignment, the CNTs are secured by depositing another lithographically defined layer on top of them. A nanowire density on the order of ten wires per micrometer yields devices with an impedance close to 50 Ω. For the measurements described in Reference [17], all of the CNTs would be metallic in the ideal case, but in practice the ensemble of tubes included some semiconducting CNTs. Nonetheless, measurements of these massively parallel devices offer insight into the fundamental physics of CNTs at RF, namely that the effects of kinetic inductance are negligible below about 200 GHz (kinetic inductance will be discussed in further detail in Chapter 5).

Like the other measurement approaches described here, this approach has its own challenges and trade-offs. For example, if the intended application is a CNT transistor, the CNTs must ideally all be semiconductors. Though this is challenging, recent advances in CNT separation suggest that this is possible.

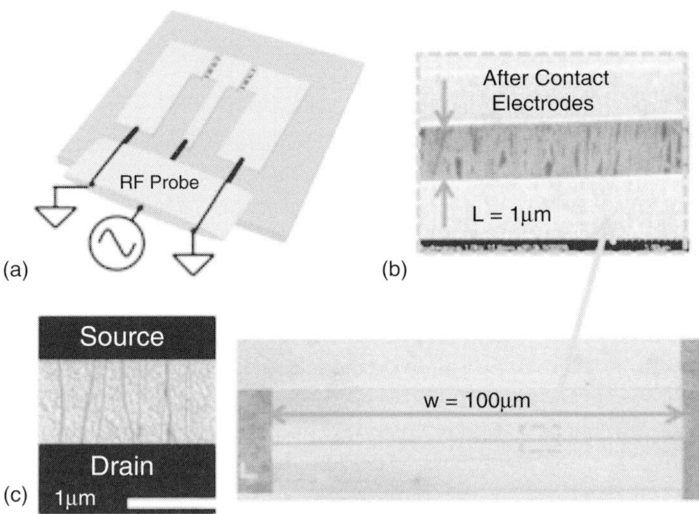

Figure 4.7. 50 Ω nanoelectronic device based on a massively parallel array of single-wall CNTs. (a) Schematic of an RF probe contacted to the CPW host structure. (b) Scanning electron microscope images of the massively parallel array aligned in the gap region (width 100 µm, length 1 µm). (c) An AFM image of several aligned CNTs. Reprinted from C. Rutherglen, D. Jain, and P. Burke, *Applied Physics Letters* 93 (2008), art. no. 083119, with permission from AIP Publishing.

Measurements made with this technique reflect the aggregate behavior of the ensemble of nanofibers. Thus, this technique is not amenable to isolating the properties of individual nanofibers or the contacts to individual nanofibers. However, one can envision using measurements of individual nanoscale building blocks to characterize and optimize single elements and contacts before integrating them into massively parallel, commercially viable devices that are compatible with bulk manufacturing.

References

[1] A. Fraser, R. Gleason, and E. W. Strid, "GHz On-Silicon-Wafer Probing Calibration Methods," *Proceedings of the 1988 Bipolar Circuits and Technology Meeting* (1988) pp. 154–157.

[2] R. B. Marks and D. F. Williams, "A General Waveguide Circuit Theory," *Journal of Research of the National Institute of Standards and Technology* 97 (1992) pp. 533–562.

[3] D. F. Williams, U. Arz, and H. Grabinski, "Characteristic-Impedance Measurement Error on Lossy Substrates," *IEEE Microwave and Wireless Components Letters* 11 (2001) pp. 299–301.

[4] L. Nougaret, G. Dambrine, S. Lepilliett, H. Happy, N. Chimot, V. Derycke, and J.-P. Bourgoin, "Gigahertz Characterization of a Single Carbon Nanotube," *Applied Physics Letters* 96 (2010) art. no. 042109.

[5] H. Happy, K. Haddadi, D. Theron, T. Lasri, and G. Dambrine, "Measurement Techniques for RF Nanoelectronic Devices," *IEEE Microwave Magazine* 15 (2014) pp. 30–39.

[6] M. C. A. M. Koolen, J. A. M. Geelen, and M. P. J. G. Versleijen, "An Improved De-embedding Technique for On-Wafer High-Frequency Characterization," in *Proceedings of the IEEE 1991 Bipolar Circuits and Technology Meeting* (1991) pp. 188–191.

[7] R. B. Marks, "A Multiline Method of Network Analyzer Calibration," *IEEE Transactions on Microwave Theory and Techniques* 39 (1991) pp. 1205–1215.

[8] C. P. Wen, "Coplanar Waveguide: A Surface Strip Transmission Line Suitable for Nonreciprocal Gyromagnetic Device Applications," *IEEE Transactions on Microwave Theory and Techniques* 17 (1969) pp. 1087–1090.

[9] T. M. Wallis, K. Kim, D. S. Filipovic, and P. Kabos, "Nanofibers for RF and Beyond," *IEEE Microwave Magazine* 12 (2011) pp. 51–61.

[10] C. Rutherglen and P. J. Burke, "Nanoelectromagnetics: Circuit and Electromagnetic Properties of Carbon Nanotubes," *Small* 5 (2009) pp. 884–906.

[11] S. Li, Z. Yu, S.-F. Yen, W. C. Tang, and P. J. Burke, "Carbon Nanotube Transistor Operation at 2.6 GHz," *Nano Letters* 4 (2004) pp. 753–756.

[12] J. M. Bethoux, H. Happy, G. Dambrine, V. Derycke, M. Goffman, and J. P. Burgoin, "An 8-GHz f_t Carbon Nanotube Field-Effect Transistor for Gigahertz Range Applications," *IEEE Electron Device Letters* 27 (2006) pp. 681–683.

[13] M. Zhang, X. Huo, P. C. H. Chan, Q. Liang, and Z. K. Tang, "Radio-Frequency Characterization for the Single-Walled Carbon Nanotubes," *Applied Physics Letters* 88 (2006) art. no. 163109.

[14] J. J. Plombon, K. P. O'Brien, F. Gstrein, V. M. Dubin, and Y. Jiao, "High Frequency Electrical Properties of Individual and Bundled Carbon Nanotubes," *Applied Physics Letters* 90 (2007) art. no. 063106.

[15] S. Vandenbrouck, K. Madjour, D. Theon, Y. Dong, Y. Li, C. M. Lieber, and C. Gaquiere, "12 GHz F_{MAX} GaN/AlN/AlGaN Nanowire MISFET," *IEEE Electron Device Letters* 30 (2009) pp. 322–324.

[16] P. Rice, T. M. Wallis, S. E. Russek, and P. Kabos, "Broadband Electrical Characterization of Multiwalled Carbon Nanotubes and Contacts," *Nano Letters* 7 (2007) pp. 1086–1090.

[17] C. Rutherglen, D. Jain, and P. Burke, "RF Resistance and Inductance of Massively Parallel Single Walled Carbon Nanotubes: Direct, Broadband Measurements and Near Perfect 50 Ω Impedance Matching," *Applied Physics Letters* 93 (2008) art. no. 083119.

[18] S. W. Hong, T. Banks, and J. A. Rogers, "Improved Density in Aligned Arrays of Single-Walled Carbon Nanotubes by Sequential Chemical Vapor Deposition on Quartz," *Advanced Materials* 22 (2010) pp. 1826–1830.

5 Modeling and Validation of RF Nanoelectronic Devices

5.1 Introduction

The development and engineering of nanoelectronic devices has been characterized by several significant technological trends. In addition to the ongoing scaling of feature sizes down to nanoscale dimensions, the need for superior performance has driven the integration of novel materials as well as the incorporation of additional device functionalities. In turn, these trends require advanced manufacturing technologies. At the same time, the clock frequencies of nanoelectronic devices have increased into the microwave and millimeter-wave range. Importantly, these trends affect not only the engineering of active devices, but also the design of interconnects between devices.

Device and interconnect scaling leads to challenges in device engineering. For example, low-dimensional systems such as nanowires manifest new and altered material properties with respect to the bulk phase. Quantum effects that may not be evident in bulk material can have a significant influence as the diameter of a wire becomes less than several tens of nanometers. In addition, the conductivity can change substantially at these scales. In metallic nanowires, this is due to a decrease of the mean free path of electrons, while in CNTs such changes reflect the ballistic nature of electron transport. For example, the input impedance of a copper antenna of length $0.47\,\lambda$ changes from about $70 - j\,8$ ohms to about $20{,}000 - j\,20{,}000$ ohms as the diameter changes from $7.5\,\mu m$ to $4\,nm$ [1]. Clearly, this presents significant challenges to scaling of copper-based RF interconnects. Additional challenges arise due to contact impedance. While contact impedance often has a negligible effect at macroscopic scales, it can impact or even govern the RF response of nanoelectronic devices. This is due in large part to the fact that the size of a nanoelectronic device and the size of a contact are often comparable. In addition, in special cases such as molecular devices, atomically small changes in the positioning of the contacts can critically influence the performance of the device. Finally, as we have noted in previous chapters, the impedance of nanoscale devices is significantly different from the $50\,\Omega$ impedance of the measurement equipment, thereby creating an inherently large impedance mismatch.

In the preceding chapters, we have demonstrated how metrology plays a crucial role in meeting these engineering challenges and enabling the understanding of material properties, circuits, interconnects, devices, and antennas at nanoscale dimensions [2]. In addition to providing insight into new nanoscale phenomena, reliable high-frequency measurements can provide a foundation for comparing results and building consensus between different research endeavors. The inherent challenges of nanoscale measurements are further augmented by the need to make measurements at practical operating frequencies in the microwave and millimeter-wave range. In this chapter, we present a crucial piece of RF metrology for nanoelectronics: computational modeling and simulations.

Dependable models are required for the design of reliable and accurate test platforms, measurement calibration and verification, as well as the extraction of quantitative circuit and material parameters from measurements. In short, without modeling and simulation, we cannot fully address the critical measurement challenges of RF nanoelectronics. We present approaches suitable for characterization of nanoscale devices, including nanowire-based interconnects and active nanotransistor devices. The focus is on the development of measurement models and methods for determination of constitutive material and device parameters. In the previous chapter, we investigated the extension of established measurement techniques to on-wafer nanoelectronic devices. Here, we use modeling to validate that approach, complementing the measurement techniques and establishing a complete measurement framework. Both full-wave, finite-element models and circuit models are used to determine the properties of nanoelectronic systems. These models are then compared with calibrated measurements in order to validate the measurement and calibration procedures. Passive, two-port test structures based on gold (Au) microbridges and platinum (Pt) nanowires are used as illustrative examples. As will become evident subsequently, the separation of the intrinsic properties of nanowires from the properties of electrical contacts presents a significant challenge. To address this problem, we present two approaches based on transmission line and circuit models. Though the modeling of any particular problem will present unique aspects, the examples presented here should provide a foundation that can be extended to address individual, specialized cases. For example, these methods can be extended to RF applications based on semiconducting nanowire devices, two-dimensional, or other emerging materials.

5.2   Modeling and Validation of Measurement Methods

5.2.1   Electromagnetic Properties of Nanoscale Conductors

We begin by addressing the basic question of transport and impedance in one-dimensional systems and related devices. One-dimensional systems such as nanowires are a fundamental building block for micro- and nanoscale systems with a broad range of applications including sensors, field effect transistors, packaging, and flexible substrates [3]–[9]. The electrical properties of such systems depend on elemental composition, size, and morphology. If the dimensions of a nanowire are on the order of the quantum mechanical wavelength of an electron, the DC resistance is on the order of $h/2e^2 \approx 12.5$ kΩ, assuming a single conduction channel (h is Planck's constant and e is the charge of an electron). If the number of conduction channels is increased, as when one accounts for spin, then this resistance is divided by the number of channels. This extreme resistance is far from the impedance of commercial, 50 Ω test equipment and such extreme impedance mismatch influences device design, modeling, and measurement. In spite of the large resistance, the operating frequency of nanoelectronic devices can extend to the terahertz regime due to their low capacitance, which is on the order of tens of attofarads to a few femtofarads. This opens the possibility of extremely high-frequency transistor applications.

We will follow the approach of Reference [10] that was developed for CNTs, but is applicable to nanowires and other low-dimensional systems. As a starting point, a simple transmission line model is used. In some cases, such as single-walled CNTs and THz systems, the model is modified to include a kinetic inductance in series with the per-unit-length resistance and per-unit-length inductance. Furthermore, a quantum capacitance may be added in series with the per-unit-length capacitance. Values of the per-unit-length inductance and per-unit-length capacitance depend on the geometry of the system and can be evaluated through analytical calculations, finite-element modeling, or measurement. The kinetic inductance per-unit-length for one-dimensional systems with a single conduction channel can be expressed as

$$L_K = \frac{h}{2e^2 v_F}, \tag{5.1}$$

where v_F is the Fermi velocity. As with the quantum resistance, if the number of channels is greater than one, then L_K is divided by the number of channels. As an example, L_K is about 16nH/μm in single-layer graphene. The quantum capacitance arises from the fact that for a low-dimensional quantum system one can add an electron to the system only if its energy is above the Fermi level. In general, the relation is complicated, but the quantum capacitance per unit length is proportional to density of states (DOS):

$$C_Q \approx e^2 DOS. \tag{5.2}$$

For a single-walled CNT the expression for C_Q is $2e^2 / (hv_F)$, which is about 100 aF/μm. The total capacitance per unit length is the series combination of the per-unit-length electrostatic capacitance, which depends on the geometry of the system, and the quantum capacitance. Note that if one considers only the kinetic inductance and quantum capacitance, the characteristic impedance of an individual, single-walled CNT is on the order of 12.5 kΩ.

In general, the effects of kinetic inductance and quantum capacitance must be handled on a case-by-case basis. When determining whether these terms must be included in the transmission line model, one must consider the distributed resistance of the microwave structure into which nanoscale building block is embedded and the relative sizes of the mean free path of the carriers and the device dimensions.

Further details about the role of quantum capacitance and inductance in transmission line models are discussed in References [10] and [11]. Lastly, in many cases the contact impedance in low-dimensional components also has to be taken into account [2], [11].

5.2.2   An Overview of Validation

The validation of high-frequency measurement and calibration methods is a critical component of quantitative metrology for RF nanoelectronics, particularly if the applications cover a broad range of operation frequencies. One cannot simply assume that established approaches for measurement and calibration of connectorized and on-wafer RF devices are valid for nanoelectronics. The validation step is all the more crucial because of the extreme impedance mismatch of nanoelectronics with respect to host structures and commercial test equipment. Matching networks can ameliorate this problem in a narrow frequency range, as discussed in Chapter 3, but for broadband measurements the device will always represent an extreme impedance load. In this chapter, we discuss the validation of measurement and calibration methods, building upon the introduction of several such measurement approaches in the previous chapters.

Modeling and simulation accomplish two main objectives related to measurement of extreme impedance loads, in general, and RF nanoelectronics, in particular. The first is the validation of the calibration methodology and the second is the extraction of circuit and material parameters from the measurements. For example, for metallic nanowires, one may need to extract the wire conductivity and the contact resistance. Validation requires the design of appropriate models and the comparison of the models' predictions to calibrated measurements. These models require reliable inputs that accurately represent devices and their constituent structures, including geometry and material parameters. In order to focus the measurement and modeling problems on the nanoscale components, in most cases the fixtures that are the part of the nanoscale test structures are de-embedded and the reference planes are moved as close as possible to the nanoscale elements of interest. If significant differences between the model predictions and the calibrated measurements are observed, then either the modeling or measurement approach must be discarded as unsuitable for characterization of nanoscale devices. For example, an established calibration procedure may be unsuitable if the underlying assumptions of the calibration process are invalid. One critical question is whether the propagating electromagnetic field corresponds to a single mode or if higher order modes are also present. As discussed in Chapter 2 single-mode propagation is a fundamental assumption of multiline TRL and other established calibration procedures. In addition, the field distribution of the fundamental mode must be investigated, as in many nanoscale devices this distribution may violate the assumption of pure TEM-mode propagation.

Once the measurement and calibration methods are validated, one can proceed to the extraction of device parameters. To extract such parameters, calibrated measurements of the device scattering parameters are compared with full-wave and circuit model simulations. For example, for conducting and semiconducting nanowires a range of the conductivities can be obtained by fitting the contact resistance and wire resistance to measured data, minimizing the deviation between the models and measurements. Note that if the model has too many unknown parameters, this approach may not have a unique solution and therefore additional measurements or methods have to be used, as discussed in the following subsection. Several modeling approaches are applicable here: full-wave finite-element models, equivalent lumped-element circuit models, and transmission line models.

5.2.3 Validation with Finite-Element Models

As a first example, we focus on the validation of multiline TRL calibration by use of full-wave, finite-element modeling (FEM), but the techniques can be generalized to other on-wafer calibration techniques. We will assume that CPWs of different lengths along with appropriate coplanar short and open circuits are used as calibration standards. The on-wafer segment of the fixtures of the nanoelectronic DUTs are required to have the same CPW geometry as the calibration structures. All calibration structures are assumed to be fabricated on the same substrate as the fixtured, nanoelectronic DUTs.

Figure 5.1(a) shows an example of a nanoelectronic DUT: a gold (Au) microbridge embedded in a two-port, CPW host structure. The center conductor of the CPW host is tapered such that it is just a two-micrometer-wide, four-micrometer-long strip at its narrowest point. The structure is fabricated on a quartz substrate by use of lithographic patterning, thin film sputtering, and liftoff. Figure 5.1(b) shows a second test structure: a bridge-free, empty device that serves as a reference device for measurements and modeling as described in this chapter. For the purpose of validation, it is instructive to start with the modeling of a simple structure like the Au microbridge. The microbridge represents a DUT with extreme impedance, albeit an extremely low impedance with respect to 50 Ω. Further, the Au bridge is continuous with the host structure, removing any complicating contact effects. The empty structure is identical except for the removal of the Au bridge and serves as a complementary DUT with an extremely high impedance. Using these two simple configurations, one can investigate both impedance extremes and test different numerical methods. Like many RF nanoelectronic devices, the reflection coefficient is high for both of these device configurations, thus providing further confidence that the validation methods may be applied to a broad range of nanoscale systems.

It is possible to model the broadband characteristics of this device by use of a number of commercial software packages, including RF finite-element-based software such as HFSS [12], CST Microwave Studio, AWR Microwave Office, COMSOL, and JCMSuite [13], as well as circuit-based software such as Spice, ANSYS Designer [14], and AWR Microwave Office [15].[1] Among these packages,

---

[1] The use of trade names is intended to provide clarity and does not constitute endorsement by NIST.

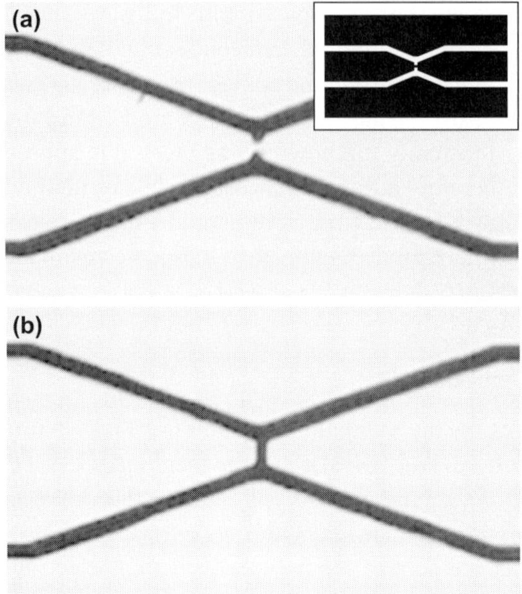

Figure 5.1.   Gold (Au) microbridge device and an empty reference device.
(a) Optical microscope image of a lithographically patterned, two-port Au device in which
a 2 μm-wide bridge connects tapered segments of a CPW center conductor. The inset
illustrates the device geometry. (b) Optical microscope image of an Au reference device
without the Au bridge.

the full-wave numerical methods such as finite-element electromagnetic solvers are
advantageous due to their three-dimensional modeling capability. FEM methods
are based on the solution of three-dimensional wave equations and enable accurate
analysis both of open and closed boundary value problems. They offer the possibility to treat inhomogeneous materials and a wide variety of shapes over a broad
frequency range. The calculated field expansion is done using polynomials of different orders and users have access and control at different stages of the solution
process. For on-wafer problems such as those treated here, it is useful to reduce the
computational overhead associated with meshing by replacing the on-wafer probes
with ideal wave ports.

The detailed steps of FEM calculations require familiarity with the modeling
software, device layout, and calculation steps. Such details are beyond the scope of
this book, but it is instructive to present results from numerical modeling of example nanoelectronic devices. Following References [16] and [17] the scattering parameters of the Au microbridge and empty DUTs were modeled by a commercial FEM
package over a frequency range of 50 MHz to 50 GHz. The results were compared
to broadband, on-wafer, multiline TRL calibration results, as shown in Fig. 5.2.
The results of the comparison show excellent agreement between the model and
the calibrated measurement can be achieved, confirming that the FEM approach

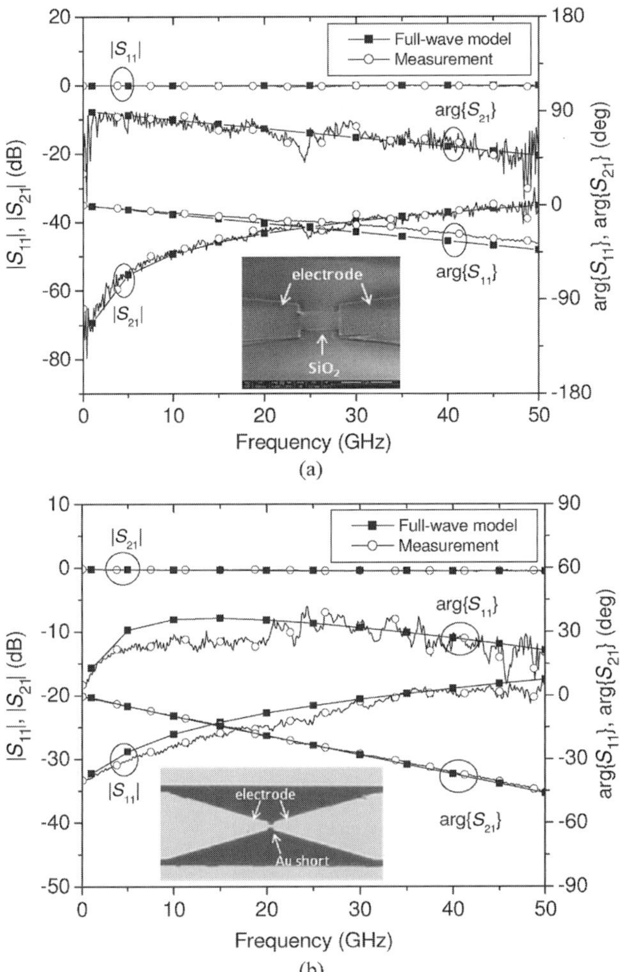

Figure 5.2. Comparison of finite-element model and measurements for a gold microbridge device and an empty reference device.
Measured and simulated scattering parameters for (a) an "empty" device and (b) a gold microbridge device (labeled "Au short"). In the empty device, a dielectric (SiO$_2$) layer has been deposited in the CPW gap. Simulations (black squares) were carried out with commercial full-wave, finite-element modeling software. Measurements (open circles) were calibrated with the on-wafer, multiline TRL method. © 2010 IEEE. Reprinted, with permission from K. Kim, T. M. Wallis, P. Rice, C.-J. Chiang, A. Imtiaz, P. Kabos, and D. S. Filipovic, *IEEE Microwave and Wireless Components Letters* 20 (2010) pp. 178–180.

is suitable for modeling of nanoelectronic devices with dimensions much smaller than the wavelength of the electromagnetic field. The agreement further suggests that there is no significant generation of higher order modes in these devices. Calculation of the electric field distribution in the vicinity of the microbridge and additional analysis shows that higher order modes transmit less than 10 percent of the incident power [18].

5.2.4   Validation with Circuit Models

Full-wave FEM solvers require significant computational resources due to large memory requirements and long computation times. Therefore, circuit models, which require less memory and have significantly shorter computation times, have their place in the toolbox of methods for high-frequency device evaluation. However, it must be noted that circuit models often require input in the form of experimentally determined or FEM-simulated parameters. Once again, the Au microbridge and the empty devices shown in Fig. 5.1 serve as the example test structure. Circuit models of the empty and nanowire devices are shown in Fig. 5.3. It is useful to introduce the parasitic gap capacitance in parallel with the contacted nanowire, as shown in the lower panel of Fig. 5.3, to represent the coupling between the two tapered waveguide segments of the host structure.

The value of the parasitic gap capacitance C_{gap} may be obtained from the comparison of the circuit model results with either the FEM simulation or the experiment. From FEM modeling, C_{gap} is 0.6 fF for the Au microbridge device, in good agreement with the value obtained by fitting calibrated measurements. In contrast, the value of

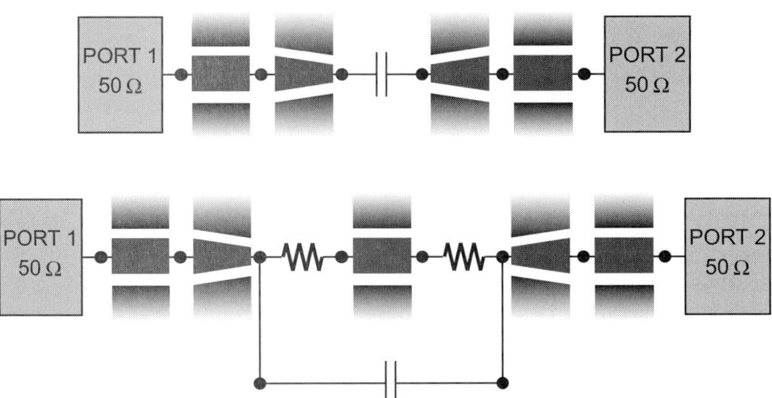

Figure 5.3.   Circuit models for an empty reference device and a gold microbridge device. Circuit models for the empty device and the microbridge device are shown in the top and bottom panels, respectively. The test platform, including the fixtures and host structure are represented by an ideal port connected to a segment of transmission line, followed by a tapered transmission line on either side of the device. The empty device is represented by a capacitive coupling, while the microbridge is represented by the same capacitive coupling in parallel with a transmission line and resistive contacts. In the case of the continuous microbridge, the resistance of the contacts is negligible. © 2011 IEEE. Adapted, with permission from K. Kim, P. Rice, T. M. Wallis, D. Gu, S. Lim, A. Imtiaz, P. Kabos, and D. S. Filipovic, *IEEE Transactions on Microwave Theory and Techniques* 59 (2011) pp. 2647–2654.

C_{gap} from calculation of the electrostatic capacitance for the given device dimensions by use of standard circuit model libraries results in an estimated value of 5.3 aF, leading to poor agreement with the measurements. The reason for this discrepancy is that the circuit model fails to account for stray fields and the resulting parasitic coupling in devices with multiple, closely spaced signal lines. This underscores the fact that for RF nanoscale devices, the impedance of the environment surrounding the nanoscale element plays an important role. For the circuit model approach, it is necessary to take care in estimating the circuit parameter values of nanoscale elements and their surrounding structural environments. Standard circuit libraries that are reliable for modeling more traditional microelectronics may be insufficient for modeling some nanoelectronic devices. When a reliable, experimentally established estimate of the parasitic capacitance C_{gap} is used, the circuit models are found to be in good agreement with experimental results for the empty device, but the agreement is poorer for the Au microbridge. Furthermore, there are significant differences between the circuit model results for different software packages [18]. In summary, circuit models may serve as a useful, complementary tool in the validation process for RF nanoelectronic devices, but such models have significant limitations.

5.3 Extracting Circuit Parameters from Measurements

5.3.1 Nanowire Device Parameters

With the validation of measurement methods demonstrated, one can proceed with the de-embedding of circuit and material parameters from broadband measurements of nanoscale devices. Without loss of generality, we will demonstrate extraction of circuit parameters for example devices that incorporate single nanowires and nanowire-like structures such as CNTs. The properties of several types of nanowire devices are summarized in Table 5.1. Note the wide variation in resistivity as well as contact resistance. For example, a Pt nanowire deposited on Au has a contact resistance of about 138 Ω, while a similar nanowire attached to Pt electrodes has a significantly higher contact resistance of 700 kΩ to 800 kΩ [24].

For nanowire structures these two quantities – the resistivity (or conductivity) of the nanowire and the contact resistance to the wire – are usually the parameters of greatest interest. The conductivity of the wire primarily reflects the specific material properties of the nanowire, though a full understanding of the device performance may require consideration of the kinetic inductance and quantum capacitance in selected cases, as described earlier and in References [2], [10], and [11]. Broadly speaking, contact resistance is a critical parameter for most nanoscale devices operating at RF and may vary strongly from device to device. One central problem of extracting these parameters is that the contact resistance is measured simultaneously with the intrinsic resistance of the device [16], [17], [20]. Additional measurements or alternative approaches are required to uniquely determine the wire conductivity and the contact resistance.

Table 5.1 Resistivity and Contact Resistance for Selected Metallic Nanowire Devices

[Citation] (Year)	[19] (2000)	[20] (2003)	[21] (2003)	[22] (2004)	[23] (2007)	[24] (2008)
NW Material	Au	Cu	Pt	Pt	Pt	Pt
NW Diameter (nm)	70	60	60	70 ± 5	60–360	80–150
Resistivity ($\mu\Omega$ cm)	4.5 (bulk: 2.5)	17.1 (bulk: 1.72)	61.5 (5.9 μm) 482 (13 μm) 545 (20 μm) (bulk: 10.8)	33 ± 5	860–3078	Not Reported
Contact Resistance	Not Reported	Not Reported	Not Reported	138 Ω	Up to 100 Ω	700 kΩ – 800 kΩ

Source: Adapted from Reference [18], with permission.

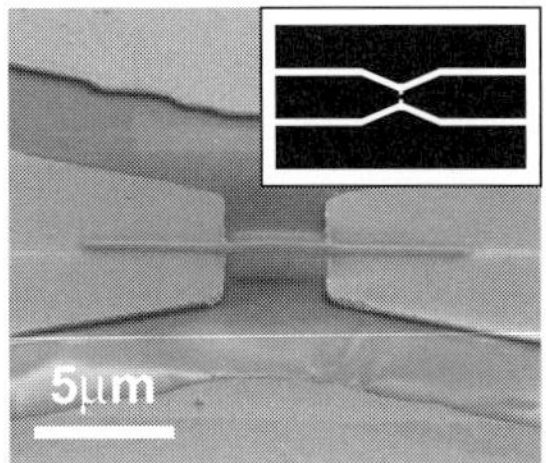

Figure 5.4.   Platinum (Pt) nanowire device.
Scanning electron microscope image of a lithographically patterned, two-port device that incorporates an oxide-supported Pt nanowire that bridges a gap in the CPW center conductor. The inset illustrates the device geometry.

The example device we will discuss here incorporates an oxide-supported Pt nanowire deposited on Au electrodes, as shown in Fig. 5.4. It is nearly identical to the Au microbridge shown in Fig. 5.1, but with the nanowire now serving as a bridge across a gap in the CPW center conductor. Pt nanowires with diameters of 150 nm and 250 nm were used. This two-port device is a useful model system as it represents a high impedance. In addition, this is an example of a system where the properties of the nanoscale element must be fully de-embedded from the on-wafer test fixtures in order to extract material and circuit parameters.

5.3.2   Full-Wave, Finite-Element Approach

As a first example of parameter extraction, we discuss estimation of the conductivity and contact resistance of the two-port Pt nanowire device by use of a full-wave, three-dimensional solver [9], [25]. The device dimensions, as determined from scanning electron microscope (SEM) images, and the wire conductivity are inputs for the simulation. To simplify the analysis, excitation ports are used in place of the probe tips. Initially, the contact between the nanowire and the electrodes is assumed to be ideal, i.e., the contact resistance $R_c = 0\ \Omega$. When the conductivity of the wire σ is set to the macroscopic, bulk Pt conductivity $\sigma = \sigma_{\text{bulk}} = 9.3 \times 10^6$ S/m, there is an enormous discrepancy between the measurement and the calculations. For example, broadband, calibrated measurements from 100 MHz to 50 GHz reveal that $|S_{21}|$ for the Pt nanowire device is between -17 dBm and -20 dBm, while the finite-element simulation with the bulk value of conductivity predicts that $|S_{21}|$ is about -2 dBm over the same frequency range. Even when the conductivity is reduced to 0.5 σ_{bulk}, this discrepancy remains. To remove this discrepancy, σ must be further reduced and a nonzero contact resistance R_c must be introduced. To introduce the contact resistance into a full-wave model, it is necessary to apply lumped-element boundary conditions. Unfortunately, full-wave, finite-element simulations do not allow a unique separation of the contact resistance and conductivity. In other words, σ and R_C can vary over a large range and still fit the experimental data. This ambiguity is illustrated in Fig. 5.5(a). When the conductivity is σ_{bulk}, agreement with the experimental data is found with $R_C = 315\ \Omega$. When the contact resistance is set to zero, agreement with the experimental data is found with $\sigma = 0.014\ \sigma_{\text{bulk}}$. In this latter case, σ may at best be interpreted as an effective conductivity parameter that reflects the combined effects of wire conductivity, contact resistance, and other parasitic effects.

This simple example serves as something of a cautionary tale. The contact resistance and the wire conductivity are in series and there is no possibility to separate them by use of a single measurement of a single device. Although the range of possible solutions is broad, the contact resistance and the wire conductivity are coupled. Specific solution pairs are shown in Fig. 5.5(b). This problem persists independently of the calibration method used to de-embed the wire properties. As with full-wave, finite-elements models, comparisons of circuit models to experimental data lead to a wide range of solution pairs for σ and R_C. Measurement strategies for separation of the contact resistance from the conductivity include measurement of wires of different length or measurements of a single device before and after treatment that systematically alters contact resistance or conductivity. These measurement strategies may be supplemented by alternative modeling approaches, such as transmission line and lumped-element models, as described in the following subsections.

5.3.3   Transmission Line Approach

Next, we outline a strategy for extraction of device parameters by use of a transmission line model. Importantly, the transmission line model requires an assumption that the characteristic impedance and propagation constant do not vary over the length of the wire. The first step in this strategy, as with several other approaches that we have described, is to translate the reference planes as close as possible to the nanowire contacts. We denote the translated S-parameter matrix

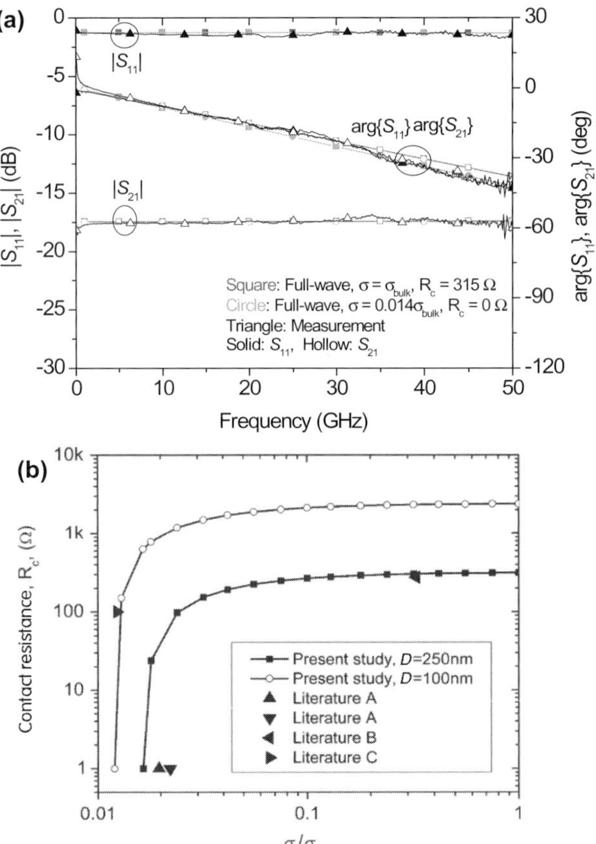

Figure 5.5.   Platinum (Pt) nanowire conductivity and contact resistance extracted by use of finite-element models.

(a) Scattering parameter magnitudes $|S_{11}|$ and $|S_{21}|$ as a function of frequency. Two solution pairs of the conductivity σ and contact resistance R_C that simulate the measured data are shown.

Reprinted from [18], with permission. (b) Sets of solution pairs for a 150 nm-diameter nanowire device and a 250 nm-diameter nanowire device. Literature A, B, and C correspond to References [21], [22], and [23], respectively. © 2011 IEEE. Reprinted, with permission from K. Kim, P. Rice, T. M. Wallis, D. Gu, S. Lim, A. Imtiaz, P. Kabos, and D. S. Filipovic, *IEEE Transactions on Microwave Theory and Techniques* 59 (2011) pp. 2647–2654.

of the nanowire device as $\boldsymbol{S}^{tot}$ and for an empty, nanowire-free device as $\boldsymbol{S}^{empty}$. Following the "empty device method" from Chapter 4, we convert these matrices to admittance matrices, $\boldsymbol{Y}^{tot}$ and $\boldsymbol{Y}^{empty}$ and then obtain the intrinsic admittance matrix for the nanowire and contacts, $\boldsymbol{Y}^{NW}$ by subtraction of $\boldsymbol{Y}^{empty}$ from $\boldsymbol{Y}^{tot}$ (alternatively, an approach based on the conversion of S-parameter matrices to impedance matrices may be developed in a similar way). It is useful to transform $\boldsymbol{Y}^{NW}$ into ABCD-matrix form: $\boldsymbol{ABCD}^{NW}$. We assume that the two contact resistances R_c are identical and in series with the nanowire. If the nanowire is represented by a transmission line of length l,

$$\boldsymbol{ABCD}^{NW} = \begin{bmatrix} 1 & R_c \\ 0 & 1 \end{bmatrix} \cdot \begin{bmatrix} \cosh(\gamma l) & Z_0\sinh(\gamma l) \\ \dfrac{1}{Z_0}\sinh(\gamma l) & \cosh(\gamma l) \end{bmatrix} \cdot \begin{bmatrix} 1 & R_c \\ 0 & 1 \end{bmatrix}, \tag{5.3}$$

where Z_0 and γ and are the characteristic impedance and propagation constant of the nanowire transmission line, respectively. To separate the conductivity and contact resistance, we require two measurements on devices with nanowires of different lengths, l_1 and l_2. Simple algebraic treatment leads to expressions for Z_0 and γ:

$$Z_0 = \sqrt{\frac{B^{NW}_i}{C^{NW}_i}} \tag{5.4}$$

and

$$\gamma_i = \frac{1}{l_i}\ln(A^{NW}_i \pm \sqrt{\left(A^{NW}_i\right)^2 - 1}. \tag{5.5}$$

A^{NW}_i, B^{NW}_i, C^{NW}_i are elements of the measured ABCD matrix $\boldsymbol{ABCD}^{NW}$ and the index $i = 1,2$ enumerates the measurements made with the nanowires of different lengths. The propagation constant is assumed to be the same for both lengths ($\gamma = \gamma_1 = \gamma_2$). Expanding the *sinh* and *cosh* functions in Equation (5.3) into Taylor series and retaining the terms to second order gives the simplified expression

$$R_c = \Re\left\{\frac{\left(l_1\right)^2\left(A^{NW}_2 - 1\right) - \left(l_2\right)^2\left(A^{NW}_1 - 1\right)}{\left(l_1\right)^2 C^{NW}_2 - \left(l_2\right)^2 C^{NW}_1}\right\}, \tag{5.6}$$

where $\Re(X)$ is the real part of X. With R_c established, it is now possible to independently determine the conductivity of the nanowire by use of a full-wave or circuit model.

It is necessary to consider the results from this approach very carefully: simple results can be deceiving. The result is sensitive to the terms in the denominator of Equation (5.6) that are usually quite small. Highly accurate measurements are required and any noise introduced into the measurements makes it almost impossible to get reliable information about the device parameters. In addition, the transmission line model accounts only for a single propagating TEM mode, but full-wave modeling reveals that there may be multiple modes. In the case of multimode propagation, it is necessary either to use the multimode calibration procedure discussed in Chapter 2 or consider the single-mode results as a first approximation to the corrected measurements. Also note that the use of redundant measurements, introduced in Chapter 3, can further decrease the statistical uncertainty of the measurement. In summary, this approach, though analytically straightforward, may be practically unreliable for extracting the properties of the wires and contacts, even if multiple devices of different lengths are measured.

5.3.4   Lumped Element Approach

Because the length scale of nanoscale devices is usually much smaller than the wavelength of the propagating microwave signal, meshing in FEM simulations presents significant challenges related to memory requirements and computation times. On the other hand, the interaction of electromagnetic waves with a nanoscale device is described well in the electrostatic limit. Therefore, it is reasonable to assume that if one moves the calibrated reference planes as close as possible to the nanoscale device, then a lumped element approach may work well. A hybrid approach that combines transmission lines and lumped elements may also be effective.

In the lumped element approach, a simple two-port "T" or "π" model can be used to represent the nanowire. To the extent that nanowire devices can be assumed to be symmetric, such circuits can be significantly simplified. A schematic of a nanowire device represented by a two-port T model in series with contact resistances is shown in Fig. 5.6. Once again, we will require measurements of two (or more) devices of different lengths, indexed by $i = 1,2$ (Z_{11} is the value of Z_1 for device of length l_1, Z_{21} is the value of Z_1 for device of length l_2, and so on). If each device is symmetric, $Z_{i1} = Z_{i2}$ and together with the shunt impedances Z_{i3} this fully characterizes each nanowire device. The ABCD matrices in terms of these impedances are

$$\mathbf{ABCD}^{NW}{}_i = \begin{bmatrix} 1 + \dfrac{Z_{i1}}{Z_{i3}} & 2Z_{i1} + \dfrac{Z_{i1}^2}{Z_{i3}} \\[2ex] \dfrac{1}{Z_{i3}} & 1 + \dfrac{Z_{i1}}{Z_{i3}} \end{bmatrix}. \tag{5.7}$$

Following a similar procedure as in the transmission line approach and replacing the nanowire transmission line matrix in Equation (5.3) with Equation (5.7), the impedances associated with this model are related to the elements of the measured ABCD matrices through

$$Z_{i3} = \frac{1}{C^{NW}{}_i} \tag{5.8}$$

and

$$Z_{i1} = \frac{A^{NW}{}_i - R_c C^{NW}{}_i - 1}{C^{NW}{}_i}. \tag{5.9}$$

Again using simple algebra, the contact resistance can be expressed as

$$R_C = \Re \left\{ \frac{l_2 C^{NW}{}_2 \left(A^{NW}{}_1 - 1 \right) - l_1 C^{NW}{}_1 \left(A^{NW}{}_2 - 1 \right)}{\left(l_2 - l_1 \right) C^{NW}{}_1 C^{NW}{}_2} \right\}. \tag{5.10}$$

In a similar way, by taking the imaginary part instead of the real part of the expression in Equation (5.10), the contact reactance is obtained. Since the wire's resistance is part of Z_{i1}, the wire conductivity can be obtained directly from the calculated

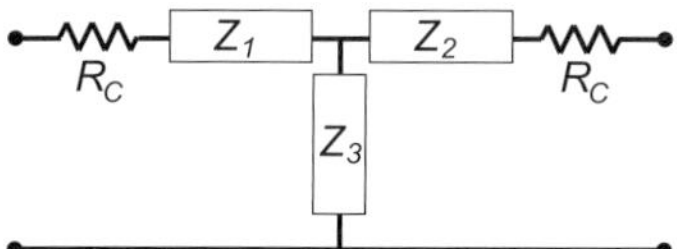

Figure 5.6.   Lumped-element model of a two-port nanoelectronic device.
The contact resistances are represented by resistors R_c at either end of the device. The nanowire (or other nanoscale element) is represented by a T model with impedances Z_1, Z_2, and Z_3.

impedances of the equivalent lumped-element T circuit in Fig. 5.6. Furthermore, validation of this approach can be done with a modified version of the circuit model in Fig. 5.3(b) in which the nanowire element is replaced by the T circuit. Validation with a full-wave model is also possible, with little change to the procedure. The full-wave model validation shows that the simple circuit model approach works well and can be used for extraction of the nanowire and contact resistance with reasonable precision.

Following the lumped element approach, the extracted values of the conductivity and the contact resistance for the Au microbridges and Pt nanowire devices are shown in Fig. 5.7(a) and Fig. 5.7(b), respectively. Two device lengths were fabricated and measured for each system: 4.0 μm and 8.0 μm. To reduce statistical contributions to uncertainties, multiple devices were measured and multiple measurements were made of each device. The Pt nanowire conductivity was calculated assuming that the wires have a circular cross section. While the contact resistance of the Au microbridge is expected to be zero, the obtained result of 0.5 Ω is reasonable and well within estimated uncertainties.

This lumped element approach assumes that there is little if any variation of the contact resistance from device to device and from contact to contact within a device. If we are only interested in isolating the properties of a nanowire or another nanoscale building block, it is desirable to design a device in such a way that the contact resistance would not come into play. One way to eliminate the contact resistance is to contact the measured nanowire to the host structure by use of capacitive coupling. Such coupling can be realized through a dielectric layer between the nanowire and the signal line or by cutting slots in the signal line and positioning the nanowire between these slots. In principle, the use of capacitive contacts requires measurement of a single device rather than multiple devices of different lengths. The disadvantage of this approach is the underlying assumption that the contact reactance can be accurately determined from knowledge of the structural form and material properties of the contacts. In addition, all dielectric layers have to be well-characterized, uniform, and free from contaminants in order to estimate dielectric constants. Little if any DC current will flow through the device. Further, with increasing separation of the nanoscale element from the host, sensitivity to nanowire properties is decreasing rapidly.

The layouts of the Au microbridge and Pt nanowire devices can be modified to have capacitive contacts. The inset in Fig. 5.8 shows an SEM image of an Au microbridge structure that has been modified with a focused ion beam such that there are slots that electrically separate the bridge element from the host CPW's

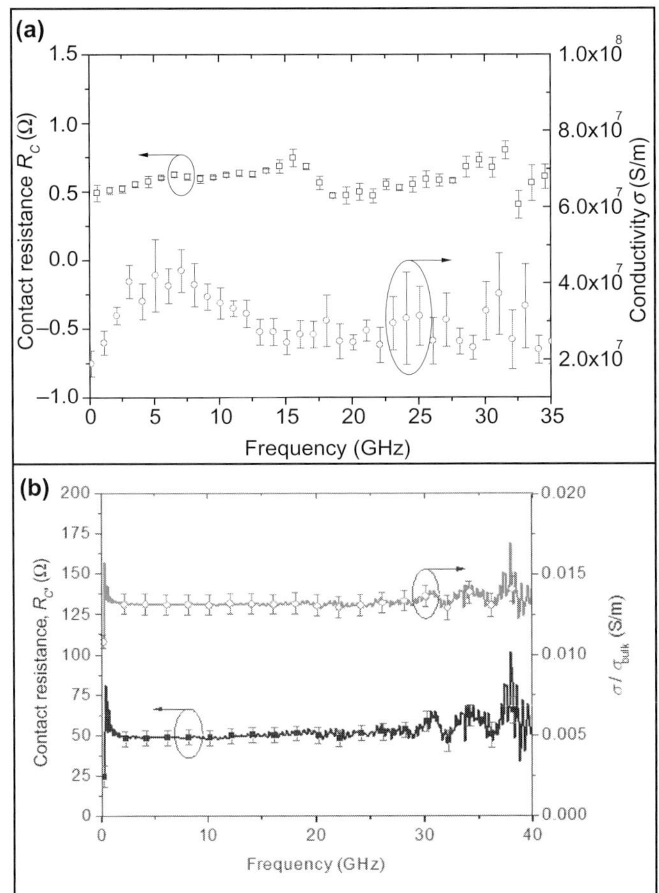

Figure 5.7.   Circuit parameters extracted by use of the lumped element approach.
A lumped element approach was used to extract the contact resistance and conductivity
for (a) Au microbridge device and (b) Pt nanowire device. © 2011 IEEE. Reprinted,
with permission from K. Kim, P. Rice, T. M. Wallis, D. Gu, S. Lim, A. Imtiaz, P. Kabos,
and D. S. Filipovic, *IEEE Transactions on Microwave Theory and Techniques* 59 (2011)
pp. 2647–2654.

center conductor. Calibrated measurements are compared with full-wave simulations in Fig. 5.8. The high level of agreement is encouraging, at least for this case
of an extremely low impedance device. In order to examine the case of extremely
high impedance devices, Pt nanowire devices were fabricated with a dielectric layer
between the wire and the conductive signal line of the host structure. For this case
the critical parameter is the thickness of the dielectric separation layer and the
length of the overlap of the nanowire with the signal line [17]. Measurements for the
capacitively coupled Pt nanowire devices yielded values of Pt nanowire conductivity consistent with the results from the lumped element approach. Simulations further reveal that the frequency-dependent amplitudes of S_{11} and S_{21} are significantly
more sensitive than the phases of S_{11} and S_{21} to changes in nanowire conductance.

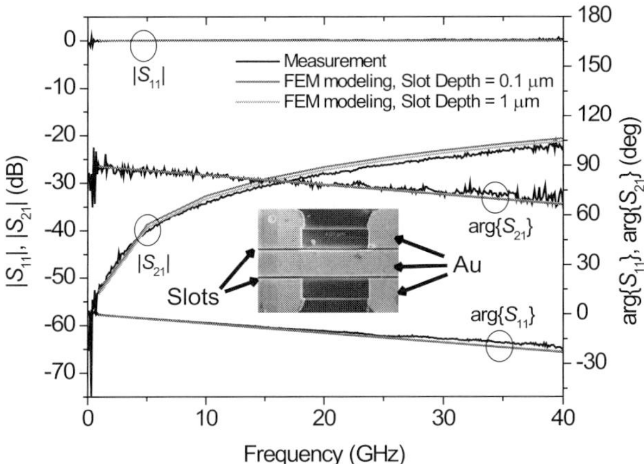

Figure 5.8. Gold (Au) microbridge device with capacitive contacts. Calibrated measurements and finite-elements simulations of the scattering parameters of an Au microbridge device. The inset shows an SEM image of the device. The gold layer has been cut by use of a focused ion beam in order to produce slots near the contact region, yielding contacts that are primarily capacitive.
(Data courtesy of D. S. Filipovic, University of Colorado, Boulder.)

5.3.5 Modeling and Parameter Extraction for CNT Devices

Historically, nanoelectronic devices based on CNTs have been of widespread interest, due to their potential applications as high-frequency transistors, nano-antennas, and interconnects. Single-walled CNTs present special problems that are not present in the more general case of nanowires. Previously we have discussed some of these problems, including high contact impedance and quantum mechanical effects. In order to extract reliable information from RF measurements of single-walled CNT devices, models must be altered by the modification of circuit and material parameters. The complex conductivity of metallic CNTs can be expressed as:

$$\sigma_{CNT}(\omega) = -j\frac{2e^2 v_F}{\pi^2 \hbar a(\omega - j\upsilon)}[S], \tag{5.11}$$

where e is the charge of electron, υ is the relaxation frequency as defined in Reference [26], ω is the angular frequency, $\hbar$ is the reduced Planck's constant, a is the CNT diameter and v_F is the Fermi velocity. It is important to notice that the units of σ_{CNT} are S, not S/m. Equation (5.11) is derived under the assumption that the thickness of the CNT wall is zero and therefore only surface currents are present. Based on this definition, the surface impedance of the CNT can be expressed as

$$Z_{CNT} = \frac{1}{2\pi a \sigma_{CNT}}\left(\frac{\Omega}{m}\right). \tag{5.12}$$

As with nanowire devices, in order to de-embed the properties of CNT devices, it is necessary in most cases to measure an empty, CNT-free device structure. The intrinsic, frequency-dependent impedance of a CNT vary widely from a few to hundreds of kΩ [27], which is consistent with theoretical predictions [28]. Note once again that the performance of CNT devices may critically depend on the quality of the electrical contacts.

So far, we have exclusively discussed nanoelectronics devices that incorporate a single CNT or other nanoscale building block. To reduce device impedance and loss, CNT bundles may be used in place of individual CNTs in interconnects as well as vertical vias in multilayered integrated circuits [29], [30], [31]. A CNT bundle may consist of as many as hundreds of CNTs. The resistance and inductance of the bundle are found to be inversely proportional to the number of CNTs [29], providing an avenue to produce device impedances that match 50 Ω test equipment. A distribution of CNT types must be assumed. For example, in References [32] and [33], the distribution is assumed to be about one-third

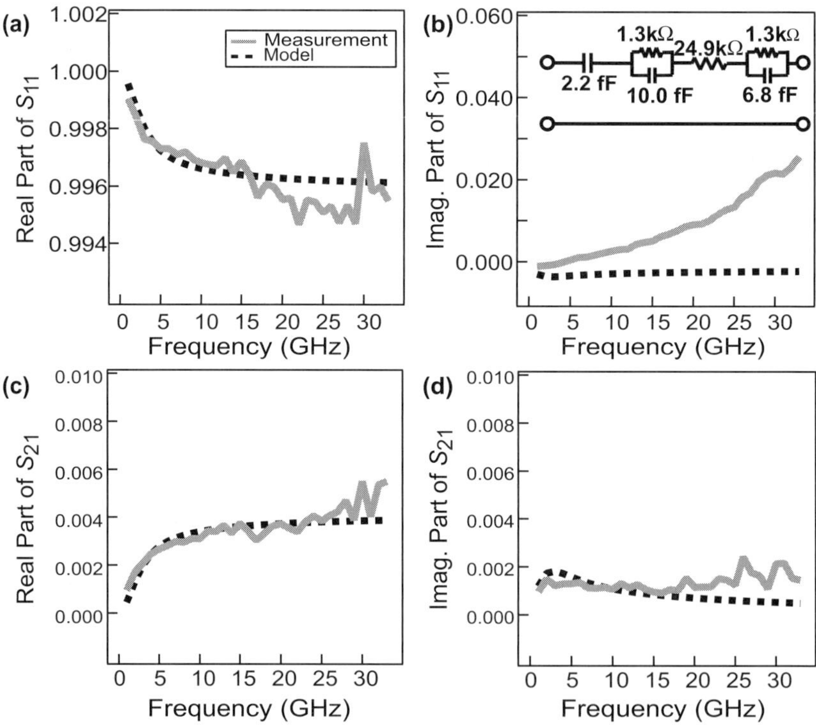

Figure 5.9.   Scattering parameters of a GaN nanowire device simulated with a genetic algorithm. Simulated scattering parameters of the circuit models of a two-port GaN nanowire device compared to calibrated measurements: S_{11} (a,b) and S_{21} (c,d). A schematic of the circuit model obtained with the genetic algorithm is shown in the inset of (b).
© 2011 IEEE. Adapted, with permission from T. M. Wallis, *2011 78th ARFTG Microwave Measurement Symposium* (2011) pp.1–5.

metallic and two-thirds semiconducting. For many applications, further improvement in electrical performance can be achieved by use of bundles made exclusively from metallic CNTs.

Measurement of CNT bundle devices may be performed by use of the same calibration and de-embedding techniques as discussed for individual nanowires and CNTs. For modeling and validation, it is assumed that each CNT has four conduction channels and a corresponding quantum resistance of 6.45 kΩ. The kinetic inductance for each CNT is about 16 nH/μm and quantum capacitance is about 100 aF/μm. Coupling between the wires is assumed to be negligible. Thus, the total resistance, inductance, and capacitance for a bundle may be determined from a simple model where all of the CNTs are in parallel. Additional, complementary approaches include multiconductor transmission line models, equivalent single-conductor transmission line models [34], [35], and equivalent multi-shell models [36]. Incorporating the quantum mechanical behavior of CNTs with a transmission line model leads to a hybrid approach [37]. Further, full-wave, finite-element models can include both the quantum mechanical properties of CNTs as well as mutual interactions between CNTs in the bundle [29], [38], [39], [40].

5.3.6   Iterative Optimization Approach

Biologically inspired, iterative optimization approaches such as genetic algorithms and neural networks provide another strategy for extraction of circuit parameters from device measurements [41]. Such approaches do not replace, but rather complement the development of and validation of models that has been described earlier. Iterative optimization approaches have been widely used before to design RF circuits [42], [43], calculate two-port error boxes in calibrated measurements [44], and generate models of nonlinear, large-signal devices [45]. They are especially well suited for generation of simple RF circuits from calibrated broadband measurements because it is not likely that the inverse problem is going to have a unique global solution. There is a multitude of different possible circuits consistent with the measured data. Iterative optimization approaches provide an automated way to explore this wide space of possible circuit models without invoking full-wave simulations.

We illustrate a basic example of and iterative approach with a genetic algorithm that generates a lumped-element model of a GaN nanowire device from calibrated scattering parameter measurements [41]. Though the nanowire device is used as an example here, iterative optimization approaches may be extended to materials characterization, such as frequency-dependent determination of parameters like complex permittivity and permeability of ferroelectric, ferromagnetic, or multiferroic materials, especially in thin film forms [46]. These methods may also be extended to interconnects, such as CNT bundle devices discussed earlier [47], as well as active devices. In the basic example discussed here, the circuit parameters are represented by a small number of cascaded elements, each representing a circuit element. For a given circuit model, the scattering parameters are calculated by generating an

ABCD matrix for each element, cascading all of the elements, and transforming the resulting matrix to a scattering parameter representation. Initially, the algorithm randomly generates a population of candidate circuit models from a pool of available elements. For the GaN nanowire devices, this pool included lumped elements such as series resistors, shunt capacitors, and transmission line elements, with possible ranges of parameter values defined based on reasonable values found in the literature [16], [48], [49].

For each generated model within the population, the scattering parameters are simulated and compared with the calibrated device measurements. The genetic approach culls the circuits from the population with the best agreement with the experiment by minimizing a cost function such as

$$c_f = \sum_{m,n} \sum_N \left(\left(\Delta_r(f_N) \right)^2 + \left(\Delta_i(f_N) \right)^2 \right), \tag{5.13}$$

where

$$\Delta_r(f_N) = \Re\left(S_{m,n}^{meas}(f_N) \right) - \Re(S_{m,n}^{sim}(f_N)), \tag{5.14}$$

and

$$\Delta_i(f_N) = \Im\left(S_{m,n}^{meas}(f_N) \right) - \Im\left(S_{m,n}^{sim}(f_N) \right), \tag{5.15}$$

where $\Im(X)$ is the imaginary part of X. In Equations (5.13) through (5.15), m and n are scattering parameter indices ($m = 1,2$; $n = 1,2$) and f_N represents the N frequencies at which the measurements were taken. S_{meas} and S_{sim} are the measured and simulated scattering parameters, respectively. The algorithm generates new circuit models by "mating" and "mutating" the circuit models with the lowest cost function c_f. This process is repeated over many cycles till the results from the generated circuit model are in a good agreement with the calibrated measurements. The process is terminated when the average cost function c_f is equal to or less than a target cost set by the user.

The simple algorithm was used to generate a model for a two-port, GaN nanowire similar to the Pt nanowire device discussed earlier. The comparison of the measured S-parameters with the simulated parameters from the circuit model is shown in Fig. 5.9. A schematic circuit model obtained for the GaN nanowire is shown in the inset of Fig. 5.9(b). The model lends itself to a simple and reasonable physical interpretation. The central 24.9 kΩ resistor may be interpreted as the intrinsic resistance of the nanowire while the combination of 1.3 kΩ resistances in parallel with the capacitances may be interpreted as contact impedances to the nanowire. That said, one should be wary of over-interpretation of the results. Though this circuit model is empirically found to be consistent with the measurements, it does not guarantee that the chosen model is globally optimal or the "best" circuit model that could be used. Still, an approach such as this provides some insight into overall device behavior. Further, it may be useful to have a model that is consistent with the measurements, independent of the physical interpretation.

References

[1] G. W. Hanson, "Fundamental Transmitting Properties of Carbon Nanotube Antennas," *IEEE Transactions on Antennas and Propagation* 53 (2005) pp. 3426–3435.

[2] C. Rutherglen and P. J. Burke, "Nanoelectromagnetics: Circuits and Electromagnetic Properties of Carbon Nanotubes," *Small* 8 (2009) pp. 884–906.

[3] U. Yogeswaran and S. Chen, "A Review on the Electrochemical Sensors and Biosensors Composed of Nanowires as Sensing Material," *Sensors* 8 (2008) pp. 290–313.

[4] S. Fiedler, M. Zwanzig, R. Schnidt, and W. Scheel, "Nanowires in Electronics Packaging." In *Nanopackaging: Nanotechnologies and Electronics Packaging*, (J. E. Morris, ed.), (Springer, 2009).

[5] A. R. Rathmell, S. M. Bergin, Y. -. Hua, Z. Li, and B. J. Wiley, "The Growth Mechanism of Copper Nanowires and Their Properties in Flexible, Transparent Conducting Films," *Advanced Materials* 22 (2010) pp. 3558–3563.

[6] S. Koo, M. D. Edelstein, Q. Li, C. A. Richter, and E. M. Vogel, "Silicon Nanowires as Enhancement-Mode Schottky Barrier Field-Effect Transistors," *Nanotechnology* 16 (2005) pp. 1482–1485.

[7] R. R. Llinas, K. D. Walton, M. Nakao, I. Hunter, and P. A. Anquetil, "Neuro-vascular Central Nervous Recording/Stimulating System: Using Nanotechnology Probes," *Journal of Nanoparticle Research* 7 (2005) pp. 111–127.

[8] A. I. Hochbaum, R. Chen, R. D. Delgado, W. Liang, E. C. Garnett, M. Najarian, A. Majumdar, and P. Yang, "Enhanced Thermoelectric Performance of Rough Silicon Nanowires," *Nature* 451 (2008) pp. 163–167.

[9] J. J. Boland, "Flexible Electronics: Within Touch of Artificial Skin," *Nature Materials* 9 (2010) pp. 790–792.

[10] P. J. Burke, "An RF Circuit Model for Carbon Nanotubes," *IEEE Transactions on Nanotechnology* 2 (2003) pp. 55–58.

[11] P. J. Burke, "Lüttinger Liquid Theory as a Model of the Gigahertz Electrical Properties of Carbon Nanotubes," *IEEE Transactions on Nanotechnology* 1 (2002) pp. 129–144.

[12] ANSYS HFSS: 3D full-wave electromagnetic field simulation. www.ansys.com/products/electronics/ansys-hfss. Accessed April 26, 2017.

[13] Burger S., et al., "JCMsuite: An Adaptive FEM Solver for Precise Simulations in Nano-optics," *Integrated Photonics and Nanophotonics Research and Applications* Conference Paper ITuE4 Optical Society of America (2008).

[14] ANSYS. Ansoft designer. www.ansys.com/. Accessed April 26, 2017.

[15] AWR microwave office: RF/microwave design software. www.awrcorp.com/products/ni-awr-design-environment/microwave-office. Accessed April 26, 2017.

[16] K. Kim, T. M. Wallis, P. Rice, C.-J. Chiang, A. Imtiaz, P. Kabos, and D. S. Filipovic, "A Framework for Broadband Characterization of Individual Nanowires," *IEEE Microwave and Wireless Component Letters* 20 (2010) pp. 178–180.

[17] K. Kim, P. Rice, T. M. Wallis, D. Gu, S. Lim, A. Imtiaz, P. Kabos, and D. S. Filipovic, "High-frequency Characterization of Contact Resistance and Conductivity of Platinum Nanowires," *IEEE Transactions on Microwave Theory and Techniques* 59 (2011) pp. 2647–2654.

[18] K. Kim, *Characterization of Carbon Nanotubes and Nanowires and Their Application*, PhD Thesis, University of Colorado (2010).

[19] P. A. Smith, C. D. Nordquist, T. N. Jackson, T. S. Mayer, B. R. Martin, J. Mbindyo, and T. E. Mallouk, "Electric-Field Assisted Assembly and Alignment of Metallic Nanowires," *Applied Physics Letters* 77 (2000) pp. 1399–1401.

[20] M. E. Toimil Molares, E. M. Hohberger, C. Scheaflein, R. H. Blick, R, Neumann, and C. Trautmann, "Electrical Characterization of Electrochemically Grown Single Copper Nanowires," *Applied Physics Letters* 82 (2003) pp. 2139–2141.

[21] J.-F. Lin, J. P. Bird, L. Rotkina, and P. A. Bennett, "Classical and Quantum Transport in Focused-Ion-Beam-Deposited Pt Nanointerconnects," *Applied Physics Letters* 82 (2003) pp. 804–805.

[22] G. De Marzil, D. Iacopino, A. J. Quinn, and G. Redmont, "Probing Intrinsic Transport Properties of Single Metal Nanowires: Direct-Write Contact Formation Using a Focused Ion Beam," *Journal of Applied Physics* 96 (2004) pp. 3458–3462.

[23] L. Penate-Quesada, J. Mitra, and P. Dawson, "Non-linear Electronic Transport in Pt Nanowires Deposited by Focused Ion Beam," *Nanotechnology* 18 (2007) pp. 215203–215207.

[24] T. Schwamb, B. R. Burg, N. C. Schirmer, and D. Poulikakos, "On the Effect of the Electrical Contact Resistance in Nanodevices," *Applied Physics Letters* 92 (2008) art. no. 243106.

[25] A. F. Mayadas and M. Shatzkes, "Electrical-Resistivity Model for Polycrystalline Films: The Case of Arbitrary Reflection at External Surfaces," *Physical Review B* 1 (1970) pp. 1382–1389.

[26] J. Hao and G. W. Hanson, "Infrared and Optical Properties of Carbon Nanotube Dipole Antennas," *IEEE Transactions on Nanotechnology* 5 (2006) pp. 766–775.

[27] P. Rice, T. M. Wallis, S. E. Russek, and P. Kabos, "Broadband Electrical Characterization of Multiwalled Carbon Nanotubes and Contacts," *Nano Letters* 7 (2007) pp. 1086–1090.

[28] S. Saluhuddin, M. Ludstrom, and S. Datta, "Transport Effects on Signal Propagation in Quantum Wires," *IEEE Transactions on Electron Devices* 52 (2005) pp. 1734–1752.

[29] N. Srivastava and K. Banerjee, "Performance Analysis of Carbon Nanotube Interconnects for VLSI Applications," *Proceedings of the IEEE/ACM International Conference on Computer-Aided Design* (2005) pp. 383–390.

[30] Y. Awano, "Carbon Nanotube Technologies for LSI via Interconnects," *IEICE Transactions on Electronics* E89-C (2006) pp. 1499–1503.

[31] M. Nihei, A. Kawabata, D. Kondo, M. Horibe, S. Sato, and Y. Awano, "Electrical Properties of Carbon Nanotube Bundles for Future via Interconnects," *Japanese Journal of Applied Physics* 44 (2005) pp. 1626–1628.

[32] A. Thess, R. Lee, P. Nikolaev, H. Dai, P. Petit, J. Robert, C. Xu, Y. H. Lee, S. G. Kim, A. G. Rinzler, D. T. Colbert, G. E. Scuseria, D. Tomanek, J. E. Fischer, and R. E. Smalley, "Crystalline Ropes of Metallic Carbon Nanotubes," *Science* 273 (1996) pp. 483–487.

[33] M. S. Dresselhaus, G. Dresselhaus, and P. Avouris, *Carbon Nanotubes: Synthesis, Structure, Properties, and Applications* (Springer, 2001).

[34] M. S. Sarto and A. Tamburrano, "Electromagnetic Analysis of Radiofrequency Signal Propagation along SWCN Bundles," *Proceedings of the Sixth IEEE Conference on Nanotechnology (IEEE-NANO 2006)* (2006) pp. 201–204.

[35] M. S. Sarto and A. Tamburrano, "Multiconductor Transmission Line Modeling of SWCNT Bundles in Common-Mode Excitation," *Proceedings of 2006 IEEE International Symposium on Electromagnetic Compatibility (EMC 2006)* (2006) pp. 466–471.

[36] M. V. Shuba, S. A. Maksimenko, and A. Lakhtakia, "Electromagnetic Wave Propagation in an Almost Circular Bundle of Closely Packed Metallic Carbon Nanotubes," *Physical Review B* 76 (2007) art. no. 155407.

[37] M. S. Sarto, A. Tamburrano, and M. D'Amore, "New Electron-Waveguide-Based Modeling for Carbon Nanotube Interconnects," *IEEE Transactions on Nanotechnology* 8 (2009) pp. 214–225.

[38] A. Naeemi, R. Sarvari, and J. D. Meindl, "Performance Comparison between Carbon Nanotube and Copper Interconnects for GSI," *IEDM Technical Digest in Electron Devices Meeting* (2004) pp. 699–702.

[39] A. Nieuwoudt and Y. Massoud, "Evaluating the Impact of Resistance in Carbon Nanotube Bundles for VLSI Interconnect Using Diameter-dependent Modeling Techniques," *IEEE Transactions on Electron Devices* 53 (2006) pp. 2460–2466.

[40] C. W. Tan and Jianmin Miao, "Modeling of Carbon Nanotube Vertical Interconnects as Transmission Lines," *Proceedings of the 2006 IEEE Conference on Emerging Technologies – Nanoelectronics* (2006) pp. 75–78.

[41] T. M. Wallis, "A Genetic Algorithm for Generating RF Circuit Models from Calibrated Broadband Measurements," *Proceedings of the 78th ARFTG Microwave Measurement Symposium* (2011) pp. 1–5.

[42] J. G. Pascual, P. F. Quesada, D. C. Rebenaque, J. L. G. Tornero, and A. A. Melcon, "A Multilayered Shielded Microwave Circuit Design Method Based on Genetic Algorithms and Neural Networks," *2006 IEEE MTT-S International Microwave Symposium Digest* (2006) pp. 1427–1430.

[43] S. Iezekiel, "Application of Evolutionary Computation Techniques to Nonlinear Microwave Circuit Analysis," *Proceedings of the 8th IEEE International Symposium on High Performance Electron Devices for Microwave and Optoelectronic Applications* (2000) pp. 230–235.

[44] A. S. Adalev, N. V. Korovkin, M. Hayakawa, and J. B. Nitsch, "Deembedding and Unterminating Microwave Fixtures with the Genetic Algorithm," *IEEE Transactions on Microwave Theory and Techniques* 54 (2006) pp. 3131–3140.

[45] J. Jargon, K. C. Gupta, D. Schreurs, K. Remley, and D. DeGroot, "A Method of Developing Frequency-Domain Models for Nonlinear Circuits Based on Large-Signal Measurements," *Proceedings of the 58th ARFTG Microwave Measurement Symposium* (2001) pp. 35–48.

[46] N. Orloff, J. Mateu, M. Murakami, I. Takeuchi, and J. C. Booth, "Broadband Characterization of Multilayer Dielectric Thin-Films," *2007 IEEE MTT-S International Microwave Symposium Digest (MTT)* (2007) pp. 1177–1180.

[47] J. J. Plombon, K. P. O'Brien, F. Gstrein, V. M. Dubin, and Y. Jiao, "High-Frequency Electrical Properties of Individual and Bundled Carbon Nanotubes," *Applied Physics Letters* 90 (2007) art. no. 063106.

[48] D. Gu, T. M. Wallis, P. Blanchard, S.-H. Lim, A. Imtiaz, K. A. Bertness, N. A. Sanford, and P. Kabos, "Deembedding Parasitic Elements of GaN Nanowire Metal Semiconductor Field Effect Transistors by Use of Microwave Measurements," *Applied Physics Letters* 98 (2011) art. no. 223109.

[49] K. Kim, T. M. Wallis, P. Rice, C.-J. Chiang, A. Imtiaz, P. Kabos, and D. Filipovic, "Modeling and Metrology of Metallic Nanowires with Application to Microwave Interconnects," *2010 IEEE MTT-S International Microwave Symposium Digest (MTT)* (2010) pp. 1292–1295.

6   Characterization of Nanofiber Devices

6.1   The Measurement Problem

Previous chapters have introduced and described a variety of measurement techniques for RF nanoelectronic devices. Here, our objective is to work through an illustrative example that highlights strategies and challenges related to implementing a specific RF nanoelectronic device measurement. To that end, this chapter will describe the broadband, two-port characterization of an individual nanofiber device (here, a "nanofiber" is broadly defined to be any individual nanotube or nanowire, or a bundle of nanotubes or nanowires). Historically, the electromagnetic characterization of individual CNTs at gigahertz frequencies was one of the first measurement challenges encountered in the relatively short history of RF nanoelectronics [1]–[4]. Interest in making accurate RF measurements of nanotubes has been driven largely by the potential uses of CNTs as high-quality interconnects in very large scale integrated circuits and as nano-antennas in communications applications [5]. In addition, such measurements were needed to investigate what influence, if any, the quantum capacitance and kinetic inductance have on the AC transport properties of CNTs [6]. The techniques that were developed for electromagnetic characterization of CNTs have subsequently been applied to additional nanomaterial systems, including semiconducting nanowires and graphene nanoribbons [7]–[9].

The twofold goals of the measurement example presented in this chapter are: (1) to obtain de-embedded complex scattering parameters for the device and (2) to extract circuit model parameters that describe the electromagnetic properties of the device, including the nanofiber and its contacts. We will consider the specific case of a semiconducting GaN nanowire, though this measurement approach generally is applicable to any nanofiber. Note that this measurement approach is not the only approach to the problem, nor is it necessarily the best approach. Nonetheless, this example measurement concretely implements the strategies that have been presented in previous chapters.

First, we introduce the nanofiber device geometry and describe approaches to fabrication. Subsequently, we discuss calibrated, on-wafer measurements of the device, including the translation of the reference planes, which allows for complete de-embedding of the scattering parameters. We present a circuit model for the device and discuss the steps needed to extract relevant circuit model parameters from the de-embedded scattering parameters. Throughout this example, we present measured and de-embedded data for the specific case of a two-port device that incorporates an individual GaN nanowire.

6.2   Device Geometry and Fabrication

In order to characterize an individual nanofiber at microwave frequencies, the nanofiber must be integrated into a microwave host device. Here, we will consider a nanofiber that is integrated into a CPW structure as illustrated in Fig. 6.1. This simple, passive device represents a practical test platform for measuring electromagnetic material properties and for optimization of contact impedance. Advanced device applications will likely be more complex, but this design presents a suitable case study for demonstration and discussion of the measurement approaches for extreme impedance devices in general and RF nanoelectronic devices in particular. In this device design, there is a small gap in the center conductor at the midpoint of the CPW. An individual nanofiber bridges this gap. The length of the gap is chosen to be between a few micrometers to tens of micrometers, depending on the length of the nanofiber under test. In this specific example, for an as-grown GaN nanowire with a length of about 10 μm, a 4 μm gap was chosen. To either side of the gap, the center conductor is tapered in order to constrain the area where the nanofiber can bridge the gap and to reduce the reflection of microwave signals that may result from an abrupt impedance mismatch. At either end of the host device, segments of CPW with a transmission line impedance of 50 Ω serve as landing sites for on-wafer, ground-signal-ground probes. Because of the high impedance of the nanofiber, a significant contribution to transmission through the device will be made by parasitic coupling across the gap in the center conductor. Thus, following the strategy laid out in Chapters 4 and 5, an empty, nanofiber-free device that is otherwise identical to the device show in Fig. 6.1 is fabricated on the same wafer.

In practice, there are several approaches to fabricating a microwave nanofiber device. One approach is to fabricate the CPW host by use of standard

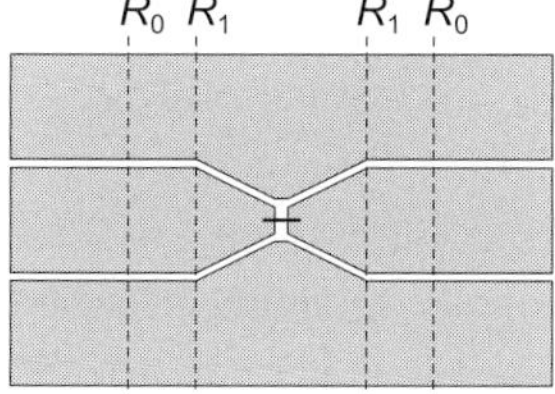

Figure 6.1.   Geometry of a nanowire device.
A CPW host device for broadband characterization of nanowire devices is illustrated (Top View). The CPW host structure is illustrated in gray, the nanowire is illustrated in black. The on-wafer, multiline thru-reflect-line (TRL) calibration establishes reference planes at positions R_0. Subsequently, the reference planes are rolled to position R_1.

photolithography techniques and then affix an individual nanofiber in the device by use of a focused ion beam (FIB) [10]. A nano-manipulator is used to transfer an individual nanofiber into the gap in the CPW host. Subsequently, the ends of the nanofiber may be fixed by metal bonds formed by use of FIB-induced deposition. These metal bonds also serve as electrical contact points between the fiber and the host structure, thus contributing to the contact impedance. The FIB-based approach enables precise control of fiber positioning and contact formation. Furthermore, this approach allows for real-time inspection of the device as FIB capabilities are typically integrated into dual-beam systems with SEMs. However, this approach is time-consuming and devices are produced one-by-one. Furthermore, the contact impedance is difficult to control and reproduce. Lastly, the ion beam may damage the nanofiber.

A higher-throughput alternative to the FIB-based approach is to align the individual fibers with dielectrophoresis. During this process, a sinusoidal AC voltage signal is applied across the CPW gap. For the GaN nanowires discussed in detail in this chapter, this signal has a frequency between 50 kHz and 100 kHz and a peak-to-peak amplitude between 10 V and 20 V [11]. As the signal is being applied, a drop of nanofibers suspended in solution is dispensed over each device site and allowed to evaporate. Due to the presence of the AC field, some of the individual nanofibers will be induced to align themselves across the gaps in the devices. The aligned nanofibers can be subsequently secured by deposition of a photolithographically defined layer on top of the contact area. A nanofiber device produced by use of this approach is shown in Fig. 6.2. Note that dielectrophoresis requires electrical contacts to the center conductor of each CPW device on the wafer. Though this approach yields multiple devices, the yield is significantly less than 100 percent and some fraction of the devices may have more than one nanofiber per gap.

Though this device geometry is simple and the fabrication steps described earlier appear straightforward, the fabrication of this two-port nanoelectronic device

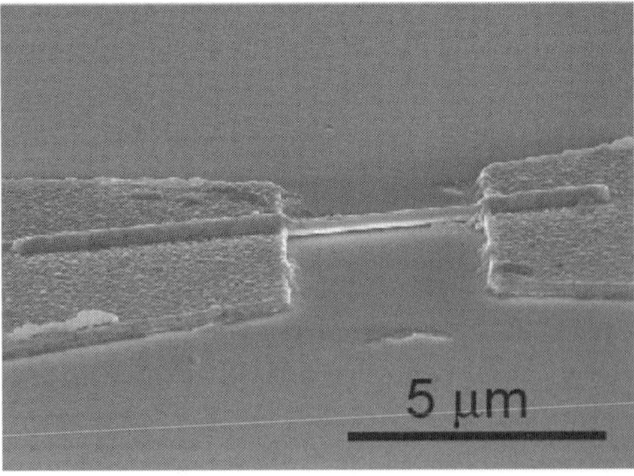

Figure 6.2.   Scanning electron microscope image of a GaN nanowire device.
The image shows an individual GaN nanowire bridging a 4 µm gap in a two-port, CPW host device.

presents several significant challenges. The first barrier is the need for high-quality nanomaterials. In the case of GaN nanowires, the wires were grown via plasma-assisted molecular beam epitaxy. The as-grown nanowires were essentially defect-free and could be removed from the growth substrate by ultrasonic agitation in isopropanol. This yielded a nanowire solution amenable to dielectrophoresis, though the use of extremely small (~µL) drops of solution placed only upon the gap in the CPW was required in order to prevent unwanted placement of stray wires in locations other than the gap in the CPW. Once the wires were placed, they were secured with appropriate metal deposition.

The next fabrication challenge is substantial: the electromagnetic properties of the resulting contact regions between the GaN nanowire and the metal host structure must be controlled. In general, the control and optimization of electrical contacts to nanomaterials is one of the most difficult steps in the fabrication of RF nanoelectronic devices. In the specific case of GaN nanowires, the fabrication of ohmic contacts to the wire requires several key steps, including appropriate choice of contact materials that may depend on the wire conductivity and carrier type (e.g., Ti and Al for n-type GaN) as well as a reactive ion etch treatment of the wires before metal deposition. Generally, contact optimization in the device development phase follows an iterative process, cycling through repeated fabrication and characterization steps. For more complex devices, such as nanowire transistors, which incorporate multiple material components and three or more contacts to the nanowire, fabrication naturally becomes increasingly challenging and complex [3], [7], [11].

Before describing the measurements themselves, it is useful to point out how strategic device design and fabrication must be developed in conjunction with the identification of measurement approaches. As a simple example, additional "empty" devices, as well as other calibration devices that will be discussed in more detail subsequently, must be fabricated along with the DUT. In a case where multiple measurement approaches are under consideration, it is conceivable that multiple sets of calibration standards may need to be included in the design. A subtler issue is that the design choice to include tapered sections of CPW will have implications for modeling the circuit and extracting model parameters, as will also be discussed in the following section.

6.3   Calibrated On-Wafer Measurements

The next step is calibrated, on-wafer measurement of the device and de-embedding of the device's two-port scattering parameters from the test platform. As outlined in earlier chapters, there are a number of well-known on-wafer calibration procedures to choose from, including SOLT, line-reflect-reflect-match (LRRM), and multiline thru-reflect-line (TRL). Here, we select multiline TRL [12] because it enables rolling of the reference planes of the measurement from their original position (labeled R_0 in Fig. 6.1) to a new location closer to the nanowire and its contact points (labeled

R_1 in Fig. 6.1). By translating the reference planes in this way, the influence of the CPW segments at either end of the device can effectively be taken out of the device measurement.

Each on-wafer calibration method requires the fabrication of a set of on-wafer calibration devices. Ideally, these calibration devices will be fabricated on the same substrate as the DUTs. For multiline TRL, the following CPW calibration devices are required: a thru (chosen here to be 0.500 mm long), a short circuit, and multiple transmission lines of differing lengths (chosen here to be 1.80 mm, 2.60 mm, 3.83 mm, and 6.10 mm). A typical wafer layout is illustrated in Fig. 6.3, including nanowire devices, empty reference devices, and a set of calibration devices. Note that we have implemented a ground-signal-ground configuration for each CPW calibration standard as well as the CPW nanowire device. Other configurations, such as ground-signal, are possible, but with corresponding trade-offs in device footprint, device performance, and measurement uncertainty.

To perform the calibrated measurements, each of the on-wafer standards is measured by use of a two-port, on-wafer probe station. Subsequently, every DUT, including at least one nanowire device and one empty device, is measured with the same probes. The measured scattering parameters of the calibrations standards are processed via the calibration algorithm in order to determine the error boxes. The error boxes are then used to de-embed the scattering parameters of the nanowire

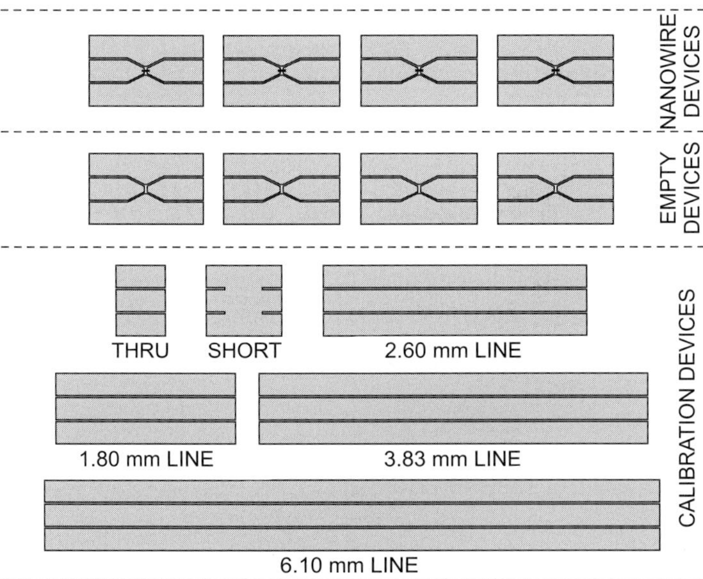

Figure 6.3.   Layout of nanowire devices and calibration devices.
Four nanowire devices under test, four empty (nanowire-free) devices, and six on-wafer calibration structures are illustrated. The six on-wafer calibration structures, including a thru, a short, and four transmission lines, are used to perform a thru-reflect-line (TRL) calibration. The empty devices are used to extract the stray capacitance, as discussed in the text.

device and empty device. In a TRL calibration, the complex scattering coefficient is also found by the calibration algorithm. In practice, it is useful to remeasure each on-wafer standard again at the conclusion of the experiment. This second set of calibration standard measurements provides an additional data set and quantitative estimate of statistical uncertainty in the calibration procedure.

The magnitudes of the as-measured, uncalibrated scattering parameters, S^M_{11} and S^M_{21}, are shown in Fig. 6.4, for both the nanowire device and the empty device. The full measured, two-port scattering parameter matrices for the nanowire device and for the empty device are represented by $S^{M\text{-}total}$ and $S^{M\text{-}empty}$, respectively. Keep in mind that the elements of the scattering parameter and admittance matrices discussed in this case study are complex valued. For simplicity, we will focus on the magnitudes of these elements throughout the discussion. Over the frequency range shown in Fig. 6.4, differences in S^M_{21} for the nanowire and empty devices are observable, though the difference decreases as the measurement frequency increases. Meanwhile little difference is discernable for S^M_{11}.

The TRL procedure yields the corresponding calibrated, two-port scattering parameters, $S^{C\text{-}total}$ and $S^{C\text{-}empty}$. By default, the reference planes of the calibrated measurement are displaced from each end of the host device by 0.250 mm (half the length of the thru standard). The location of these reference planes are labeled as R_0 in Fig. 6.1. Because the TRL algorithm extracts the complex propagation constant in addition to the error boxes, it is possible to roll the reference planes to a new position at the boundary between the straight and tapered sections of the host CPW, labeled as R_1 in Fig. 6.1.

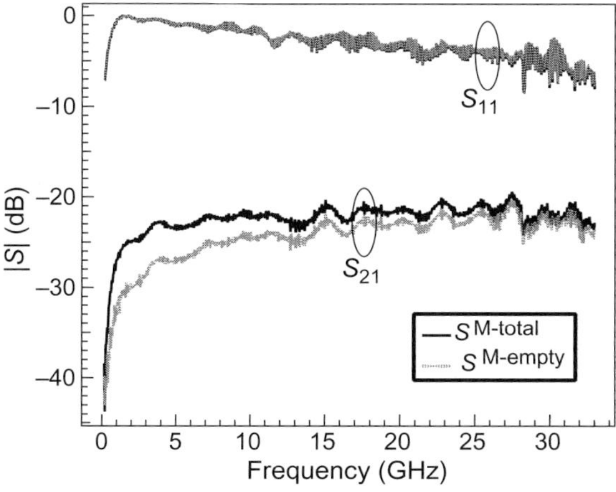

Figure 6.4. Raw S-parameter measurement of GaN nanowire device.
The magnitudes of the raw, uncalibrated scattering parameters, S^M_{11} and S^M_{21}, are shown as a function of frequency, as measured with a VNA. Measurements of a nanowire device, $S^{M\text{-}total}$ (solid black curve), and an empty reference device, $S^{M\text{-}empty}$ (dashed gray curve), are shown.

A few algebraic manipulations are required to transform a calibrated scattering parameter matrix S^C with reference planes at R_0 to a calibrated scattering parameter matrix $S^{C'}$ with reference planes at R_1. The first step is to transform S^C into a transmission matrix T^C. The propagation constant γ is known from the TRL algorithm. The distance between R_0 and R_1 is L. A transmission matrix T^L for a line with propagation constant γ and length L can be constructed as follows:

$$T^L = \begin{bmatrix} e^{-\gamma L} & 0 \\ 0 & e^{\gamma L} \end{bmatrix}. \tag{6.1}$$

A new transmission matrix with reference planes at R_1 can be formed from

$$T^{c'} = (T^L)^{-1}\, T^C\, T^L. \tag{6.2}$$

As a final step, the transmission matrix $T^{C'}$ is transformed back to a scattering parameter matrix $S^{C'}$. Note that the above analysis assumes that all transmission matrices share the same reference impedance. If the transmission matrices do not share the same reference impedance, then additional algebraic steps will be required to transform the impedances.

The magnitudes of the calibrated, transformed scattering parameters for the nanowire and empty device with the reference planes rolled to the new position R_1, S^{total} and S^{empty}, are shown in Fig. 6.5. As with the uncalibrated scattering parameters, clear differences are discernable in S_{21} but not in S_{11}. Note that the frequency-dependent ripples in the measurements shown in Fig. 6.4 have been eliminated. The de-embedded scattering parameters shown in Fig. 6.5 are now smooth curves as a function of frequency.

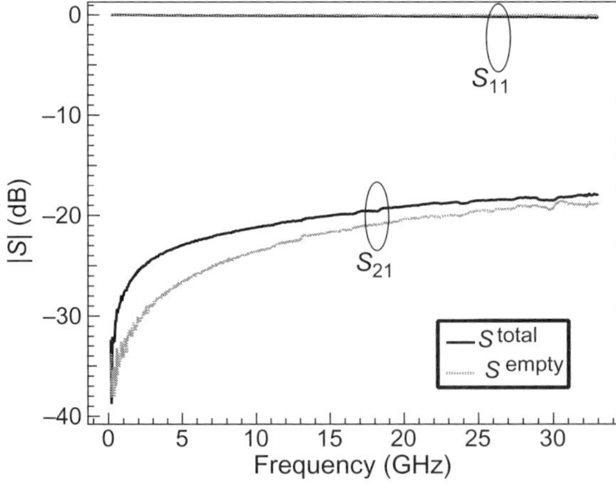

Figure 6.5. De-embedded S-parameter measurement of GaN nanowire device. The magnitudes of the de-embedded, calibrated scattering parameters, S_{11} and S_{21}, with the reference planes translated to position R_1, are shown as a function of frequency. Measurements of a nanowire device, S^{total} (solid black curve) and an empty reference device, S^{empty} (dashed gray curve) are shown.

Ultimately, the objective is to de-embed the properties of the nanowire and its contacts from the properties of the host structure. In particular, we seek to determine S^{nw}, the complex scattering parameter matrix of the nanowire and its contacts. The location of reference plane R_1 implies that S^{nw} also includes the influence of the tapered CPW sections. The de-embedding is more easily accomplished if the scattering parameter matrices are transformed to an impedance matrix representation. S^{total} and S^{empty} are transformed to impedance matrices Y^{total} and Y^{empty}, which are related to Y^{nw}, the impedance matrix of the nanowire and its contacts, in the following way:

$$Y^{total} = Y^{nw} + Y^{empty}. \tag{6.3}$$

Y^{nw} is readily determined by use of Equation 6.3. Figure 6.6 shows the magnitudes of elements of the impedance matrices Y^{total}, Y^{empty}, and Y^{nw}. Comparing Y^{total} and Y^{empty}, clear differences are discernable in both Y_{11} and Y_{21}. Note that the linear dependence of Y^{empty}_{21} on frequency is consistent with purely capacitive coupling across the empty gap. As a final step, the complex matrix Y^{nw} may be converted to S^{nw}, as shown in Fig. 6.6(d).

6.4 Uncertainty Analysis

Uncertainty analysis is a vital step in establishing reliable metrology tools for emerging application areas. Unfortunately, this step has often been overlooked during RF and microwave measurements of nanoelectronic devices. Consider the measurements of the magnitude of the admittance parameters of the GaN nanowire devices, Y^{nw}_{11} and Y^{nw}_{21}, shown in Fig. 6.6. It is clear that the magnitude of Y^{nw}_{21} is quite small with a maximum value of about 1 mS. A reliable uncertainty analysis is particularly important in the measurement of such small quantities where the value of the uncertainty may represent a significant fraction of the value of the measurand. In this section, we will estimate the uncertainty budget for the measurement of the GaN nanowire devices.

Before discussing the specific case of the two-port nanowire device measurement, it is useful to review several core concepts of uncertainty analysis. Suppose there are N different components of uncertainty in a measurement process. Let u_i be the standard uncertainty for component i ($i = 1, 2, ..., N$) and let u_i be equal to the positive square root of the estimated variance. If the components of uncertainty are uncorrelated, one simple way to estimate the total standard uncertainty is

$$u_{total} = \sqrt{u_1^2 + u_2^2 + ... + u_N^2}. \tag{6.4}$$

In describing contributions of different components of uncertainty, it is useful to distinguish between methods used to estimate the uncertainty values [13]. Uncertainties determined from the statistical analysis of multiple, repeated measurements are referred to as "Type A" uncertainties. For example, Type A uncertainties

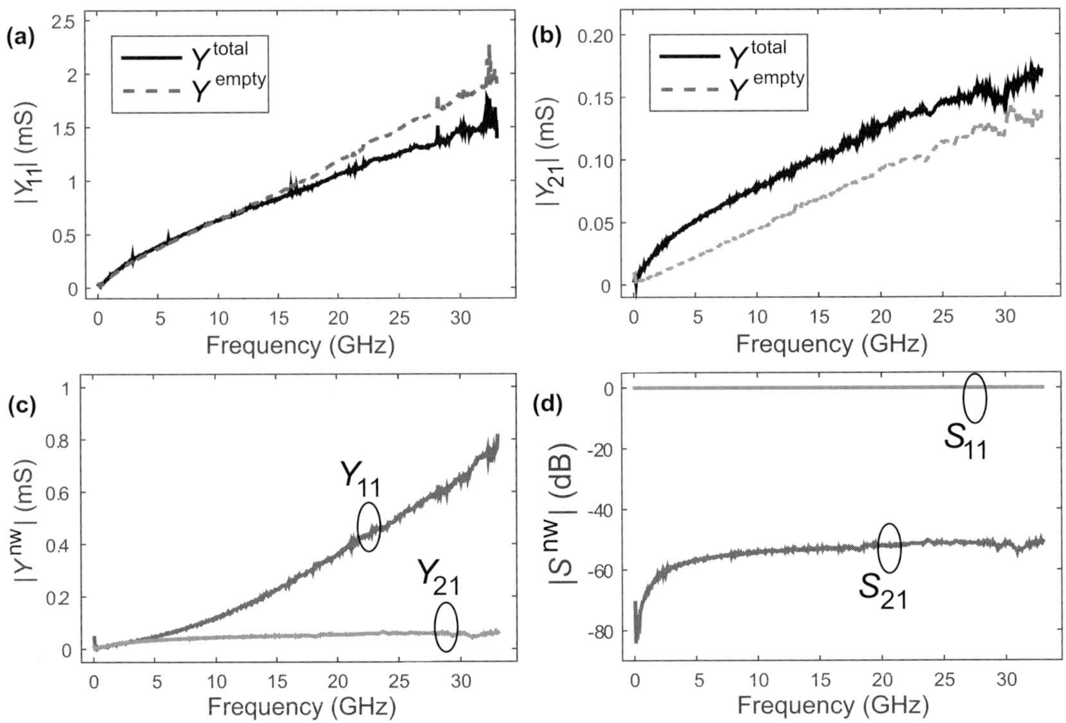

Figure 6.6. Calibrated measurement of GaN nanowire scattering parameters.
(a) and (b) Magnitudes of the admittance matrix elements Y_{11} and Y_{21} of the matrices $\mathbf{Y}^{total}$ and $\mathbf{Y}^{empty}$. (c) Magnitudes of the admittance matrix elements Y_{11} and Y_{21} of the matrix $\mathbf{Y}^{nw}$. (d) Magnitudes of the scattering matrix elements S_{11} and S_{21} of the matrix $\mathbf{S}^{nw}$.

may be calculated from repeated measurements of the same system as well as from measurements of multiple, nominally identical, systems. Uncertainties determined from nonstatistical methods are referred to as "Type B" uncertainties. Examples of Type B uncertainties include uncertainties calculated from physical models or uncertainties taken from manufacturer specifications.

Returning to the example of the two-port nanowire devices, the measurement process rests on the calibrated, on-wafer measurement of two scattering parameter matrices, $\mathbf{S}^{total}$ and $\mathbf{S}^{empty}$. Let ΔS^{total}_{ij} and ΔS^{empty}_{ij} be the uncertainties in the matrix element ij of these scattering parameter matrices. As the scattering parameters are transformed to an admittance representation, the uncertainties will propagate in the following way:

$$\Delta Y^{total}_{ij} = \sqrt{\sum_{k,l}\left(\frac{\partial Y^{total}_{ij}}{\partial S^{total}_{kl}}\Delta S^{total}_{kl}\right)^2} \tag{6.5}$$

and

$$\Delta Y^{empty}_{ij} = \sqrt{\frac{(\Delta Y^{A-empty}_{ij})^2}{N_{empty}} + \sum_{k,l}\left(\frac{\partial Y^{empty}_{ij}}{\partial S^{empty}_{kl}}\Delta S^{empty}_{kl}\right)^2}. \tag{6.6}$$

The partial derivatives of the form $\delta Y/\delta S$ are determined from the equations that prescribe the transformation from scattering parameters to admittance parameters [14]. Note the inclusion of the Type A uncertainty term $\Delta Y^{A-empty}_{ij}$, in Equation (6.6). This term is included because imperfections in the fabrication process and substrate inhomogeneity may lead to variations of geometric and electromagnetic properties from device to device. The value of $\Delta Y^{A-empty}_{ij}$ is estimated from statistical analysis of N_{empty} repeated measurements of empty devices. Once ΔY^{total}_{ij} and ΔY^{empty}_{ij} are known, the total uncertainty in the magnitude of the elements of Y^{nw} is given by

$$\Delta Y^{nw}_{ij} = \sqrt{(\Delta Y^{empty}_{ij})^2 + (\Delta Y^{total}_{ij})^2 + (\Delta Y^{B}_{ij})^2 + \frac{(\Delta Y^{A-nw}_{ij})^2}{N_{repeat}}}. \tag{6.7}$$

Further variations from measurement to measurement may result from lack of repeatability in the position of the GSG probes on the device. To account for this component of the uncertainty, the measurement of a given nanowire device is repeated N_{repeat} times and the Type A uncertainty term ΔY^{A-nw}_{ij} is determined from statistical analysis of the resulting measurements. A single Type B uncertainty term, ΔY^{B-nw}_{ij}, is included to account for systematic sources of uncertainty in the measurement process, including the validity (or lack of validity) of Equation 6.3.

A typical uncertainty budget for the magnitude of the admittance matrix elements Y^{nw}_{ij} of a two-port nanowire device, based on the estimation process described earlier, is shown in Table 6.1. The values of ΔS^{total}_{ij}, ΔS^{empty}_{ij}, and

Table 6.1 Typical Uncertainty Budget for the Magnitude of the Admittance Matrix Elements Y^{nw}_{ij} of a Two-Port Nanowire Device

Uncertainty Component	k=2 Value at 10 GHz	k=2 Value at 35 GHz
ΔS^{total}	0.005	0.005
ΔS^{empty}	0.005	0.005
$\Delta Y^{A-empty}$	0.005 mS	0.025 mS
ΔY^{total}	0.050 mS	0.050 mS
ΔY^{empty}	0.050 mS	0.055 mS
ΔY^{A-nw}	0.015 mS	0.030 mS
ΔY^{B}	0.050 mS	0.050 mS
ΔY^{nw}	0.085 mS	0.090 mS

$\Delta Y^{B\text{-}nw}_{ij}$ are estimated from the literature in References [15], [16]. Note that the Type A uncertainty values, $\Delta Y^{A\text{-}empty}_{ij}$ and $\Delta Y^{A\text{-}nw}_{ij}$, vary the strongest between the two frequencies shown in the table. The coverage factor k is two for the uncertainty components shown in the table, i.e., they correspond to twice the standard uncertainty. When the complete uncertainty calculation is carried out via Equation 6.4, the estimated uncertainty ΔY^{nw}_{ij} is about 0.085 mS at 10 GHz and 0.090 mS at 30 GHz. Comparison to Fig. 6.6 reveals that these uncertainty estimates represent a significant fraction of the value of Y^{nw}_{11}, namely about 75 percent of Y^{nw}_{11} at 10 GHz and about 15 percent of Y^{nw}_{11} at 30 GHz. Further comparison to Fig. 6.6 reveals that these uncertainty estimates exceed the value of Y^{nw}_{21} throughout the measured frequency range. The significant uncertainties of this measurement technique imply that any circuit or material parameters extracted from these measurements should be treated as estimates at best. The largest contribution to the total uncertainty comes from the uncertainty in the calibrated, on-wafer measurement of scattering parameter matrices, as propagated via Equations 6.5 and 6.6. Estimation and minimization of uncertainties in calibrated on-wafer measurements are ongoing challenges, but do provide a way forward for improving both on-wafer measurements in general as well as RF nanoelectronic device measurements in particular. On-wafer implementation of the extreme impedance techniques introduced in Chapter 3 may also provide an avenue to finding a measurement approach with lower uncertainties.

6.5   Extraction of Parameters from Circuit Models

In some cases, the calibrated measurement of the de-embedded scattering parameters of the device, S^{nw}, may be sufficient. However, in a research and development environment, additional information about the nanoelectronic device may be sought. In the case of the simple, prototype GaN nanowire device discussed in this chapter, a number of questions may be asked. What are the intrinsic electromagnetic material properties of the nanowire? Are the contacts Schottky-like or ohmic? Are there strong differences between the contacts at either end of the nanowire? Are there strong variations in the scattering parameters from device to device? In order to address such questions, we will develop a circuit model for the device and extract corresponding circuit parameters.

As discussed in Chapter 5, circuit models of RF nanoelectronic devices often combine both distributed and lumped elements. Nanofibers with different physical properties may be better represented by alternative choices within a circuit model. For example, a metallic nanowire may be represented by a simple, lumped-element resistor, provided that its length is much shorter than the wavelength in the experiment. For some nanomaterials, such as graphene and CNTs, being measured at high enough frequencies, quantum phenomena such as kinetic inductance and quantum capacitance may make a significant contribution. In such cases, appropriate circuit elements must be added to the model.

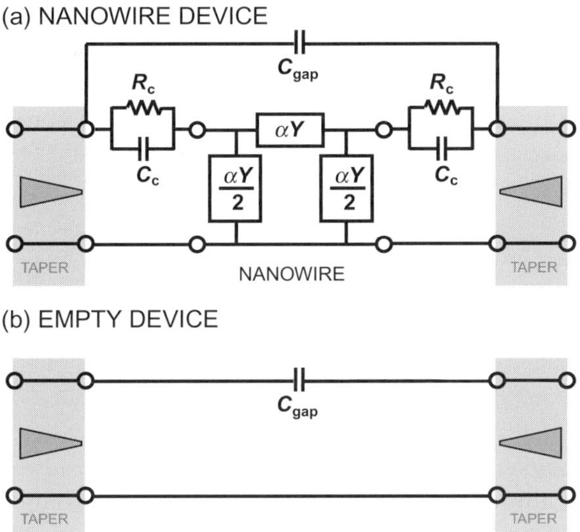

Figure 6.7. Circuit models for a nanowire device and an empty device.
(a) Circuit model for a nanowire device, including contact resistance (R_c) and electrostatic contact capacitance (C_c), as well as a π-network representing the nanowire. (b) Circuit model for the empty, nanowire-free device. Both models include the tapered section of the CPW host as well as the parasitic capacitance C_{gap}.

Figure 6.7 shows possible circuit models of the two-port GaN nanowire device, with and without the nanowire and its contacts. In this model, the contact resistance R_c, electrostatic contact capacitance C_c, and parasitic capacitance C_{gap} are represented by lumped elements while the nanowire itself and the tapered segments to either side of the wire are modeled as distributed elements. The contact capacitance C_c may make a significant contribution in certain nanofiber devices, such as CNTs. Here, in the case of annealed GaN nanowires with ohmic contacts, $(\omega\, C_c)^{-1} \gg R_c$. Thus, C_c will be ignored for the remainder of the discussion. The tapered segments are modeled as a finite series of CPW transmission lines with successively narrower center conductor widths. The models of the tapered segments are parameterized by an effective length, l_{taper}. Finally, the GaN nanowire is represented by a π-network. The admittance of each element of the π-network is proportional to the characteristic complex admittance Y of an equivalent transmission line

$$Y = \sqrt{\frac{G_{nw} + j\omega C_{nw}}{R_{nw} + j\omega L_{nw}}} \tag{6.8}$$

where R_{nw}, L_{nw}, G_{nw}, and C_{nw} are the per-unit-length series resistance, series inductance, shunt conductance, and shunt capacitance, respectively. We require that if one of the two ports of the π-network is loaded with a characteristic impedance $Z = 1/Y$, the total impedance at the unloaded port will be Z. This requirement leads to a value of the constant $\alpha = 2\,/\,\sqrt{5}$. In the specific example of the GaN nanowire

Table 6.2 Typical Circuit Parameters for a Two-Port GaN Nanowire Device

Parameter	Value
C_{gap}	690 aF
l_{taper}	130 μm
R_{nw}	58 kΩ / μm
C_{nw}	6500 aF / μm
R_c	3.6 kΩ

device, G_{nw} and L_{nw} are negligible, i.e., $R_{nw} \gg \omega L_{nw}$ and $\omega C_{nw} \gg G_{nw}$. The complex scattering parameters of the model network can be calculated with a commercial software package. Subsequently, the unknown parameters can be adjusted to fit the model response to the calibrated, measured scattering parameter data.

The model has five unknown parameters: R_c, C_{gap}, l_{taper}, R_{nw}, and C_{nw}. Two of these parameters, C_{gap} and l_{taper}, may be determined by fitting the modeled scattering parameters of the empty device to the measured scattering parameters. This initial fitting procedure requires that the model of the nanowire device be modified to represent an empty device by removing the π-network that represents the nanowire as well as the lumped elements representing the contacts (R_c and C_c). Once C_{gap} and l_{taper} are known, the remaining unknown parameters, R_c and the ratio R_{nw} / C_{nw}, may be determined by fitting the modeled scattering parameters for the full nanowire device to the measured scattering parameters. R_{nw} may then be determined by enforcing the condition that the DC resistance equal the sum of the wire resistance and twice the contact resistance. As we have seen in other cases, unique determination of contact and nanowire properties requires additional measurements, in this case in the form of a DC resistance measurement. Alternatively, the measured nanowire device data may be fit with R_c fixed to a reasonable value [17]–[19], leaving the ratio R_{nw} / C_{nw} as the only remaining fitting parameter. Typical values for R_c, C_{gap}, l_{taper}, R_{nw}, and C_{nw} for a two-port GaN nanowire device [9] are summarized in Table 6.2.

References

[1]   Z. Yu and P. J. Burke, "Microwave Transport in Metallic Single-Walled Carbon Nanotubes," *Nano Letters* 5 (2005) pp. 1403–1406.

[2]   M. Zhang, X. Huo, P. C. H. Chan, Q. Liang, and Z. K. Tang, "Radio-Frequency Characterization for the Single-Walled Carbon Nanotubes," *Applied Physics Letters* 88 (2006) art. no. 163109.

[3]   J. M. Bethoux, H. Happy, G. Dambrine, V. Derycke, M. Goffman, and J. P. Burgoin, "An 8-GHz f_t Carbon Nanotube Field-Effect Transistor for Gigahertz Range Applications," *IEEE Electron Device Letters* 27 (2006) pp. 681–683.

[4] J. J. Plombon, K. P. O'Brien, F. Gstrein, V. M. Dubin, and Y. Jiao, "High-Frequency Electrical Properties of Individual and Bundled Carbon Nanotubes," *Applied Physics Letters* 90 (2007) art. no. 063106.

[5] P. Russer, "Nanoelectronics-Based Integrated Antennas," *IEEE Microwave Magazine* 11 (2010) pp. 58–71.

[6] C. Rutherglen and P. J. Burke, "Nanoelectromagnetics: Circuit and Electromagnetic Properties of Carbon Nanotubes," *Small* 5 (2009) pp. 884–906.

[7] S. Vandenbrouck, K. Madjour, D. Théron, Y. Dong, Y. Li, C. M. Lieber, and C. Gaquiere, "12 GHz F_{MAX} GaN/AlN/AlGaN Nanowire MISFET," *IEEE Electron Device Letters* 30 (2009) pp. 322–324.

[8] S. Salahuddin, M. Lundstrom, and S. Datta, "Transport Effects on Signal Propagation in Quantum Wires," *IEEE Transactions on Electron Devices* 52 (2005) pp. 1734–1742.

[9] T. Mitch Wallis, Dazhen Gu, Atif Imtiaz, Christopher S. Smith, Chin-Jen Chiang, Pavel Kabos, Paul T. Blanchard, Norman A. Sanford, and Kris A. Bertness, "Electrical Characterization of Photoconductive GaN Nanowires from 50 MHz to 33 GHz," *IEEE Transactions on Nanotechnology* 10 (2011) pp. 832–834.

[10] P. Rice, T. M. Wallis, S. E. Russek, and P. Kabos, "Broadband Electrical Characterization of Multiwalled Carbon Nanotubes and Contacts," *Nano Letters* 7 (2007) pp. 1086–1090.

[11] P. T. Blanchard, K. A. Bertness, T. E. Harvey, L. M. Mansfield, A. W. Sanders, and N. A. Sanford, "MESFETS Made from Individual GaN Nanowires," *IEEE Transactions on Nanotechnology* 7 (2008) pp. 760–765.

[12] R. B. Marks, "A Multiline Method of Network Analyzer Calibration," *IEEE Transactions on Microwave Theory and Techniques* 39 (1991) pp. 1205–1215.

[13] ISO, *Guide to the Expression of Uncertainty in Measurement* (International Organization for Standardization, 1993).

[14] D. M. Pozar, *Microwave Engineering* (Addison-Wesley, 1993), p. 235.

[15] D. F. Williams, C. M. Wang, and U. Arz, "An Optimal Multiline TRL Calibration Algorithm," *2003 IEEE MTT-S International Microwave Symposium Digest* 1–3 (2003) pp. 1819–1822.

[16] K. Kim, T. Mitch Wallis, P. Rice, C. Chiang, A. Imtiaz, P. Kabos and D. S. Filipovic, "A Framework for Broadband Characterization of Individual Nanowires," *IEEE Microwave and Wireless Component Letters*, 20 (2010) pp. 178–180.

[17] L. M. Mansfield, K. A. Bertness, P. T. Blanchard, T. E. Harvey, A. W. Sanders, and N. A. Sanford, "GaN Nanowire Carrier Concentration Calculated from Light and Dark Resistance Measurements," *Journal of Electronic Materials* 38 (2009) pp. 495–504.

[18] N. A. Sanford, P. T. Blanchard, K. A. Bertness, L. Mansfield, J. B. Schlager, A. W. Sanders, A. Roshko, B. B. Burton, and S. M. George, "Steady-State and Transient Photoconductivity in c-axis GaN Nanowires Grown by Nitrogen-Plasma-Assisted Molecular Beam Epitaxy," *Journal of Applied Physics* 107 (2010) art. no. 034318.

[19] N. A. Sanford, L. H. Robins, P. T. Blanchard, K. Soria, B. Klein, B. S. Eller, K. A. Bertness, J. B. Schlager, and A. W. Sanders, "Studies of Photoconductivity and Field Effect Transistor Behavior in Examining Drift Mobility, Surface Depletion, and Transient Effects in Si-doped GaN Nanowires in Vacuum and Air," *Journal of Applied Physics* 113 (2013) art. no. 174306.

7 Instrumentation for Near-Field Scanning Microwave Microscopy

7.1 Introduction

In the preceding chapters, we have focused on broadband, calibrated measurements of nanoelectronic devices. In particular, we have described techniques for the measurement of calibrated, complex scattering parameters and the subsequent extraction of circuit model parameters. In order to facilitate the ongoing development of novel, RF nanoelectronic devices, it is highly desirable to complement scattering parameter measurements with local, intra-device measurements. Furthermore, nondestructive, spatially localized characterization of nanomaterials and other nanoelectronic building blocks is critical for engineering of RF nanoelectronics. Thus, in this chapter, we introduce broadband, near-field probes, especially those integrated with scanning probe microscopes. Here, we consider the practical implementations of such scanning probe systems.

In designing an NSMM system, several critical questions must be considered. Will the probe be implemented with a resonant or nonresonant microwave circuit? What type of microwave probe will be used: a sharpened metal tip, a planar structure such as a stripline, or perhaps a resonant cavity with a sub-wavelength aperture? What distance-following mechanism will be used to maintain a constant separation between the probe and the sample under test (SUT)? Depending on how these questions are addressed, any of a wide variety of NSMM designs may be engineered. In addition, the instrumentation directly impacts calibration techniques, which will be described in detail in the following chapter, as well as the underlying physical models and theory of operation for NSMMs. Before proceeding to the detailed discussion of contemporary approaches to NSMM instrumentation, we will briefly review the historical development of near-field microwave probing.

7.2 Historical Development

In an ideal, classical optical microscope, it has long been known that the resolution is limited by diffraction. The diffraction limit, also known as the Abbe limit, is on the order of λ, where λ is the wavelength of the probing illumination. More generally, it is extremely difficult to resolve sub-wavelength features with far-field systems in which the probe-sample distance r is much larger than both λ and the size of the illuminating source D. As a result, improvements in the resolution of far-field microscopes have historically relied on the use of smaller and smaller wavelengths, pushing into the extreme-ultraviolet regime and below. Things are quite different in the near field. In particular, in the near-field regime, evanescent waves make a significant contribution to the total field and enable sub-wavelength resolution. Thus, a near-field probe may be implemented by devising an illuminating or field-focusing source of dimension D that illuminates at wavelength λ and is positioned a very short distance r from the probe, with $r << \lambda$.

An early proposal to implement a near-field probe was made by Edward H. Synge in 1928 [1]. Synge proposed to create a probe by opening a sub-wavelength aperture in an otherwise opaque barrier that was positioned close to the sample of interest such that it was within the near field. An image could then be generated by scanning the aperture back and forth above the sample. Synge's probe concept won endorsement from Albert Einstein himself [2], but the instrumentation limits of his day prevented Synge from implementing his idea. Note that there is an important trade-off required for implementation of such a near-field probe. In the far field, all points on an object within the field of view may be imaged simultaneously by use of far-field optics. In the near field, an image can only be obtained by serially scanning the probe over the object and making a separate measurement at each probe position. Thus, the near-field strategy requires additional instrumentation for scanning and longer times for image acquisition. Experimental work in near-field microscopy did not emerge until the late 1960s and early 1970s. Bryant and Gunn demonstrated a probe-based microscope for measurement of the local resistivity of a semiconductor crystal with spatial resolution on the order of 1 mm [3]. Their system consisted of a sharpened probe integrated with a bridge-based impedance detector. In order to maximize the sensitivity of their system, the electrical length of the probe signal path was required to be equal to an integral number of half-wavelengths at the operation frequency (450 MHz), effectively creating a resonant circuit. Later, Ash and Nicholls produced an aperture-based microscope that was able to measure the local relative permittivity of dielectrics with 0.5 mm spatial resolution [4]. The central element of their apparatus was a 10 GHz resonator with a sub-wavelength aperture, 1.5 mm in diameter. With the aperture positioned directly above a dielectric sample, they measured shifts in both the quality factor Q and resonant frequency f_0 of the resonator as a function of local permittivity. Measurements of these two variables have since become a hallmark of NSMM measurement techniques.

Over the course of the intervening decades, numerous and varied implementations of NSMM systems and related tools have been reported [5],[6]. For example, several NSMM designs have been developed that combine microwave compatibility with the spatial resolution of vacuum scanning tunneling microscopy [7]–[9], pushing NSMM toward the atomic scale. Engineering and design of NSMM probes has also been an area of particular interest. Custom probe designs include fully microfabricated coaxial probes [10], field-focusing probes sculpted with a focused ion beam [11], and metal-coated nanowire probes [12], as shown in Fig. 7.1. These

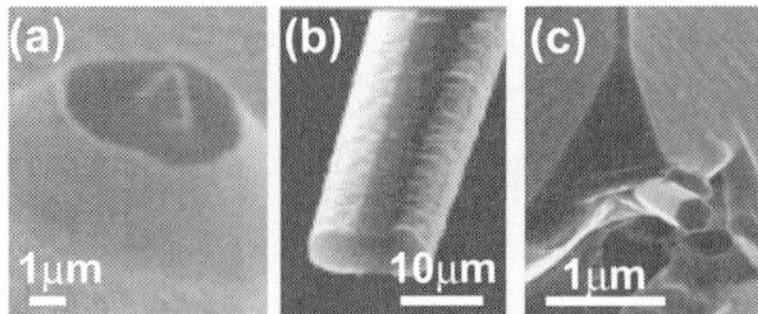

Figure 7.1.   Specialized probes for near-field scanning microwave microscopy.
(a) Scanning electron microscope (SEM) image of the tip of a microfabricated coaxial probe. All components of the probe, including the cantilever body, the coplanar waveguide signal path, and the coaxial antenna probe, were constructed by following microfabrication techniques developed for microelectromechanical systems. Adapted with permission from Y. Q. Wang, A. D. Bettermann, and D. W. van der Weide, *Journal of Vacuum Science and Technology B* 25 (2007) pp. 813–816. © 2007, American Vacuum Society.
(b) SEM image of a balanced stripline resonator probe. The stripline was formed by depositing Al layers on either side of a tapered quartz bar. The tapered end of the bar was then further shaped by use of a focused ion beam in order to further confine the fields near the probe tip. Adapted from V. V. Talanov, A. Scherz, R. L. Moreland, and A. R. Schwarz, *Applied Physics Letters* 88 (2006) art. no. 134106, with permission from AIP Publishing.
(c) SEM image of a GaN nanowire-based probe. The defect-free, mechanically robust nanowire was inserted into the apex of a silicon microcantilever by use of a focused ion beam and a nanomanipulator. In order to form a continuous microwave signal path, a thin metal layer was subsequently deposited on the cantilever by use of atomic layer deposition [12]. © IOP Publishing. Adapted with permission. All rights reserved.

examples, along with the NSMM systems discussed in this chapter, represent but a small fraction of published implementations. While an exhaustive study of NSMM instrumentation is beyond the scope of this book, we will cite several representative systems that exemplify different design strategies and choices. Note that as a result of the ongoing development of NSMM instrumentation, several different NSMM systems are now commercially available.

7.3   Probe and Sample Motion

7.3.1   Distance-Following Mechanisms

In general, NSMM measurements depend on the local impedance presented to the probe tip by the SUT. This impedance depends not only on the sample's electromagnetic material properties, but also on the geometries and relative position of the probe and the sample. Thus, the distance between the probe tip and the sample, z, is a critical parameter for NSMMs. In some imaging modes, the probe tip is kept in direct contact with the sample ($z = 0$ nm), while in other modes z may be on the order of 1 nm. A distance- or height-following mechanism is integrated into most scanning probe systems in order to maintain a constant value of z while the tip is raster-scanned across the sample surface. In certain situations, it may be desirable to make height-dependent NSMM measurements over several orders of magnitude in z, spanning from 1 nm to as much as a few millimeters. At the outset of height-dependent measurements, the distance-following mechanism is used to establish an initial tip-sample distance. Subsequently, the height-dependent measurement may be carried out by increasing the tip-sample distance until the probe interaction with the sample is negligible.

Over the last several decades, as scanning probe microscopes have become workhorses for nanotechnology research and development, many distance-following strategies have been developed. Several distance-following mechanisms are shown in Fig. 7.2. For example, in the scanning tunneling microscope (STM) a metal tip is positioned within a nanometer or less of the sample [13]. When a potential difference is present between the tip and sample, a tunneling current flows between the tip and the sample due to quantum mechanical tunneling of electrons. In order to maintain this small distance, a feedback loop is used to adjust z in order to maintain a constant tunneling current between the tip and the sample, as illustrated in Fig. 7.2(a). STM-type feedback has been implemented in NSMMs [7], but such systems are limited to conducting and semiconducting samples. In order to extend STM to dielectric materials, RF STMs have been developed. In place of the DC tunneling current, the feedback loops in RF STMs maintain a constant amplitude of an odd harmonic signal that is generated by nonlinear behavior of the tunneling current [8], [14]. While STM provides lateral spatial resolution on the atomic scale, extremely clean surfaces are required. This requires an ultra-high vacuum environment as well as in situ sample cleaning capabilities.

Distance-following strategies based on AFMs allow access to a wide variety of samples. In an AFM, the probe tip is integrated into a microcantilever beam. A simple approach to distance-following in this configuration is contact mode AFM, in which the probe is in direct contact with the sample, leading to a deflection of the microcantilever. Then, a feedback loop is used to adjust z in order to maintain a constant beam deflection, as shown in Fig. 7.2(b). One common approach for monitoring the deflection of the cantilever beam is the so-called beam-bounce approach, in which a laser is focused on a spot near the free end of the beam. The beam is reflected off the cantilever onto a quadrant photodetector. By monitoring the difference in intensities between different quadrants of the detector, both the beam deflection and torsion may be monitored. Interferometric optical detection or capacitive motion detection may also be used to monitor the deflection of the cantilever. As an alternative to contact mode AFM, dynamic, noncontact approaches have been developed. The earliest noncontact AFM modes drive resonant vibrations of the cantilever [15]. Changes in the vibration frequency and amplitude may be directly correlated to surface topography. Further innovations in noncontact AFM modified this approach to incorporate deliberate tapping of the surface with the probe tip [16]. To date, most AFM-based NSMM systems have utilized contact mode, but noncontact NSMM is highly desirable as it has the potential to reduce mechanical wear of both the probe tip and the sample.

Design choices for distance-following in an NSMM or related probe microscope are often dictated by the intended application. As an example, consider an experiment in which optical illumination is introduced in order to excite the sample, such as the optical illumination of a photoconductive system [17], [18]. In general, any external stimulus that leads to change of the local impedance can be detected by NSMM. Here, the excitation of additional carriers in the photoconductive material will be detected. Since photoconductive samples may be sensitive to stray optical illumination from beam-bounce detection instrumentation, alternative, light-free approaches such as tuning-fork-based feedback may be necessary for distance following [17]. A schematic of a tuning-fork-based distance-following system is shown in Fig. 7.2(c). A quartz tuning fork is placed in mechanical contact with the probe tip. This may be done by use of a clamp or by directly integrating the probe tip with one of the tines of the tuning fork. The fundamental-mode resonance frequency of a typical quartz tuning fork is about 32 kHz and has a quality factor on the order of 1000. During operation, the fork vibrates at resonance, excited by ambient energy sources or in some cases, driven by an actuator. As the probe tip is brought within a few nanometers of sample, probe-sample interactions damp the vibrations. A feedback loop is used to adjust the probe-sample distance to maintain a constant vibration frequency or amplitude. A lock-in technique may be used to improve the sensitivity of the feedback system to small signals.

The distance-following strategy has significant ramifications for NSMM. The distance-following feedback loop provides a mechanism to visualize the sample geometry. By convention, the resulting images are almost always referred to as

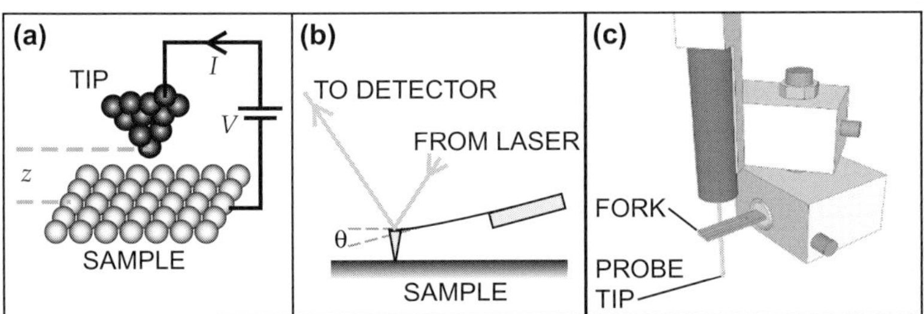

Figure 7.2.   Examples of distance-following mechanisms.
(a) Schematic of an STM junction. An atomically sharp tip is placed a small distance z from a conducting sample. In constant current mode, a fixed potential difference V is applied between the tip and sample and a feedback loop adjusts z to maintain a constant tunneling current I. (b) Schematic of a contact-mode AFM. When the probe tip is in contact with the sample, the cantilever beam is deflected at an angle θ, which is measured by use of a laser that reflects off of the cantilever and onto a split or quadrant photodetector. The feedback loop adjusts the cantilever height to maintain a constant deflection. (c) Schematic of shear-force detection, including a tuning fork in contact with a sharp NSMM probe. A feedback loop is used to adjust the probe-sample distance to maintain a constant vibration frequency or amplitude of the tuning fork.
Reprinted from J. C. Weber, J. B. Schlager, N. A. Sanford, A. Imtiaz, T. M. Wallis, L. M. Mansfield, K. J. Coakley, K. A. Bertness, P. Kabos, and V. M. Bright, *Review of Scientific Instruments* 83 (2012) art. no. 083702, with permission from AIP Publishing.

"topographic images," regardless of which distance-following strategy has been implemented. However, as each of these strategies relies on a different physical interaction with the sample, subtle differences exist between different types of topographic images. Further complications are introduced due to the fact that apparent dimensions in topographic images may be functions of additional parameters, including scan speed, scan direction, and any electric potential difference between the probe and the sample. A detailed understanding of topographic imaging mechanisms is particularly important in the interpretation and analysis of NSMM images. As we point out elsewhere in this book, the capacitive contribution to an NSMM image is highly dependent on the tip-sample geometry as well as the electromagnetic material properties of the sample. Thus, any attempt to isolate the material properties from the overall measurement requires knowledge of the sample geometry. A topographic image, often obtained simultaneously with the microwave signal image(s), is usually the primary source of knowledge about the apparent sample geometry. Furthermore, extraction of material properties required accurate modeling of the tip-sample system, including parasitic capacitance, as will be described in detail in Chapter 9.

The theory of Tersoff and Hamann, which builds upon previous work by Bardeen, provides one simplified approach to interpretation of STM topographic images. In the Bardeen formalism [19], the tunneling current I is given by

$$I = \frac{4\pi e}{\hbar} \sum_{\mu,v} f\left(E_\mu\right)\left[1 - f\left(E_v + eV\right)\right]\left|M_{\mu v}\right|^2 \delta\left(E_\mu - E_v\right), \tag{7.1}$$

where $f(E)$ is the Fermi function, $M_{\mu v}$ are the tunneling matrix elements between the states of the tip (Ψ_μ) and the sample (Ψ_v), E_μ is the energy of the states Ψ_μ in the absence of tunneling, V is the potential difference between the tip and the sample, e is the electron charge, and h is Planck's constant. Assuming the electronic structure of the tip apex may be represented by an s-wave, Tersoff and Hamann have shown that Equation (7.1) implies [20]

$$\frac{dI}{dV} \propto \rho_S\left(\mathbf{r}, E_F + eV\right), \tag{7.2}$$

where ρ_S is the sample local DOS, $\mathbf{r}$ is the position of the probe tip apex, and E_F is the Fermi energy of the sample. In other words, contrast in STM topographic images arises from variations in the local electronic structure, namely the local DOS evaluated at the probe position. Though local electronic structure and geometric arrangement of matter in the sample are related, they are not identical. In fact, certain adsorbed species that reduce the local DOS appear as depressions in STM topographic images, though the species is in fact adsorbed on top of the surface.

In an AFM, a combination of distance-dependent forces governs the interaction between the probe tip and the sample. When the probe tip is extremely close to the sample, Pauli repulsion dominates. However, as the probe is retracted from the sample surface, attractive van der Waals-type forces quickly become the dominant interaction mechanism. This combination of attractive and repulsive forces can be described by a Lennard-Jones potential. A typical form is

$$V(z) = \phi\left[\left(\frac{z_0}{z}\right)^{12} - 2\left(\frac{z_0}{z}\right)^{6}\right], \tag{7.3}$$

where ϕ and z_0 are positive constants. An example of a Lennard-Jones potential is shown in Fig. 7.3. In contact-mode AFM, the probe–sample interaction is repulsive, while in noncontact mode the interaction is attractive. For other imaging modes, such as intermittent contact mode, the interaction may be more complex.

Ultimately, the probe–sample interaction may be sensitive to a number of factors, including chemical interactions between the tip and the sample or the presence of a static electric or magnetic field in the junction. While simple models such as the Tersoff-Hamann STM theory may be modified to provide a more detailed picture of the probe–sample interaction, a more expedient approach is to characterize the probe-sample behavior empirically through measurements. In an STM-like system, the measurements would likely take the form of height-dependent current spectroscopy, while in an AFM-like system, the measurements would likely take the form of height-dependent force spectroscopy, also known as a "force-distance curve." Another pragmatic approach to the problem is to engineer areas into the sample that are known to be free of topographic features. In Chapter 14, we will describe how the presence of flat regions in an NSMM image may be leveraged to empirically de-convolve material properties from topographic cross-talk.

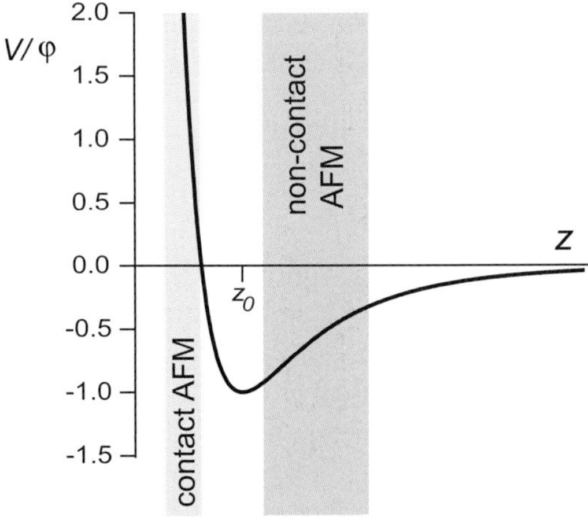

Figure 7.3.   Lennard-Jones potential.
The interaction between a cantilever probe, located a distance z above the sample, may be described by a Lennard-Jones potential. Shaded regions correspond to "contact AFM," in which the forces are purely repulsive, and "noncontact AFM," in which attractive forces dominate.

7.3.2 Probe and Sample Positioning

While one may often refer to "retracting the tip" or "scanning the tip," in the case of NSMM systems, as well as other scanned probes that incorporate microwave circuits, in practice it is usually best to keep the probe fixed and move the sample with respect to the tip. This is due to the fact that the probe is usually connected directly to the microwave signal path. This path often incorporates cabling or other elements that are sensitive to bending or other mechanical disturbances. Repeated, large-scale motion of the microwave hardware can thus introduce instability in the measured magnitude and phase. As a result, many NSMMs contain two sets of positioning elements. The first is for small-scale, fine positioning up to a few micrometers. This small-scale set of positioners is usually implemented by use of piezoelectric transducers and may move the tip or the sample, depending on the implementation. Closed-loop scanning stages implement a feedback loop to maintain constant lateral fine positions, often by use of a capacitive sensor. The fine motion generates the scanning of the probe relative to the tip. The second set of positioning is for large-scale, coarse motion over length scales from a few micrometers to a few millimeters. This large-scale positioner is often implemented by use of a stepper motor and usually moves only the sample in order to improve stability and reduce uncertainty and drift in microwave measurements.

7.4 Microwave Probes and Circuits

7.4.1 Aperture Probes versus Tip Probes

Recall that Synge's original near-field microscope concept relied upon an aperture in an opaque screen, placed at a small distance above the SUT. When microwave radiation is incident upon such an aperture of sub-wavelength dimensions, the aperture spatially confines the microwave field, effectively focusing the field to sub-wavelength dimensions and enabling sub-wavelength resolution in the near field. Nearly two decades after Synge's proposal, Bethe developed a more complete, general analysis of diffraction of electromagnetic radiation by sub-wavelength holes [21]. As practical implementations of the near-field microscope concept were developed, an alternative approach to spatial confinement was realized. It was recognized that a microwave signal path terminating in a sharp, conducting point could serve as an aperture-free alternative. Imagine such a probe, having a conical shape, but terminated at its terminal apex by a hemisphere. If this hemisphere, the probe tip, has sub-wavelength electrical dimensions, then it may also serve as a field-focusing element in the near-field regime. Both aperture-based and tip-based NSMM instruments have been realized, though tip-based designs are currently predominant.

The apparatus developed by Ash and Nicholls in 1972 is an example of a microscope that incorporates an aperture [4]. A schematic of their instrument is shown in Fig. 7.4. At resonance, the open resonator operates at a frequency f_0 and has a beam radius of w_0. An opaque diaphragm blocks the transmission of the beam onto the

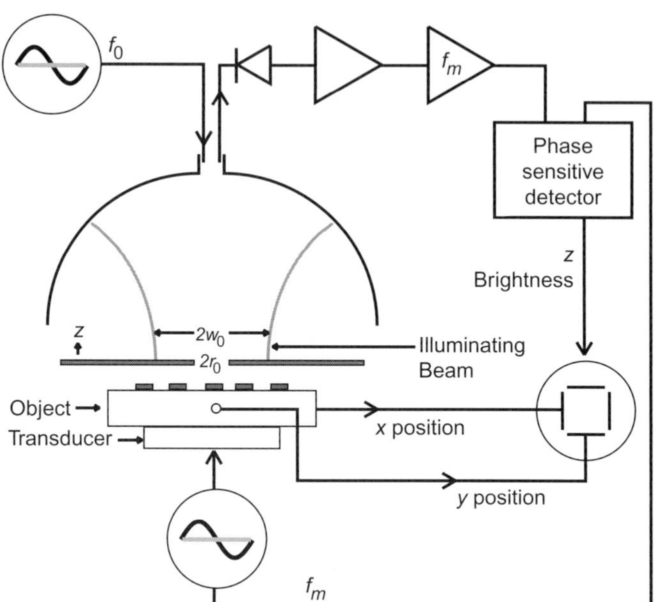

Figure 7.4.   A near-field scanning microwave microscope with an aperture probe. A schematic diagram of the aperture-based microscope developed by Ash and Nicholls in 1972. The design builds upon the original ideas of E. H. Synge, first published in 1928. Reprinted by permission from Macmillan Publishers Ltd: Nature. E. A. Ash and G. Nicholls, *Nature* 237 (1972) pp. 510–512. © 1972.

sample, save for a small aperture of radius r_0. The sample object is scanned in a raster pattern under the aperture and the reflected signal is mapped as a function of position to produce an image. In order to improve sensitivity, the sample was vibrated at a modulation frequency f_m and a lock-in technique was implemented with a phase sensitive detector. Furthermore, a low-noise amplifier and an amplifier tuned to f_m are incorporated into the receiving system. In the original work [4], the resonator operated at a frequency f_0 of 10 GHz, corresponding to a resonant wavelength λ_0 of 3 cm, and the aperture diameter r_0 was 0.75 mm. Given the relatively large spatial distances involved, neither precise distance following nor control of the aperture-sample distance along z were required. This simple, groundbreaking system was shown to resolve individual lines in a metallic grating with a line width of $\lambda_0/60$. The system was also able to resolve contrast between a region with a relative permittivity of 2.58 and a neighboring region with a relative permittivity of 2.24.

Tip-based NSMM designs provide an alternative to aperture-based designs. As many contemporary AFMs and STMs utilize sharpened points as probes, tip-based NSMMs can be readily adapted from existing scanning probe instruments. One approach to a tip-based NSMM design is to modify STM instrumentation to incorporate a microwave signal path [7], [8], [22]. STM relies on quantum mechanical tunneling of electrons from a needle-like probe to a bulk sample, thus making it a natural platform for developing an NSMM. Note that STM-based NSMM is not appropriate for all applications, as it requires a conducting sample and usually requires a vacuum environment.

A schematic of an STM-based NSMM is shown in Fig. 7.5 [7]. From an electrical point of view, there are three major components to this NSMM design. First, the probe tip is integrated into a $\lambda/2$ coaxial, transmission-line resonator. As we will discuss in this chapter, the use of a resonant circuit increases the sensitivity of an NSMM, though at the cost of being limited to a finite number of operating frequencies, namely the resonant frequency and higher harmonics. If a hollow capillary tube is used in place of the solid center conductor within the resonator, the tube may serve as a socket into which sharpened metal probe tips are inserted [23]. This provides a simple mechanism for probe replacement, should a tip become inadvertently bent or dulled by use. The resonant cavity is effectively terminated by the capacitively coupled, RF port of a bias tee. The bias tee serves as an interface between the resonator and the other two major electrical components: the microwave electronics and the DC tunneling current circuit.

The microwave electronics include a source, which supplies a signal to the resonator at the resonant frequency f_0. The signal reflected from the resonator is transmitted to a diode detector by use of a directional coupler. The output of the diode detector serves as the input to a frequency-following feedback circuit that locks

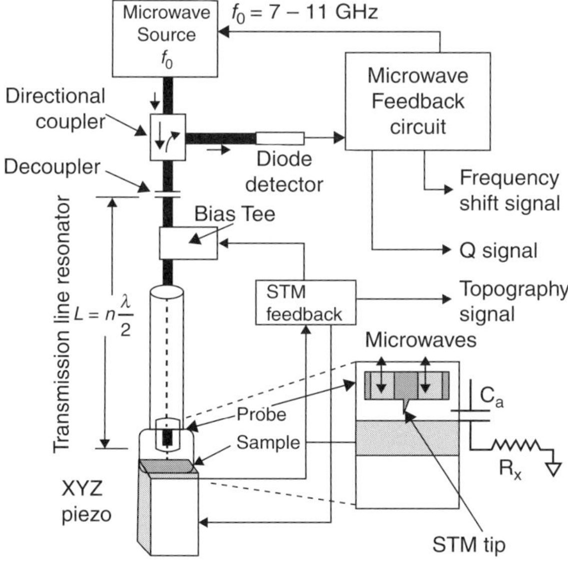

Figure 7.5. A near-field scanning microwave microscope design based on an STM. The bias tee serves as a junction between the three main electrical components of the microscope: the resonant probe, the DC STM electronics, and the microwave electronics. Reprinted from A. Imtiaz and S. M. Anlage, "A Novel STM-Assisted Microwave Microscope with Capacitance and Loss Imaging Capability," *Ultramicroscopy* 94 (2003) pp. 209–216. © 2003 with permission from Elsevier.

the source frequency to f_0. In order to lock onto the resonant frequency, a low-frequency (~3 kHz) signal f_{mod} is used to frequency modulate the signal from the microwave source. The low-frequency signal also serves as a reference for a lock-in amplifier. The output of the lock-in amplifier is then time-integrated and fed back to the frequency control of the microwave source. The time-integrated signal also serves as a measurement of the resonator frequency shift [24]. A second lock-in amplifier, referenced to twice f_{mod}, is used to extract the quality factor Q, though a calibration procedure is required to quantitatively convert the output of the second lock-in to Q [25]. Thus, the feedback circuit is designed to output the shift in the resonant frequency and the resonator quality factor. Both of these measurands are sensitive to changes in the tip-sample impedance and, in turn, the local electromagnetic properties of the sample. As in conventional STM, the DC tunneling current circuit incorporates a feedback loop that serves as the distance-following mechanism. Specifically, in constant current mode, the DC tip-sample bias voltage is fixed and the feedback loop adjusts the tip-sample distance such that a constant tunneling current is maintained. For bias voltages in the range between millivolts and volts, corresponding tunneling currents typically are in the range between picoamps and nanoamps. The varying tip-sample distance is the topography signal. Thus, three types of images can be simultaneously obtained by this type of NSMM: a topographic image, a frequency shift image, and a quality factor image.

7.4.2 Resonant Probes versus Nonresonant Probes

The preceding NSMM implementations, shown in Fig. 7.4 and Fig. 7.5, are examples of resonant probes. Resonant probes are formed by integrating the field-focusing feature (aperture or tip) into a resonant structure, such as a cylindrical cavity or a resonant microwave circuit. The motivation for this approach is clear: by integrating the probe into a resonant structure, the signal-to-noise ratio of the NSMM is improved by a factor of Q. On the other hand, a critical limitation of resonant NSMMs is that they necessarily operate at or near f_0 or, in some cases, harmonic multiples of f_0. Some leeway in the operating frequency can be obtained by introducing a tunable resonant structure, but this may require compromising the resonator's Q and, in turn, the signal-to-noise ratio of the NSMM. Applications of resonant NSMMs include measurements of permittivity and loss tangent in bulk solids [26] and liquids [27], as well as measurements of surface resistance in metallic thin films [28].

When the probe tip is positioned near an SUT, the sample presents a load to the microwave resonator circuit, leading to perturbations of both the resonant frequency f_0 and quality factor Q. The magnitude of these perturbations will depend on the local, electromagnetic material properties of the sample, the sample geometry, and the vertical distance between the probe and the sample z. The dependence upon z suggests the most straightforward approach to resonant NSMM measurements: a height-dependent measurement of f_0 and Q. Initial, unperturbed values of f_0 and Q may be established by making a measurement while the resonant NSMM's probe tip is far from the sample. Although height-dependent NSMM measurements are at times referred to as "approach curves," in practice it is often useful to begin the measurement with the probe at its closest approach distance and then make measurements of f_0 and Q as the tip is retracted from the sample surface. Note that this requires measurement of the full spectral response at each height, in order to get a reliable fit and determination of f_0 and Q. This can be time consuming, and requires a high degree of mechanical stability as well as an accurate determination of the height z. Detailed methods to extract material properties from height-dependent NSMM measurements will be discussed in Chapter 9.

Another resonant NSMM implementation is shown in Fig. 7.6 [26]. The system is based on a quarter-wavelength coaxial resonator ($f_0 \sim 1$ GHz and $Q \sim 1000$) that is operated in transmission mode. A probe tip is mounted on the center conductor and protrudes through a small opening in the resonator. For this system, the probe consists of a metal wire, 50 μm to 100 μm in diameter, with the probe tip sharpened to submicrometer dimensions. In this implementation, a sapphire ring coated with a thin metal layer encircles the tip wire at the aperture in order to shield the resonator from far-field propagating components. The relative permittivity and loss tangent of single-crystal dielectrics have been extracted from measurements made

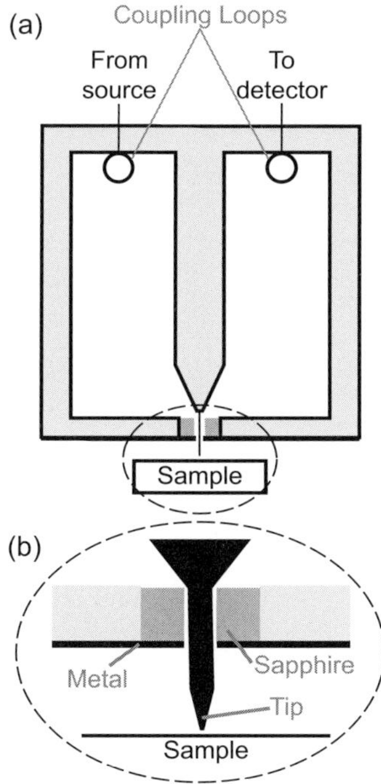

Figure 7.6. A near-field scanning microwave microscope design based on a coaxial resonator. (a) A schematic of the coaxial resonator, which is operated in transmission mode. (b) A closer view of the probe tip, which is connected to the center conductor of the resonator and extends toward the sample through an opening in the bottom of the resonator. Adapted from C. Gao and X.-D. Xiang, *Review of Scientific Instruments* 69 (1998) pp. 3846–3851, with permission from AIP Publishing.

with this type of NSMM. The extraction relies upon a quasi-static model of the probe–sample interaction that is based on the method of images and yields values of material parameters that are in good agreement with values obtained by bulk characterization techniques [26], [29].

Yet another implementation of a resonant NSMM is based on a two-dimensional, stripline resonator in place of the resonant cavity [30]. To fabricate the probe, a half-wavelength stripline resonator ($f_0 \sim 1$ GHz) is patterned on a dielectric substrate, as shown in Fig. 7.7(a). One end of the resonator is capacitively coupled to the RF source. The other end of the resonator is tapered and terminates in a sharpened point that extends beyond the end of the dielectric substrate, forming the probe. A second dielectric layer is added above the patterned resonator and ground planes are added to the top and bottom to complete the structure, as shown in Fig. 7.7(b). In Reference [30], the NSMM is operated in reflection mode. The reflected RF signal is separated from the incident signal by use of a three-port circulator and measured by use of a crystal detector. Furthermore, the sensitivity of the instrument is increased by modulating the sample position at a low frequency (100 Hz) and the use of a lock-in technique. Relative to cavity, coaxial, and rectangular waveguide resonators, stripline resonators offer the advantage of low cost and compact footprint. However, as with many planar transmission structures, striplines may radiate into free space, potentially causing unwanted effects depending on the experimental conditions.

Beyond the examples described earlier, there are numerous alternative implementations of resonant NSMM systems. In addition to providing improved

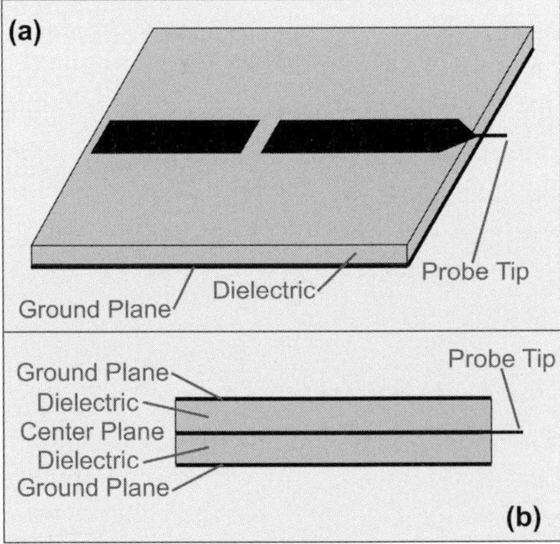

Figure 7.7.   A near-field scanning microwave microscope design based on a planar stripline resonator. (a) A schematic revealing the center plane of the probe structure, including the stripline resonator integrated with a probe tip. The feedline is capacitively coupled to the resonator. (b) Side view of the probe structure after a second set of dielectric and ground plane layers has been added on top of the center plane.
Adapted from M. Tabib-Azar, D.-P. Su, A. Pohar, S. R. LeClair, and G. Ponchak, *Review of Scientific Instruments* 70 (1999) pp. 1725–1729, with permission from AIP Publishing.

signal-to-noise, the resonant elements of these NSMM systems are generally well understood and their electrical performance and electromagnetic field structure can be precisely engineered. Furthermore, modeling of these systems is often simplified by the use of well-established, lumped-element models for resonant cavities and circuits. In recent years, a number of NSMM designs have been realized without the inclusion of an engineered resonant structure in the microwave circuit. These nonresonant NSMMs follow many of the same operational principles as resonant NSMMs and display many of the same advantages. From an instrumentation and design perspective, nonresonant NSMMs may be easier to implement in that they require only a microwave signal path to the probe tip with relatively low loss, without the need for a specific resonant structure or circuit. This advantage is particularly important if an existing scanning probe microscope platform is retrofitted or modified to work in an NSMM mode.

When a one-port microwave network is incorporated into a scanning probe microscope, the reflection coefficient will be frequency dependent. This frequency dependence will be determined not only by the load impedance, but also by the physical implementation of the signal path, including cable connections, adapters, and the transition from a guided-wave structure to the probe tip. The cumulative effect of these interfaces and the associated impedance mismatches is that sharp mimima in the reflection coefficient exist at selected frequencies. An example of such a minimum is shown by the solid curve in Fig. 7.8. This minimum resembles a resonance and an effective "resonant" frequency f_0 and effective quality factor Q may be assigned to the curve by fitting the curve with an appropriate function, e.g., a Lorentzian function. However, it should be noted that these minima represent frequencies at which the microwave network is well-matched to the source impedance rather than a true resonance. As with a resonant NSMM, when the probe tip of a nonresonant NSMM is brought near an SUT, the shape of the local minimum will be altered, leading to shifts in f_0 and Q. The sign and magnitude of these shifts will depend sensitively upon the local sample impedance.

In practice, there will be multiple minima in the reflection coefficient that will work as operating frequencies for the NSMM. For example, Imtiaz et al. imaged a semiconductor reference sample at 2.3 GHz, 5.0 GHz, 9.6 GHz, 12.6 GHz, and 17.9 GHz with a nonresonant, commercial NSMM and observed frequency-dependent contrast in the resulting images [31]. The NSMM sensitivity may vary strongly for different operating frequencies and some local minima may be unsuitable as NSMM operating points. In a nonresonant system, the most effective operating frequencies are often found by trial and error. To optimize the NSMM sensitivity to the reflection coefficient, a tunable phase shifter may be inserted into the microwave signal path, between the source and the probe. With the probe out of contact, the phase shifter can be tuned to sharpen the local minimum in the reflection coefficient.

A full description of the position-dependent shifts in frequency and quality factor requires a complete measurement of the reflection coefficient as a function of frequency at each point of interest (techniques for extracting quantitative material parameters from such measurements are described in Chapter 9). Such an approach

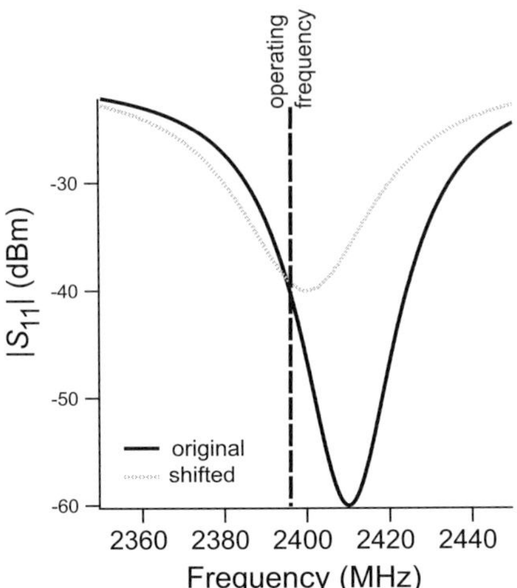

Figure 7.8.   Operating conditions for an NSMM.
A local minimum in the magnitude of the reflection coefficient ($|S_{11}|$) at 2.41 GHz is illustrated. This original peak (black curve) may be shifted during operation to a new position (gray curve) due to vertical movement of the probe or lateral movement to a different area of the sample. If the system is operated at the minimum (here, 2.41 GHz), then the measured changes are ambiguous: based on $|S_{11}|$ measurements at a single frequency, the user will be unable to distinguish shifts to lower frequency from shifts to higher frequency. To overcome this, an off-peak operating frequency is chosen, as illustrated by the dashed, vertical line.

is comprehensive, but impractical for most applications due to time and stability constraints. A practical alternative is to select a single operating frequency that is near, but not equal to f_0, as illustrated in Fig. 7.8. The magnitude and phase of the reflection coefficient are then measured at each probe position to generate images. Recall that the local minima that are used to establish the operating frequency occur when the microwave network is well-matched to the source impedance. Thus, barring a priori knowledge about whether the local impedance will result in better or poorer impedance matching, the direction of the shift in f_0 will initially be unknown for a given NSMM and a given sample. However, by operating at frequency offset from f_0 (on one side of the "resonance" curve), the contrast is unambiguous. In order to quantitatively map shifts in f_0 and Q to material parameters, it is necessary to phenomenologically validate a tip-sample model by use of a known reference or calibration sample, which will be introduced in Chapter 8.

7.5   Other Aspects of Near-Field Scanning Microwave Microscope Instrumentation

So far, we have focused on two aspects of NSMM design and implementation in some detail: mechanical positioning and microwave electronics. Though these two aspects are particularly important for NSMM, a successful instrument requires engineering several additional subsystems, including environmental control, noise reduction, and thermal control. Fortunately, many of the required techniques and methods have been previously developed for other scanning probe microscopes and microwave measurement systems. Environmental and noise issues have been addressed to varying degrees in the instrumentation examples discussed earlier. Here, we will briefly review three additional, specialized scanning probe microscope examples. Readers who are developing their own, customized NSMM systems are encouraged to consult the vast literature on this topic for a more comprehensive review of scanning probe microscope designs.

In Reference [8], Kemiktarak et al. described an implementation of an RF scanning tunneling microscope (RFSTM) that operates at cryogenic temperatures at megahertz frequencies and enables sensitive measurement of high-frequency mechanical motion. The system utilizes a resonant circuit design, incorporating the resistive tunnel junction with an LC tank circuit to produce a resonant frequency of about 115 MHz. Specifically, the probe tip is fixed on a printed circuit board, which also includes the tank circuit, a bias tee, and a directional coupler. In this way, these components are necessarily located close to the tunnel junction and within the cryostat, thus reducing both signal loss and thermal noise. The low-temperature environment is further leveraged by incorporating a cryogenic amplifier into the RF output path. The sensitivity of the RFSTM to mechanical motion was demonstrated by measuring the eigenfrequencies of a micromechanical resonator between 1.0 MHz and 3.0 MHz. Furthermore, topographic imaging based on measurement of the RF reflection coefficient enabled a 100-fold improvement in scan speed relative to conventional constant current STM imaging.

High-speed imaging with scanning probe microscopes is a tantalizing prospect. Because of its subsurface imaging capability and compatibility with liquid environments, AFM-based NSMM is a promising tool for measurements of biologically important systems. However, progress in this area has been slow. Recently, AFM imaging at a rate of five frames per second has been demonstrated, capturing the motion of a myosin V molecule walking on an actin filament [32]. In order to achieve relatively high frame rates, specialized scanning techniques were implemented, including active damping of the scanner to reduce mechanical noise and optimization of the feedback electronics to enable high-speed operation. In addition, the size of the AFM cantilever was reduced in order to increase the mechanical resonance frequency. While such advances in high-speed scanning have yet to be adopted in an NSMM design, many potential applications of high-speed NSMM exist, including real time observation of subsurface processes occurring beneath membranes.

Subsurface imaging of liquids and other soft matter with NSMM can be furthered by customized sample containment. Custom sample cells have long been used for containing and imaging liquids within the vacuum environment of electron microscopes. Recently, Tselev et al. developed a silicon-based cell for imaging processes in liquids with NSMM [33]. The liquid is encapsulated within a silicon well that is covered by a dielectric membrane (50 nm-thick silicon nitride or 8 nm-thick silicon dioxide (SiO_2)). The NSMM probe tip scans along the flat surface of the membrane. Since the membrane is a dielectric, it functions as a microwave-transparent window. In this configuration, electrochemical processes were imaged with an estimated lateral resolution of 250 nm. In separate experiments, yeast cells immersed in glycerol were imaged and the NSMM probe-depth was estimated to be a few hundred nanometers.

The field of near-field scanning microwave microscopy has grown rapidly in recent years and now incorporates a wide variety of microscopy techniques and instruments. The field of scanning probe microscopy is broader still. Here, we have illustrated several core choices required for effective design of an NSMM system by use of a small number of examples. Naturally, as the field moves forward, additional designs and imaging modes will emerge.

References

[1] E. H. Synge, "A Suggested Method for Extending Microscopic Resolution into the Ultra-microscopic Region," *Philosophical Magazine Series 7*, 6, No. 35, (1928) p. 356.

[2] M. Berry, E. Wolf, N. Bloembergen, N. Erez, and D. Greenberger, *Progress in Optics volume 50* (Elsevier, 2007) pp. 145–148.

[3] C. Bryant and J. Gunn, "Noncontact Technique for the Local Measurement of Semiconductor Resistivity," *Review of Scientific Instruments* 36 (1965) pp. 1614–1617.

[4] E. A. Ash and G. Nicholls, "Super-resolution Aperture Scanning Microscope," *Nature* 237 (1972) pp. 510–512.

[5] A. Imtiaz, T. M. Wallis, and P. Kabos, "Near-field Scanning Microwave Microscopy," *IEEE Microwave Magazine* 15 (2014) pp. 52–64.

[6] B. T. Rosner and D. W. van der Weide, "High-Frequency Near-Field Micrsocopy," *Review of Scientific Instruments* 73 (2002) pp. 2505–2525.

[7] A. Imtiaz and S. M. Anlage, "A Novel STM-Assisted Microwave Microscope with Capacitance and Loss Imaging Capability," *Ultramicroscopy* 94 (2003) pp. 209–216.

[8] U. Kemiktarak, T. Ndukum, K. C. Schwab, and K. L. Ekinci, "Radio-Frequency Scanning Tunneling Microscopy," *Nature* 450 (2007) pp. 85–88.

[9] J. Lee, C. J. Long, H. Yang, X.-D. Xiang, and I. Takeuchi, "Atomic Resolution Imaging at 2.5 GHz Using Near-Field Microwave Microscopy," *Applied Physics Letters* 97 (2010) art. no. 183111.

[10] Y. Q. Wang, A. D. Bettermann, and D. W. van der Weide, "Process for Scanning Near-Field Microwave Microscope Probes with Integrated Ultratall Coaxial Tips," *Journal of Vacuum Science and Technology B* 25 (2007) pp. 813–816.

[11] V. V. Talanov, A. Scherz, R. L. Moreland, and A. R. Schwarz, "A Near-Field Scanned Microwave Probe for Spatially Localized Electrical Metrology," *Applied Physics Letters* 88 (2006) art. no. 134106.

[12] J. C. Weber, P. T. Blanchard, A. W. Sanders, J. C. Gertsch, S. M. George, S. Berweger, A. Imtiaz, K. J. Coakley, T. M. Wallis, K. A. Bertness, N. A. Sanford, P. Kabos, and V. M. Bright, "GaN Nanowire Coated with Atomic Layer Deposition of Tungsten: A Probe for Near-Field Scanning Microwave Microscopy," *Nanotechnology* 25 (2014) art. no. 415502.

[13] G. Binnig and H. Rohrer, "Scanning Tunneling Microscopy," *Surface Science* 126 (1983) pp. 1–3.

[14] G. P. Kochanski, "Nonlinear Alternating-Current Tunneling Microscopy," *Physical Review Letters* 62 (1989) pp. 2285–2288.

[15] Y. Martin, C. C. Williams, and H. K. Wickramasinghe, "Atomic Force Microscope Force Mapping and Profiling on a Sub 100-Å Scale," *Journal of Applied Physics* 61 (1987) pp. 4723–4729.

[16] Q. Zhong, D. Inniss, K. Kjoller, and V. B. Elings, "Fractured Polymer/Silica Fiber Surface Studied by Tapping Mode Atomic Force Microscopy," *Surface Science Letters* 290 (1993) p. L688.

[17] J. C. Weber, J. B. Schlager, N. A. Sanford, A. Imtiaz, T. M. Wallis, L. M. Mansfield, K. J. Coakley, K. A. Bertness, P. Kabos, and V. M. Bright, "A Near-Field Scanning Microwave Microscope for Characterization of Inhomogeneous Photovoltaics," *Review of Scientific Instruments* 83 (2012) art. no. 083702.

[18] A. Hovsepyan, A. Babajanyan, T. Sargsyan, H. Melikyan, S. Kim, J. Kim, K. Lee, and B. Friedman, "Direct Imaging of Photoconductivity of Solar Cells Using a Near-Field Scanning Microwave Microprobe," *Journal of Applied Physics* 106 (2009) art. no. 114901.

[19] J. Tersoff and D. R. Hamann, "Theory of the Scanning Tunneling Microscope," *Physical Review B* 31 (1985) pp. 805–813.

[20] J. Tersoff and D. R. Hamann, "Theory and Application for the Scanning Tunneling Microscope," *Physical Review Letters* 50 (1983) pp. 1998–2001.

[21] H. A. Bethe, "Theory of Diffraction by Small Holes," *Physical Review* 66 (1944) pp. 163–182.

[22] M. Farina, D. Mencarelli, A. Di Donato, G. Venanzoni, and A. Morini, "Calibration Protocol for Broadband Near-Field Microwave Microscopy," *IEEE Transactions on Microwave Theory and Techniques* 59 (2011) pp. 2769–2776.

[23] A. Imtiaz, Quantitative Materials Contrast at High Spatial Resolution with a Novel Near-Field Scanning Microwave Microscope, Ph.D. Dissertation, University of Maryland (2005).

[24] D. E. Steinhauer, C. P. Vlahacos, S. K. Dutta, F. C. Wellstood, and S. M. Anlage, "Surface Resistance Imaging with a Scanning Near-Field Microwave Microscope," *Applied Physics Letters* 71 (1997) pp. 1746–1738.

[25] D. E. Steinhauer, C. P. Vlahacos, S. K. Dutta, B. J. Feenstra, F. C. Wellstood, and S. M. Anlage, "Quantitative Imaging of Sheet Resistance with a Scanning Near-Field Microwave Microscope," *Applied Physics Letters* 72 (1998) pp. 861–863.

[26] C. Gao and X.-D. Xiang, "Quantitative Microwave Near-Field Microscopy of Dielectric Properties," *Review of Scientific Instruments* 69 (1998) pp. 3846–3851.

[27] A. P. Gregory, J. F. Blackburn, K. Lees, R. N. Clarke, T. E. Hodgetts, S. M. Hanham, and N. Klein, "A Near-Field Scanning Microwave Microscope for Measurement of the Permittivity and Loss of High-Loss Materials," *2014 84th ARFTG Microwave Measurement Symposium* (2014) pp. 1–8.

[28] J. Kim, K. Lee, B. Friedman, and D. Cha, "Near-Field Scanning Microwave Microscopy Using a Dielectric Resonator," *Applied Physics Letters* 83 (2003) pp. 1032–1034.

[29] C. Gao, T. Wei, F. Duewer, Y. Lu, and X.-D. Xiang, "High Spatial Resolution Quantitative Microwave Impedance Microscopy by a Scanning Tip Microwave Near-Field Microscope," *Applied Physics Letters* 71 (1997) pp. 1872–1874.

[30] M. Tabib-Azar, D.-P. Su, A. Pohar, S. R. LeClair, and G. Ponchak, "0.4 μm Spatial Resolution with 1 GHz (l = 30 cm) Evanescent Microwave Probe," *Review of Scientific Instruments* 70 (1999) pp. 1725–1729.

[31] A. Imtiaz, T. M. Wallis, S.-H. Lim, H. Tanbakuchi, H.-P. Huber, A. Hornung, P. Hinterdorfer, J. Smoliner, F. Kienberger, and P. Kabos, "Frequency-Selective Contrast on Variably Doped p-type Silicon with a Scanning Microwave Microscope," *Journal of Applied Physics* 111 (2012) art. no. 093727.

[32] N. Kodera, D. Yamamoto, R. Ishikawa, and T. Ando, "Video Imaging of Walking Myosin V by High-Speed Atomic Force Microscopy," *Nature* 468 (2010) pp. 72–76.

[33] A. Tselev, J. Velmurugan, A.V. Ievlev, S.V. Kalinin, and A. Kolmakov, "Seeing through Walls at the Nanoscale: Microwave Microscopy of Enclosed Objects and Processes in Liquids," *ACS Nano* 10 (2016) pp. 3562–3570.

8   Probe-Based Measurement Systems

8.1   An Overview of Probe-Based Measurement Systems

A critical component of the NSMM is the broadband probe. Thus, a fundamental understanding of the probe's near-field interaction with investigated materials and devices is necessary for interpretation of NSMM measurements. In this chapter, we will discuss the fundamental concepts and modeling of probe-based measurement systems. We will place particular emphasis on near-field probes that are similar to AFM cantilevers, including models for local effects around the tip apex, as well as the parasitic effects of the cantilever's body. This introduction should serve as a foundation for understanding the application of these measurement systems to semiquantitative and quantitative characterization of both devices and materials. In addition, we shall introduce techniques for calibration of broadband scanning probe systems that enable quantitative characterization of device and material properties.

In recent years a number of variants of NSMMs have been reported. They can be distinguished based on the principles of their design. The designs include instruments based on transmission lines, [1]–[4], waveguides [5], [6], resonant cavities [4], [7] and other scanning probe microscope architectures [8]–[10]. A more complete summary of these techniques can be found in review articles [11], [12] and in the preceding chapter. In general, microwave microscopes utilize either resonant or nonresonant probes. In the case of resonant probes, changes in resonant frequency and quality factor are measured either as a sample is brought toward the probe tip or as an inhomogeneous sample is scanned laterally beneath the tip. The changes in resonant frequency and quality factor are related to the local electromagnetic properties of the sample by use of models and calibration techniques. In the case of nonresonant probes, the probe consists either of an electrically small aperture or a protruding, electrically small antenna integrated into a one- or two-port microwave network. The probe-sample coupling is characterized through position-dependent measurements of the complex reflection coefficient or complex transmission coefficient. For both resonant and nonresonant probes, in order to extract material properties of the sample, it is necessary to calculate the detailed field configuration in the probe-sample region as a function of the probe geometry, probe-to-sample distance, and properties of the probed specimen.

In this chapter, we address the existing state-of-the-art models of the tip–sample interaction. This interaction has direct consequences for application and calibration of probe-based systems. We take an experimental approach based on measurement of calibration artifacts combined, where possible, with VNA calibration procedures. In order to extract characteristic parameters from such measurements, the tip–sample interaction must be modeled and a corresponding inverse problem must be solved. As no unifying theoretical approach exists at present, we will review several different methods. Due to the near-field nature of this interaction, several different modeling approaches may be used, independently or in combination: lumped-element circuits, transmission line circuits, or finite-element modeling. To date, such approaches have been successful when applied to microscopes operating in reflection mode. The situation is more complicated and much less understood for transmission mode operation, lacking established procedures for calibration and device parameter extraction.

We will start with a description of the simplest models. We begin with tip-sample capacitance models, including the coupling capacitance between the tip and the sample in cases where there is a gap between the tip and the sample surface. We discuss models of the most common tip shapes and their influence on the interpretation of measurements. In the next step, we approach the microwave probing problem from the electromagnetics point of view, implementing solutions based on Maxwell's equations. From there, we describe existing calibration procedures for NSMMs that are being adopted for quantitative characterization of devices. In Chapter 9, we will expand on this discussion to describe particular applications to electromagnetic characterization of materials.

8.2   Simple Tip-Sample Models

8.2.1   General Considerations

Tip-sample models are indispensable for quantitative characterization of material properties. As with the modeling of microwave devices, these models can come in many forms, including analytical models, circuit models, and finite-element-based calculations. These techniques are powerful, but can't be generalized. Therefore, one has to treat each measurement individually. Models of tip–sample interactions may be developed by use of the same commercial software programs that are used for modeling of microwave devices, albeit with many of the same restrictions and complications that were described in discussion of broadband modeling of nanoelectronic devices. In the case of near-field microscopy, the situation is slightly simplified. Due to the near-field nature of the interaction, modeling can be done without invoking full microwave solvers. Rather, the modeling can be done at a single frequency with electro- and magnetostatic packages. Even in this simplified case, each particular measurement has to be considered individually with little possibility of generalization. Therefore, we will focus on simpler lumped-element models and quasi-analytical approaches that have proven to be useful for the development of calibration approaches and, in turn, quantitative (or semiquantitative) characterization.

In order to get quantitative information about the device properties and introduce reasonable calibration procedures, it is necessary to capture all of the important contributions that influence the measurement in the tip-sample model. Thus, in NSMM experiments, the tip–sample interaction model embodies critical aspects of the experimental configuration. For near-field probes that are similar to AFM cantilevers, the underlying electromagnetic interaction with the sample is more or less the same for all NSMM experimental configurations, independent of whether the modeling of these interactions is based on a resonant cavity, a transmission line, or other implementation. However, the nature of probe-sample coupling and parasitic coupling within the system may vary strongly from implementation to implementation. For example, if the probe tip is not in mechanical contact with the sample (e.g., if the tip is a height h above the sample), then the electrostatic interaction of the tip with the sample, which we will refer to as the "coupling capacitance" or "coupling impedance," must be included in addition to the sample impedance that arises from interactions within the material. In a lumped-element picture, this likely will take the form of one or more capacitors. These capacitances contribute to the measured response and incorporate the influence of probe and sample geometry, as discussed later in this chapter. In addition, there are configuration-dependent, parasitic interactions of the tip with the sample that must be included in the model.

To summarize, there are three major elements in a tip-sample model of a (non-contact) NSMM: the parasitic impedance, the coupling capacitance, and the impedance characterizing the sample itself. We will address all these impedances as we proceed through this chapter. We start with the coupling capacitance. The coupling capacitance is critical to calibration of NSMMs, but the coupling capacitance is only one part of the interaction. Full calibration requires taking into account all the components of the measurement path, including those components outside of the tip-sample model such as any resonator structure. Finally, because the parasitic capacitance is common to all of the models, we will discuss it separately when addressing specific calibration techniques.

8.2.2   Coupling Capacitance: Parallel Plate Model

The natural first approximation is to model the coupling capacitance as a parallel plate capacitor with the two electrodes formed by the tip and the sample surface, respectively [13]. This approach, although somewhat crude, has been used extensively with surprisingly good results. This applies especially for cases where the objective is understanding the basic physics of a system, obtaining preliminary approximations of material parameters, or semiquantitative estimates of relative contrast within an image. More precise models are needed to obtain the absolute values of the measured parameters.

The coupling capacitance between the tip apex and sample, approximating the system as a parallel plate capacitor is

$$C_{apex-pp} = \frac{\pi \varepsilon_0 R^2}{h},$$ (8.1)

where R is the effective tip radius and h is the tip-sample surface distance. This approach can be improved by including a correction for the fringing capacitance. If the tip apex can be considered to be a disc-shaped terminus of a cylinder, the fringing stray capacitance of the disc can be expressed as [14]

$$C_{apex-strdsc} = 2\pi \varepsilon R \cdot ln\left(\frac{2\pi e R}{h}\right).$$ (8.2)

The authors of Reference [14] found empirically that the correction term

$$C_{apex-strdsc} = 2\varepsilon R \cdot ln\left(\frac{8\pi R}{eh}\right) + \frac{\varepsilon d}{\pi} \cdot \left(ln\left(\frac{h}{8\pi R}\right)\right)^2$$ (8.3)

provides accuracy within about 2 percent when R/h is on the order of unity with increased accuracy at larger R/h ratios. e is the base of natural logarithm. The resulting capacitance is the sum

$$C_{apex} = C_{apex-pp} + C_{apex-strdsc}.$$ (8.4)

Clearly, the parallel plate approach can work well only for conducting samples. Its application for dielectric materials is limited and can be used only as a first approximation.

8.2.3   Coupling Capacitance: Spherical and Conical Tip Shapes

The comparison of NSMM models with experimental results clearly shows that better agreement between model and experiment can be obtained by approximating the tip apex as a sphere. In this case, one can use the method of images to model the tip–surface interaction and the corresponding coupling capacitance. The method of images is well known from electrostatics and it can be applied to a broad range of materials [15]. In References [10], [16]–[18] the method of images was used to derive the apex coupling capacitance for isotropic materials and in Reference [19] the method was extended to anisotropic dielectrics. Indeed, for many applications it is possible to get good agreement between experiments and simple analytical expressions derived by use of the method of images. For example, in the cases of a tip over a metallic or dielectric surface, the apex coupling capacitance can be expressed as

$$C_{apex-m/d} = 4\pi \varepsilon_0 R \sinh(\alpha) \sum_{n=1}^{\infty} A^{n-1} \left(\sinh(n\alpha)\right)^{-1},$$ (8.5)

where $A = 1$ for a metal and $A = \left(\dfrac{\varepsilon_r - 1}{\varepsilon_r + 1}\right)$ for a dielectric. The parameter $\alpha = \cosh^{-1}(1 + a')$ where $a' = h/R$. This approach substantially improves agreement with experimental results compared to the parallel plate model. For the interested reader, more sophisticated approaches based on full solution of the electrostatic problem and finite-element methods can be found in References [20]–[23].

Further refinements to the method-of-images include corrections that more accurately capture the shape of the tip. This is a complex problem that can be treated correctly and completely only by use of numerical methods. In many applications, the apex is modeled as a sphere and the rest of the tip is modeled as a cone described by its taper angle θ and its length L. The coupling capacitance of a combination of the spherical apex with the conical tip is given by a logarithmic expression. For a tip over a metallic surface the apex coupling capacitance can be expressed as [24]

$$C_{apex} = 2\pi\varepsilon_0 RK' \ln\left(1 + \frac{R(1 - \sin\theta)}{h}\right), \tag{8.6}$$

where the constant K' has to be determined empirically from experimental measurement of a known sample.

In many common experimental situations, the sample configuration is a thin dielectric film deposited on a highly conductive or metallic substrate. The apex coupling capacitance for a conical tip over a dielectric film backed by a metal is obtained by modification of Equation (8.6). Specifically, this is done by replacing h with $h + \dfrac{d}{\varepsilon_r}$, leading to [25]

$$C_{apex}(h,d) = 2\pi\varepsilon_0 R\ln\left(1 + \frac{R(1 - \sin\theta)}{h + \dfrac{d}{\varepsilon_r}}\right), \tag{8.7}$$

where ε_r is the relative permittivity of the dielectric and d is the thickness of the film. Sometimes it is useful to use this expression to obtain the derivative of the coupling capacitance with respect to the tip distance from the surface:

$$C_{apex}'(h,d) = 2\pi\varepsilon_0 \cdot \frac{R^2(1 - \sin\theta)}{\left(h + \dfrac{d}{\varepsilon_r}\right)\left(h + \dfrac{d}{\varepsilon_r} + R\sin\theta\right)}. \tag{8.8}$$

In References [26]–[29], a more detailed expression for a thin film over metallic surface is derived, once again assuming a conical tip geometry:

$$C_{cone}(h,d) = \frac{-2\pi\varepsilon_0}{(\ln(\tan\theta/2))^2}\left[\begin{array}{l} \ln\left[L\left(\dfrac{d}{\varepsilon_r}+h+R(1-\sin\theta)\right)^{-1}\right]+\left(\dfrac{d}{\varepsilon_r}+h+R(1-\sin\theta)\right) \\[2mm] +R(1-\sin\theta)+\dfrac{R\cos^2(\theta)}{\sin\theta}\ln\left(\dfrac{d}{\varepsilon_r}+h+R(1-\sin\theta)\right) \end{array}\right].$$

$$(8.9)$$

The corresponding derivative with respect to the tip distance from the surface is

$$C_{cone}'(h,d) = \frac{2\pi\varepsilon_0}{(\ln(\tan\theta/2))^2}\left[\ln\left[L\left(\frac{d}{\varepsilon_r}+h+R(1-\sin\theta)\right)^{-1}\right]-1+\frac{R\cos^2(\theta)/\sin\theta}{\left(\dfrac{d}{\varepsilon_r}+h+R(1-\sin\theta)\right)}\right]. \qquad (8.10)$$

Note that these capacitances may have an additional constant term that is independent of h and therefore is not explicitly given in the foregoing.

Another interesting, empirical approach was introduced in Reference [30], in which the authors used high-frequency, finite-element calculations that were subsequently fitted with analytical formulas. The characteristic impedance of the tip-sample system expressed in terms of propagation constant γ and reference impedance Z_0 was found to be [30]

$$Z_{in} = \frac{1-\Gamma^2+2j\Gamma\sin(\gamma)}{1+\Gamma^2-2j\Gamma\cos(\gamma)}Z_0. \qquad (8.11)$$

From Equation (8.11), it follows that the coupling capacitance is

$$C_{cpl} = \frac{2\Gamma\cos(\gamma)-(1+\Gamma^2)}{2\omega Z_0\Gamma\sin(\gamma)} = \frac{2\Re(S_{11})-(1+|S_{11}|^2)}{2\omega Z_0\Im(S_{11})}, \qquad (8.12)$$

where Γ is the reflection coefficient and ω is the NSMM operating frequency. The operators $\Re$ and $\Im$ denote the real and imaginary parts of a complex number, respectively.

8.2.4   Coupling Capacitance: Elementary Antenna Approach

As stated earlier, the use of electrostatics to describe NSMM tip–sample interactions is successful because of the near-field nature of the problem. Although the electrostatic approach is effective, it is useful to also discuss the simplest dynamic, electromagnetic solutions to the problem of tip–sample interactions. It is indeed possible to apply full electromagnetic finite-element solvers to the problem. The calculated results provide field profiles and the corresponding dependence of the impedance on the distance from the surface. However, due to computational constraints, the use of numerical methods is limited to the region closest to the tip. Though such simulations accommodate arbitrary tip shapes, other parts of the probe structure, such as a cantilever beam, have to be neglected. Here, we will focus on analytical and semi-analytical solutions that could be applied by the interested reader to directly obtain quantitative NSMM results without the need for numerical solutions. Ideally, the numerical methods serve as an important and efficient way to check the validity of the analytical approach. We will not describe the numerical solutions of the problem at hand, but in many instances the analytical asymptotic solutions have been validated by numerical methods.

In the electromagnetic wave solution of the coupling capacitance, we will assume that the tip can be modeled as an elementary dipole or as a small wire antenna with length much smaller than the electromagnetic wavelength. In addition, the problem is assumed to be confined to the near field such that the distance of the elementary dipole from the surface is also much smaller than the electromagnetic wavelength. The general solutions are usually derived in the form of differential-integral equations with simplification to asymptotic cases representing the far and near field. Here we will focus on results for the elementary vertical electric dipole (VED). The reader will find other configurations discussed in the referenced literature.

The problem of an elementary electric dipole radiating above a high-loss half space was originally formulated by Sommerfeld [31], [32]. The problem is treated through the solution of Maxwell's equations in a cylindrical coordinate system by use of the Hertz vector potentials with the appropriate boundary conditions (see Chapter 2). The solution leads to so-called Sommerfeld integrals that are directly solvable only for very special cases, such as those described in References [33]–[36]. Notably, Lindell and Alanen introduced an elegant approach [37], [38], known as exact image theory that introduced a way to implement the imaging theory within the Sommerfeld integral framework.

In the VED model, the important parameter is the impedance of the antenna above a conducting or lossy surface. The real part of this impedance represents the loss due to tip–sample interactions. For conducting samples, the inclusion of the losses due to such a resistive load is important to consider for quantitative assessment. The imaginary part of the impedance relates directly to the coupling capacitance. The treatments of Wait [39], [40] and Lindell and Alanen [41] determined the impedance change of a VED and other elementary antennas in the presence of a conducting half-plane and derived simple asymptotic forms for practical use, albeit with some limitations. Reference [42] compared the numerical integration of constitutive equations with limiting cases. In particular, the far-field and near-field limits for all the configurations of elementary electric and magnetic dipoles have been determined in simple functional forms. Because the near-field limit can be used for characterization of the tip-sample coupling, we will explicitly give the result for a VED in the limit of close proximity of the dipole to a lossy surface.

In the limit $\alpha N \ll 1$, where the interaction distance of the dipole over the sample is much smaller than the wavelength of the incident electromagnetic field λ, the change of the impedance of a VED over a conducting surface with permittivity $\varepsilon_1 = \varepsilon_{r1}\varepsilon_0$ and conductivity σ_1 can be expressed as [42]

$$\frac{Z - Z_0}{R_0} \cong \frac{3(N+1)}{4}\left[f_1(D) + \left(\frac{N+1}{2}\right)^2 f_2(D)\right] + j\left\{\frac{3(1+A)\cdot D}{\alpha^3}\right\}e^{-A}, \quad (8.13)$$

where Z_0 is the impedance of the same dipole located in free space and R_0 is the free space resistance of the antenna, while the parameters $N^2 = (\varepsilon_1 / \varepsilon_0) - j(\sigma_1 / \omega\varepsilon_0), \alpha = 2h(2\pi / \lambda), A = \alpha\sqrt{N^2 - 1}$ and $D = \dfrac{N^2 - 1}{N^2 + 1}$. The functions are defined as

$$f_1(x) = x + \left[1 - x(1+x)\right]\beta + \frac{1}{3}(1+x)(1-x^2)\beta^2, \quad (8.14)$$

$$f_2(x) = -\frac{x}{3} + \left[1 + x(1-x)\right]\beta + \left[1 - x(1+x)(2-x)\right]\beta^2 + \frac{1}{3}\left[1 + x(1-x)(2-x^2)\right]\beta^3,$$
$$-\frac{1}{5}(1-x)(1-x^2)^2\beta^4$$
$$(8.15)$$

where $\beta = \dfrac{N-1}{N+1}$.

For the case of VED over a perfectly conducting plane the change of the impedance can be expressed as [40]

$$\frac{Z - Z_0}{R_0} \cong \frac{1}{N} f_3(\frac{2\pi}{\lambda}\cdot\alpha), \quad (8.16)$$

where

$$f_3(x) = 3[x^{-2}(1 + jx)e^{-jx} - Ei(-jx)] \quad (8.17)$$

and the exponential integral is defined as

$$Ei(-jx) = -\int_x^\infty \frac{e^{-jy}}{y}\,dy. \quad (8.18)$$

8.3 Calibration Procedures for Microwave Scanning Probe Microscopes

8.3.1 Calibration of Near-Field Scanning Microwave Microscopes Operating in Reflection Mode

Quantitative and reproducible measurements of intrinsic material properties can be done only through calibration of the measurement systems, including broadband scanning probe microscopes. As many existing commercial and home-made NSMMs operate in contact mode, most of the existing calibration approaches were derived and implemented assuming that the tip of the cantilever is in direct contact with the sample surface. The procedures described as follows can be extended to noncontact mode operation.

One approach to quantitative impedance measurements is to introduce a calibration procedure that is based on calibration artifacts. One of the first of such procedures for local, quantitative impedance measurements with a resonant, network-analyzer-based NSMM was introduced in Reference [43]. In that work, the calibration artifacts comprised a series of different-sized, square, metallic patches deposited on a SiO_2 film that was in turn supported by low resistivity silicon. During the calibration procedure the probe is in contact with the calibration artifact. A sequence of measurements is made in which the probe is first positioned on one or more patches and then on the bare, SiO_2 film. The differential capacitance is then defined as the difference between the capacitance of the metallic patch and the capacitance of the SiO_2 thin film. The relative capacitance was empirically found to be proportional to the change in the NSMM reflection coefficient. Independently, the perturbation of the reflection coefficient of the NSMM is calculated in the presence of known capacitive loads represented by the metallic patches and the SiO_2 film [44]. Thus, by measurement of a series of patch capacitors, it is possible to quantify the relationship between the measured responses of the NSMM system to each capacitive load. In turn, this gives the required calibration coefficient that relates the measured change in the reflection coefficient to the known capacitance of the artifact. Further, by use of a combination of FEM simulations and measurements over different thicknesses of the oxide layer, the effective tip radius is obtained. Knowledge of the effective tip radius is critical for the transfer of the calibration to an arbitrary sample. An additional result from this work is that the capacitance between the body of the cantilever (or other probe-supporting structure) and the sample surface can be treated as a constant. This observation has been used in all subsequent calibration approaches that are being discussed.

Building upon the work in Reference [43], another calibration artifact was designed that incorporated a set of circular Au/Ti microcapacitors deposited on a SiO_2 staircase [45]. In addition, a bare SiO_2 staircase without metallic patches was incorporated into the same sample. A sketch of the measurement system, including the calibration artifact, is shown in Fig. 8.1 along with a simple capacitance model. The measurement system shown in Fig. 8.1(a) operates as follows: the microwave signal is delivered to the tip from a VNA. The VNA frequency is swept in a narrow range around a local minimum in the reflection coefficient corresponding to the most optimal impedance match to the resonator. Here, the resonator comprises the cantilever, probe tip, half wavelength coaxial resonator, and a 50 Ω shunt resistor. In the next step, an operating frequency is selected as indicated in Fig. 8.1(c). Position-dependent measurements of the reflection coefficient are

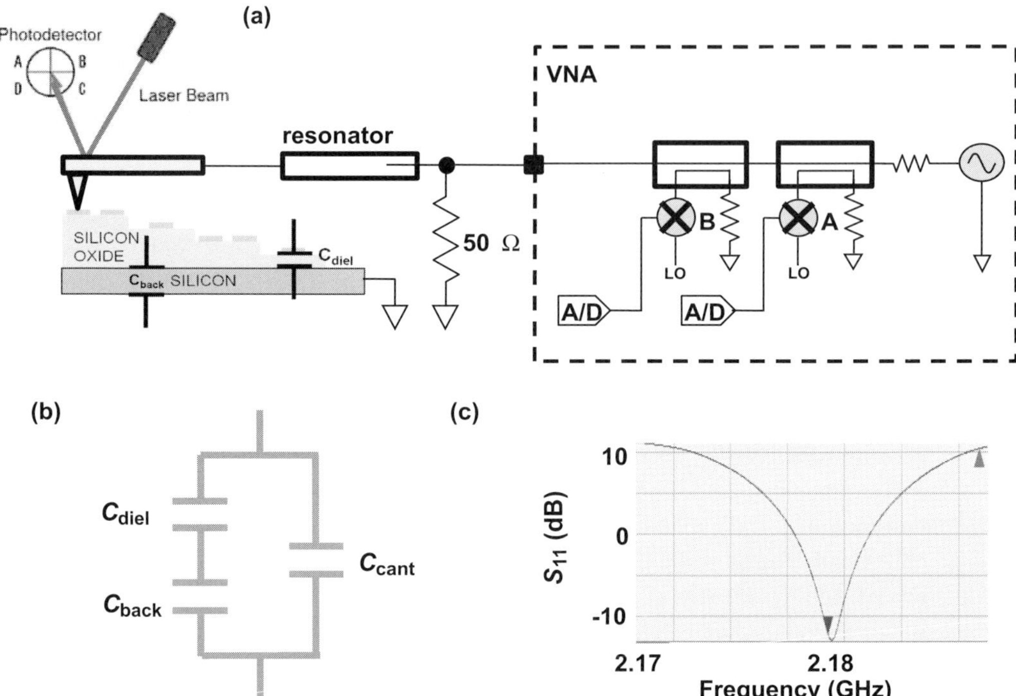

Figure 8.1. An AFM-based NSMM.
(a) Schematic of an NSMM and related test equipment, including the VNA, half-wavelength resonator, cantilever, calibration sample, and beam-bounce detection. (b) A simple circuit model of the probe interaction with the calibration sample. (c) An example of a local minimum in the reflection coefficient, with the operating frequency (downward pointing triangle) offset slightly from the local minimum.
Adapted from H. P. Huber et al., *Review of Scientific Instruments* 81 (2010) art. no. 113701, with permission from AIP Publishing.

carried out at this selected frequency. The topography is measured simultaneously with the amplitude and phase of the complex reflection coefficient S_{11}, defined in the usual way:

$$S_{11} = (Z_L - Z_0)/(Z_L + Z_0), \tag{8.19}$$

where Z_L is the load impedance and $Z_0 = 50\ \Omega$ (the reference impedance). Typical NSMM images are shown in Fig. 8.2 (metal capacitors) and Fig. 8.3 (bare oxide staircase).

The capacitors on the reference sample are modeled as ideal, parallel plate capacitors with a uniform SiO_2 dielectric sandwiched between the circular metal disc and a corresponding circular region on the supporting Si substrate. Equations (8.2) or (8.3) may be used to correct for the fringing capacitance, but for this system the effect of the fringing capacitance can be neglected. The total capacitance is modeled as the parallel plate capacitance C_{diel} in series with an additional capacitance

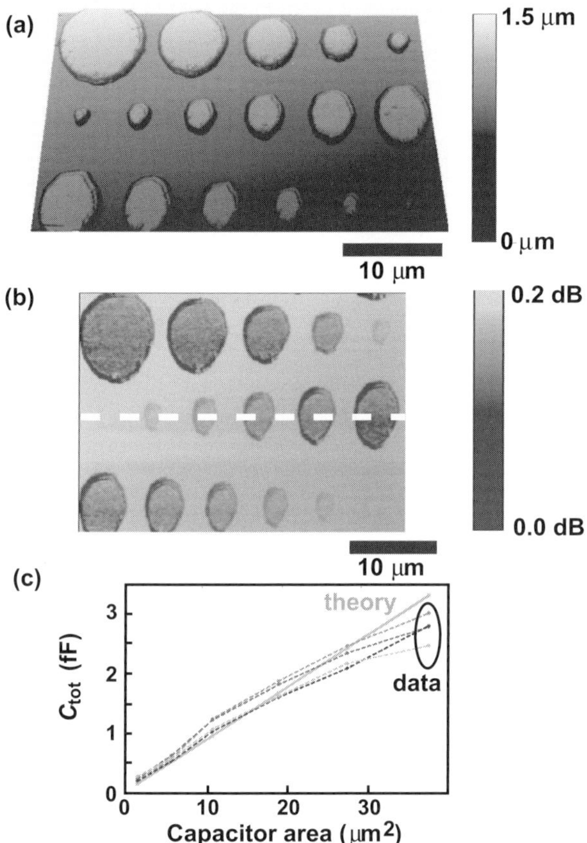

Figure 8.2. Measurement of microcapacitors with an NSMM. (a) Topography and (b) amplitude of the reflection coefficient imaged for a series of oxide-supported, gold microcapacitors. (c) Total capacitance of a series of capacitors, obtained by the calibration procedure outlined in the text. Line cuts taken along the dashed line in (b) provide the raw input for the calibration procedure. Theoretical calculations based on Equations (8.20) and (8.21) are shown as a solid line while data sets are shown as dashed lines. Adapted from H. P. Huber et al., *Review of Scientific Instruments* 81 (2010) art. no. 113701, with permission from AIP Publishing.

between the Si/SiO$_2$ interface and the sample holder, C_{back}. Thus, the total capacitance C_{tot} is

$$\frac{1}{C_{tot}} = \frac{1}{C_{diel}} + \frac{1}{C_{back}} \tag{8.20}$$

with

$$C_{diel} = \frac{\varepsilon A}{d}, \tag{8.21}$$

where A is the pad area, d is the thickness of the dielectric, and ε is the permittivity of the SiO_2. The parameters A and d may be obtained directly from the topography images, provided that the positioning elements in the scanning probe microscope have been calibrated.

There is an additional parasitic capacitance between the body of the cantilever and the sample surface, which is in parallel to C_{tot}. The parasitic capacitance can be effectively removed by making a differential measurement:

$$\Delta S_{11} = S_{11}^{Au} - S_{11}^{Ox}, \tag{8.22}$$

where S_{11}^{Au} is the reflection coefficient on the Au/Ti pad and S_{11}^{Ox} is the reflection coefficient on an adjacent dielectric oxide surface. It is reasonable to assume that the parasitic capacitance is constant across these two positions, as the change in height from the pad to the oxide is much smaller than the total distance from the sample surface to the cantilever body.

Having accounted for the parasitic capacitance, the magnitude of the relative reflection coefficient $|\Delta S_{11}|$ can be related to the total capacitance

$$C_{tot} = \alpha |\Delta S_{11}|, \tag{8.23}$$

where α is a calibration constant in fF/dB. The calculated capacitances for the metal capacitors are shown in Fig. 8.2(c). From the measurements in Fig. 8.2, the fitted calibration constant is found to be $\alpha = 1.5$ fF/dB and the background, interfacial capacitance $C_{back} = 2$ fF. Once again, in order to measure the permittivity of an arbitrary sample, one additional parameter must be found: the effective tip radius. In the present approach, the effective tip radius is obtained from the measurement of the bare dielectric staircase, shown in Fig. 8.3 (another common method for determination of the effective tip radius is discussed later in this chapter and is based on AFM approach curves). We will return to the topic of capacitance calibration based on microcapacitor artifacts in Chapter 11.

Though the use of a differential measurement (Equation (8.22)) nominally removes the effects of parasitic capacitances, an alternative is to estimate the parasitic capacitance by retracting the cantilever to a certain distance that is a few micrometers above the sample. The retraction is performed in series of small steps in height h and the reflection coefficient is measured at each step. At a certain distance from the surface, the dependence of the extracted capacitance becomes linear with distance. The slope of the extracted capacitance as a function of height then represents the contribution of the stray capacitance. This can be expressed as

$$C_{cant} = k_{cant}(h - h_0), \tag{8.24}$$

where the constants k_{cant} and h_0 are determined by fitting the linear region of the measured, height-dependent curve with Equation (8.24). The corrected capacitance over the dielectric steps, which accounts for the effects of C_{cant} in parallel with C_{tot}, is plotted in Fig. 8.3(b). The effective tip radius may then be determined

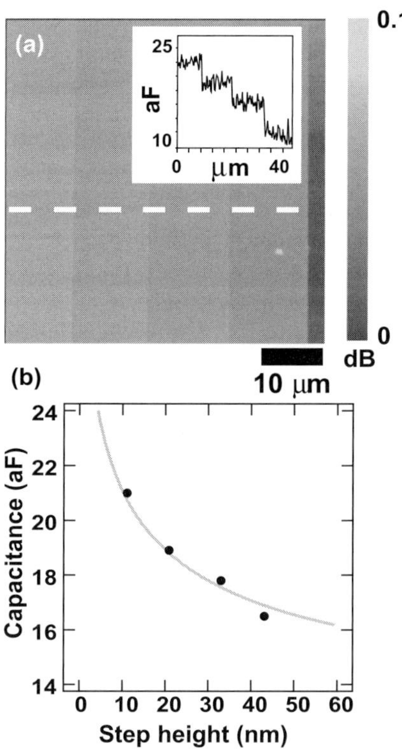

Figure 8.3. Measurement of a SiO$_2$ staircase with an NSMM [45].
(a) Amplitude of the NSMM reflection coefficient imaged for a bare SiO$_2$ staircase. The inset shows a line cut along the dashed white line, converted to capacitance following the procedure described in the text. (b) Average capacitance on each step (black dots) compared to the trend predicted by Equation (8.7) (solid gray line).
Adapted from H. P. Huber et al., *Review of Scientific Instruments* 81 (2010) art. no. 113701, with permission from AIP Publishing.

from Equation (8.7), where contact corresponds to $h = 0$. Additional corrections may be made for fringe capacitance contributions in the form of Equation (8.2) or Equation (8.3).

Though this calibration procedure can be applied in many areas, it does not address measurements of loss or non-capacitive reactance. In addition, the procedure requires a number of micro-fabricated capacitance standards that may contribute additional uncertainty to the measurement. Introducing an alternative approach based on one-port calibration of network analyzers reduces the number of required standards and provides full quantitative determination of both resistive and reactive parts of the local impedance [46]. As earlier, this procedure assumes that the NSMM operates in contact mode. As in one-port calibration of guided-wave systems, the definition of the calibration reference plane is an important consideration (see Chapter 2). Here, it is possible to define the reference plane either at the probe tip or at another location in the signal path, as will be discussed later.

From Chapter 2, recall that in a three-term error model, the measured, raw reflection coefficient S_{11m} is related to the corrected reflection coefficient S_{11} through

$$S_{11} = \frac{S_{11m} - e_{00}}{e_{10}e_{01} + e_{11}(S_{11m} - e_{00})}, \tag{8.25}$$

where the error coefficients are defined as follows: e_{00} is directivity, the product $e_{10}e_{01}$ is tracking, and e_{11} is the port match. From the measurements of three electrically distinct standards with known reflection coefficients, it is possible to calculate the complex error coefficients. Once the values of the error coefficients are known, the real and imaginary parts of the impedance at the reference plane can be determined from the corrected reflection coefficient via Equation (8.19). The reference impedance Z_0 can be chosen arbitrarily. In Reference [46], the reference impedance was chosen to be 10 kΩ and three capacitors from the calibration artifacts described earlier were chosen as standards. This approach provides calibrated, local measurements of complex impedance by use of NSMMs, but it has a relatively large uncertainty, particularly for small capacitance values, and the procedure is not transferable from the calibration substrate to an arbitrary sample. This relatively large uncertainty arises from the difficulty in estimation of the change in fringing electromagnetic fields around the tip when moved from calibration standards to the SUT.

The problem of transfer from the calibration substrate to an arbitrary sample was solved by Gramse et al. [29]. Their approach builds on the initial calibration ideas of Hoffman et al. in Reference [46] and Farina et al. in Reference [47]. In the latter reference, one central idea was the variation of the probe-sample distance to create different calibration "standards." The Gramse et al. approach works in situ directly on the SUT and does not require a special calibration sample [29]. The particular innovation in the approach is to simultaneously perform an NSMM measurement with a low-frequency electrostatic capacitance measurement as the tip approaches the sample. Though no longer necessary for calibration, well-characterized samples and substrates remain useful for validation of the procedure and for systematic study of effective tip radii on different classes of samples. The calibration procedure presented in Reference [29] works in typical NSMM operating frequency ranges (about 1 GHz to 26 GHz) and takes into account the full test platform response, including the VNA, cabling, and the specific geometry of the tip–sample interaction area. Because this approach may become a widely used procedure for NSMM calibration, we describe the steps of this method in greater detail subsequently. Our description will assume a measurement system similar to that shown in Fig. 8.1(a), but it is straightforward to adapt the procedure to other configurations.

In the first step, the measured, complex reflection coefficient is converted into a complex impedance (admittance). Next, a one-port calibration is applied [46], following Equation (8.25). The reference plane is chosen to be right behind the cantilever chip shown in Fig. 8.1(a). This choice of reference plane means that the measured DUT includes the body of the cantilever and the tip-sample system. As discussed in Chapter 2, this choice is arbitrary, but has direct consequences for the implementation and interpretation of the de-embedding procedure. This choice of reference plane was necessitated by complications encountered in Reference [46] arising from the dependence of the stray capacitance of the cantilever chip upon the probe-sample distance. These complications may also reflect the fact that the probe tip does not fulfill the condition for single mode propagation at the reference plane. Strictly speaking, this precludes a definition of the reference plane at the probe tip. Note that this problem is significantly mitigated for conducting or highly planar samples.

The mapping of the raw measured S_{11m} onto corrected S_{11} is the same as in the previous case, with three complex error coefficients that have to be determined from the measurement of at least three reference samples with known impedances. However, this requires the positioning of these reference samples at the calibration reference plane. For the given choice of the reference plane, this represents a challenge. However, this problem is also addressed by utilizing a height-dependent measurement of the reflection coefficient in conjunction with an electrostatic force measurement. Specifically, the amplitude and phase of the reflection coefficient S_{11m} is measured as a function of the tip-sample distance h at a fixed, selected frequency. For metallic and purely dielectric substrates, the character of the impedance change is purely capacitive, which allows the simultaneous, low-frequency measurement of the electrostatic force. As in electrostatic force microscopy (EFM), the force is proportional to $dC\,/\,dh$ [23]–[24]:

$$F_{es}(h) = \frac{1}{4}\frac{dC}{dh}V_0^2 \cos(2\omega t) \rightarrow \frac{dC}{dh} = \frac{2F_{es,2\omega}}{V_0^2}, \tag{8.26}$$

where V_0 is the amplitude of low-frequency modulation voltage and ω is the modulation frequency. Integrating the measured $F_{es}(h)$ curves gives the desired tip-surface capacitance and the corresponding admittance $Y(h) = j2\pi fC(h)$. This admittance acts as the reference sample in place of a calibration artifact and serves as the input into the mapping equation from the measured to the corrected reflection coefficient. Since this creates an overdetermined system (assuming that the measurements are made at more than three heights), an optimization algorithm is used to solve Equation (8.25) for the error coefficients. It is important to recognize that dC/dh in Equation (8.26) may have to be adjusted by a small offset to account for the stray capacitance that is not detected by EFM. This EFM-based calibration procedure has been verified across multiple operating frequencies [29].

Having calibrated the S_{11} reflection coefficients, the S_{11} images can be converted into the real and imaginary parts of impedance or admittance. Once again, to extract parameters of interest from calibrated reflection coefficient measurements of an arbitrary sample, it is necessary to know the effective tip geometry. Fortunately, the approach curve measurements enable the extraction of the tip geometry such

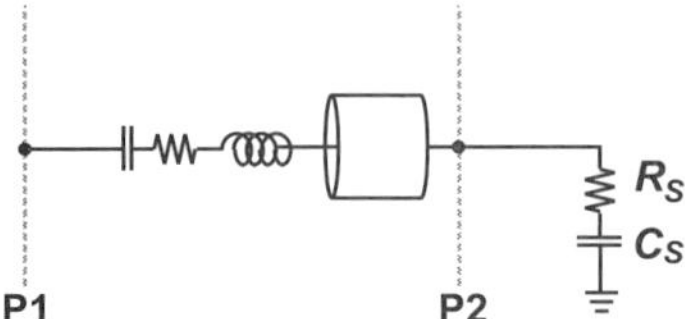

Figure 8.4. Critically coupled resonator and sample impedance.
Reference plane P1 is at the input to the resonator and reference plane P2 is at the tip of the probe. The sample resistance R_S and sample capacitance C_S contribute to the overall sample impedance Z_S.
© [2012] IEEE. Adapted, with permission from Jonathan D. Chisum and Zoya Popović, *IEEE Transactions on Microwave Theory and Techniques* 60 (2012) pp. 2605–2615.

as radius R, cone angle θ, and cone height L by comparison of the calibrated measurements to finite-element simulations. In some cases, the cone height and cone angle may be fixed to manufacturers' nominal values, thus reducing the set of fitting parameters in the determination of tip geometry and parasitic capacitance. These parameters are in general considered to be effective dimensions, absorbing the influence of any imperfect approximations or unknown electromagnetic field interactions.

Finally, a thorough characterization of an NSMM system has been described in Reference [48], including the introduction of a calibration approach for spatially resolved, subsurface measurements as a function of material and sample depth. Although the spatial resolution was in the micrometer range the results are in principle scalable to nanometer dimensions. The approach was designed for a critically coupled resonator system. As with many NSMMs, the probe's interaction with the sample leads to changes in the resonance frequency and the quality factor, which manifest as changes in the complex reflection coefficient. A schematic circuit model of the critically coupled resonator and the sample impedance is shown in Fig. 8.4. The schematic also shows the reference plane at the input to the resonator (P1) and at the base of the cantilever (P2). At the P1 reference plane, a standard coaxial or waveguide calibration is performed. To translate to the next reference plane P2, it is necessary to de-embed the resonator between P1 and P2. This can be accomplished through a frequency-swept measurement of the reflection coefficient when the tip is far from the surface of the sample. From this frequency sweep, the circuit parameters of the resonator in the vicinity of the operating frequency can be established. Having established these circuit parameters, the resonator can be de-embedded and the reference plane can be moved to P2 to get the calibrated probe-sample coupling impedance.

8.3.2   Calibration of an Interferometric Scanning Microwave Microscope

Naturally, variations in RF and microwave scanning probe microscope design necessitate modifications of the calibration procedures. The remaining sections discuss how to approach calibration for specific cases that vary from the NSMM system we have treated previously.

In Reference [49], the authors introduced an interferometric scanning microwave microscope as a means to improve the sensitivity of microwave microscopes. As we discussed in Chapter 3, interferometric approaches are an effective strategy for measurement of high-impedance devices that display a large mismatch with 50 Ω test equipment. The measurement resolution of a VNA can be estimated from the relative variation

$$\frac{\Delta Z_{DUT}}{Z_{DUT}} = \left[\frac{(Z_{DUT} + Z_{ref})^2}{2 Z_{DUT} Z_{ref}} \right] \Delta S_{11DUT}, \tag{8.27}$$

where Z_{DUT} is the DUT impedance, ΔZ_{DUT} represents the change in DUT impedance due to the change of the device properties during the scan, and Z_{ref} is the reference impedance, usually 50 Ω. Clearly, the uncertainty in the measurements degrades as Z_{DUT} increases. An early approach to reduction of this uncertainty was to introduce a comparator circuit and amplifier between the tip and the VNA port. This improved the signal-to-noise ratio for large impedances, albeit with some limitations. Further improvement was realized by including an adjustable interferometer between the VNA and the tip. This has several advantages, including compatibility with an increased range of impedances and a broader range of operating frequencies.

The interferometer connection to the VNA is shown in Fig. 8.5. This is once again a single port measurement. The signal from the source is split and directed through an interferometer that resembles a Mach-Zehnder configuration. The interferometer comprises a coaxial power divider and two hybrid couplers, one of which is connected to an attenuator. The signal from the source at port 1 of the VNA (a_I in Fig. 8.5) is split in the power divider and one arm is fed to the tip of the microscope (a_{inc}). The reflected signal from the tip-sample system (a_{ref}) is combined with the second part of the signal passing through the attenuator in the second coupler. The resulting interferometric signal is then amplified and analyzed in the receiver in port 1 of the VNA. The attenuator is adjusted to cancel the signal a_3 in front of the amplifier through destructive interference between the two branches of the interferometer. The variable attenuator ensures that complete cancellation through interference can be obtained for any impedance values.

This case represents an easily implemented improvement to the signal-to-noise ratio of NSMMs. However, use of the interferometer configuration requires a modified calibration procedure; a traditional, single port calibration procedure is not sufficient. Although the whole assembly of the interferometer and the tip-sample junction is connected to a single port of the VNA, the calibration has to be done between the port 1 source of the VNA through the assembly to the receiver of the port 1. To circumvent this, a calibration procedure based on a modified one-port error model was developed for interferometric NSMM. A set of oxide microcapacitors similar to those described earlier may serve as reference standards for the interferometer calibration.

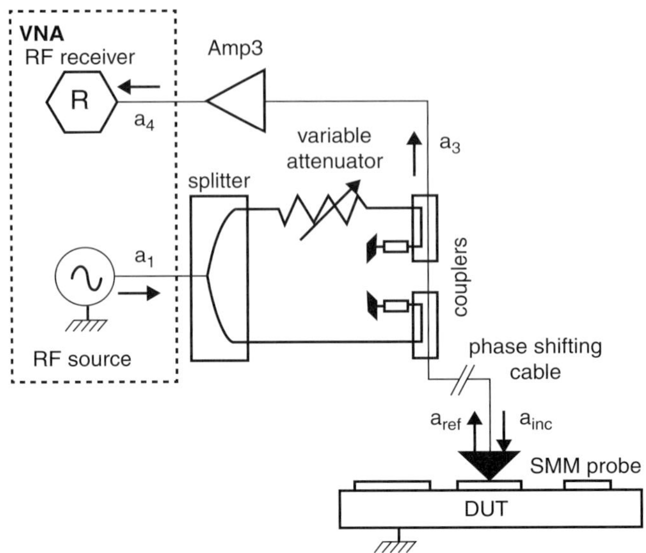

Figure 8.5.   Interferometric scanning microwave microscope.
Schematic of the circuit and test platform for implementing an interferometric scanning
microwave microscope NSMM. The interferometric measurement is made by use
of a VNA. Reprinted from T. Dargent, K. Haddadi, T. Lasri, N. Clement, D. Ducatteau,
B. Legrand, H. Tanbakuchi, and D. Theron, *Review of Scientific Instruments* 84 (2013) art.
no. 123705, with permission from AIP Publishing.

The calibration is carried out as follows. The reference load with reflection coefficient S_{11ref} is obtained by positioning the tip onto a thick dielectric that serves to establish the reference impedance. The attenuator is set to minimize the interference signal such that at a given frequency the transmission signal (a_3) is close to zero, thus ensuring high measurement sensitivity for impedances around the reference impedance. As shown in Fig. 8.5, a_4 represents the incoming signal from the DUT into the VNA. One can consider this as a signal entering a second port, an effective "port 2," loaded with the reflection coefficient of SUT. Thus, it is possible to represent the effect of the interferometer as a matrix complex scattering parameters resembling a two-port error box. It is possible to relate the measured reflection coefficient in terms of the scattering parameters of the interferometer through Equation (8.25), provided that the e_{ij} coefficients now represent the S_{ij} parameters of the interferometer with indices equal to 0 replaced by 1 and indices equal to 1 replaced by 2. S_{11} represents the calibration impedance reflection coefficient. These parameters are known as "interferometer transition coefficients." The interferometer transition coefficients are determined from measurements of three known standards chosen from a calibration kit. Increasing the number of calibration standards leads to an overdetermined system and, in turn, a more robust calibration with reduced statistical uncertainty. If the calibration standards are all pure capacitors, it is possible to reduce the complexity of numerical optimization by assuming that only the phase shift of the reflection coefficient has to be taken into account.

8.3.3   Time-Domain Approaches in Scanning Microwave Microscopy

An alternative approach to NSMM data processing is to move post-processing steps from the frequency domain to the time domain [50]. The underlying idea that motivates moving to the time domain is that in broadband near-field microscopy both near- and far-field interactions exist simultaneously. The objective is to disentangle these contributions based on their separation in time. Broadly speaking, the near-field interactions at the tip are instantaneous and the far-field echo is delayed by a certain amount of time that depends on distance from the source to the reference plane. This time separation allows partial disentangling of the signals and also can be used as a filter to reduce the influence of either set of interactions from the images. Notably, using this approach for de-noising raw measured images does not require ultra-wideband measurements, though such measurements may be required in the case that quantitative information about a selected physical or circuit parameter of the system is required.

Practical implementation of this procedure begins with measurement of the reflection coefficient over a limited frequency range. The frequency dependence of the reflection coefficient must be measured at each point of the scanned image area. The measured frequency-domain responses are linearly combined to a time-domain signal via [50]

$$s(t) = \sum_{f_i} \Re\left[K(f_i)S(f_i)e^{j2\pi f_i} \right], \tag{8.28}$$

where f_i are the measurement frequencies, $S(f)$ is the reflection coefficient as measured in the given frequency band, and $K(f)$ is a weighting function. For equidistant frequency points within the selected bandwidth, Equation (8.28) becomes a standard finite inverse Fourier transform. An example of an NSMM data set transformed to the time domain is shown in Fig. 8.6. The time-domain response shown in Fig. 8.6 allows windowing of the local and nonlocal interactions. With proper selection of the windows, it is possible to de-noise the image by cutting out the local or nonlocal interactions and transforming back to the frequency domain. In Reference [50], the authors used as $K(f)$ in Equation (8.28) a Kaiser-Bessel windowing function, but other windowing functions are possible.

An example application of this approach to images of living myotubes is shown in Fig. 8.7. Figure 8.7(a) is a microwave image taken by using an STM-based NSMM and Fig. 8.7(b) is the same data after post-processing in the time domain. The de-noising in the image is clearly observable, revealing details of the image not visible in the STM image, enabling the identification of additional features as extracellular fluids and saline crystals in the post-processed image. This approach is compatible with and complementary to the NSMM calibration procedures described in the preceding sections. The calibration of the microscope complemented by

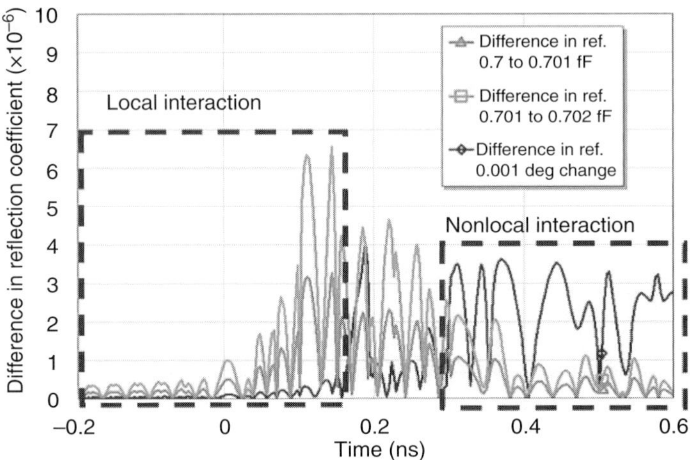

Figure 8.6.   Time-domain measurements with an NSMM.
The differences in the reflection coefficient measured by use of NSMM as the tip-sample coupling capacitance is changed from 0.700 fF to 0.701 fF then to 0.702 fF. The data has been transformed from the frequency domain to the time domain. Time frames during which local and nonlocal interactions dominate are indicated (dashed line windows). Reprinted from Reference [50], with permission from the Royal Society of Chemistry.

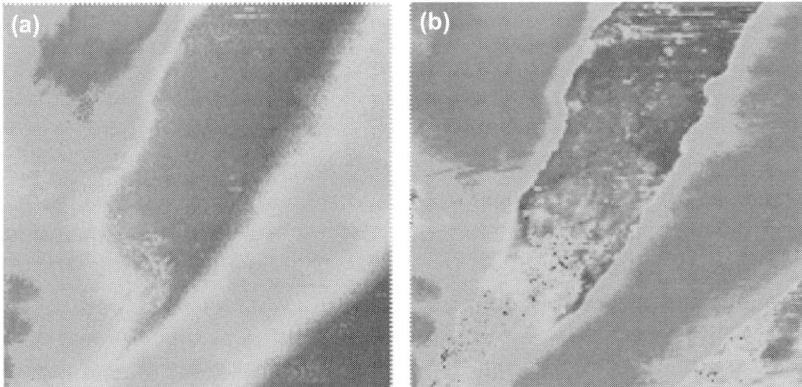

Figure 8.7.   De-noising of microwave images by use of time-domain processing.
(a) As-measured reflection coefficient amplitude image of living myotubes acquired with an STM-based NSMM. (b) The same image, following time-domain post-processing of the data. Reprinted from Reference [50], with permission from the Royal Society of Chemistry.

time-domain post-processing provides a solid foundation for quantitative measurement of material and device properties. Note that this post-processing approach has to be distinguished from direct time-domain measurements of the materials and devices. Such direct measurements require either direct impulse measurement techniques or time-domain, tip-enhanced scattering techniques.

8.3.4   Calibration of Evanescent Microwave Magnetic Probes

For some applications it is useful to trade spatial resolution for sensitivity to a specific field parameter. To that end, another alternative is to use a magnetic dipole as the sensing element in the probe tip instead of an electric dipole. In fact, some of the earliest experimental sub-wavelength microwave techniques were applied to magnetic materials. They were introduced in the late 1950s and were based on a sub-wavelength aperture in a waveguide. The objective was to map inhomogeneity in thin magnetic films [51], [52].

A novel variant of magnetic-dipole-based, near-field microscope was introduced in Reference [53] for characterization of metal samples properties with conductivity resolution of $5\times10^{-3}\,\sigma$. The probe is based on a $\lambda/4$ resonator built as part of a CPW transmission line ending with a small loop. When in close proximity to a sample, the loop couples to the sample through the coupling capacitance and loading impedance of the sample. These interactions are manifest as changes in the admittance of the probe circuit. Because the coupling to the resonator is through the reactive channel, the admittance (or impedance) calibration procedure for a magnetic-dipole probe is similar to the impedance calibration of electric-dipole probes.

However, the challenge of calibrating the DC magnetic field of the electromagnet remains. For both magnetic-dipole- and electric-dipole-based NSMMs, the standard approach for characterization of the dynamic response of the magnetic materials is through the measurement of the eigen-modes of the magnetic system, represented in the linear regime and microwave frequency range through ferromagnetic resonance (FMR). Measurements using FMR are at times performed with a static magnetic field applied to the sample while the frequency is swept. More often, due to the difficulty of de-embedding of frequency-dependent elements in the test platform, the measurements are made at a constant frequency while the biasing magnetic field is swept. The latter approach is easier to calibrate, because the sample response reflects physical changes in the material as a function of the magnetic field. Therefore, with the frequency fixed, the intensity of the resonance response is measured relative to the nonresonant response. Given that determination of the DC magnetic resonance field(s) is usually the primary objective, the calibration procedure is reduced to calibration of the local magnetic field. In some cases, this can be done using nuclear magnetic resonance (NMR), but in most cases and particularly for on-chip measurements a different approach has to be used.

A more accessible field-calibration method is to use a reference sample with a well-known dynamic magnetization response. The natural choice of a reference material for this purpose is single crystal yttrium iron garnet (YIG), due to its well-known properties, low intrinsic linewidth, and well-established FMR response. The well-known, resonant response of single-crystal YIG can be used to calibrate the DC magnetic field of electromagnets incorporated into NSMM systems. The configuration of the microscope electromagnet is shown in Fig. 8.8(a). The sample is positioned on the top of the magnet and therefore it is difficult to calibrate the correct field value at the position of the sample from calculations or simulations. To

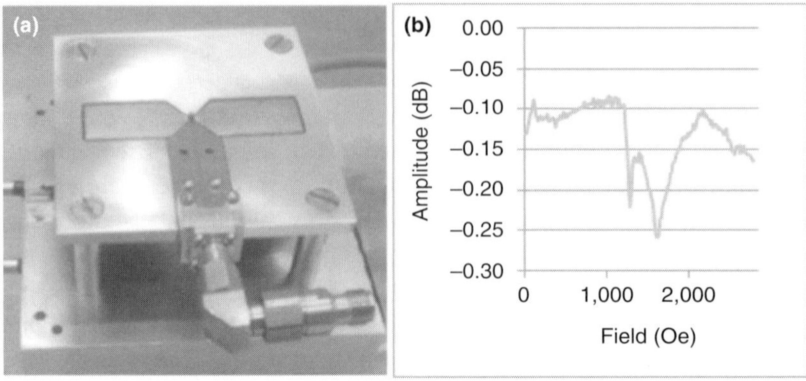

Figure 8.8.   Near-field scanning microwave measurements of a YIG film.
(a) Photograph of an electromagnet and RF signal path integrated into the sample holder
for use with a commercial broadband scanning microwave microscope. (b) Amplitude of the
local, reflected signal from a YIG thin film measured with an NSMM operating at a constant
frequency. On the Field axis, note that 1,000 Oe corresponds to 80 kA/m in SI units.

mitigate this problem, a thin epitaxial YIG film is used as a reference sample. The
field-swept FMR response is locally measured to calibrate the field value, as shown
in Fig. 8.8(b). In addition to providing a convenient approach to field calibration,
this measurement also demonstrates that magnetization dynamics can be measured
with an electric dipole NSMM. The position of the resonances in the field-swept
FMR measurement at constant applied RF frequency determines the DC resonant
fields. As the field-dependent resonance frequencies for YIG are well known, the
position of the dip in the curve calibrates the field value for a given probe position.
Also note that each resonance dip corresponds to an excited spin wave mode in the
magnetic sample. Therefore, the measured dip(s) represent dynamic excitations of
the measured sample. Thus, by selecting the corresponding DC field of an excited
mode and scanning over the sample, one can visualize the spin wave pattern for a
selected mode [54].

8.3.5   Evanescent Microwave Microscopes in Transmission Mode

Recently, some efforts in scanning microwave microscopy have moved to the implementation of microscopes in transmission mode rather than reflection mode.
This trend is particularly important for nondestructive subsurface tomography of
defects and interfaces. An early report of such an effort appeared in Reference [55],
in which the feasibility of using an NSMM in transmission mode was investigated
by use of three-dimensional, finite-element modeling. A sensitivity analysis was
carried out by varying different parameters of the system including the measurement frequency and observing the resultant change in the complex transmission
coefficient. The models are supported by preliminary measurements on a Si substrate with spatially varying dopant concentrations. The investigation concluded

that measurement of transmission provides improved sensitivity with respect to existing reflection-mode NSMMs, especially for the phase. This improved sensitivity has important experimental implications, but at present it comes at the price of quantitative metrology.

A well-established calibration procedure for a transmission mode NSMM does not presently exist. A further complication is that the traditional, two-port VNA calibration procedures are not directly transferable to this case. One significant challenge is the establishment of the reference planes. On one side of the sample, the reference plane can be chosen in a similar way to the case of a single port calibration at the probe tip. On the other side, the plane can be chosen to be at a designated port, probe, or second electric dipole (antenna). However, the device now includes not only the tip-sample assembly of the first port, but also the transmission coupling for the second port as well as the complex interaction of the microwaves throughout the thickness of sample. Thus, the total DUT is a complicated, multifaceted system and it is extremely challenging to de-embed the material properties from the total DUT measurement. Furthermore, neither the probe tip nor the second antenna fulfills the conditions for single mode propagation at the reference plane.

Moving forward, the establishment of the necessary error model and error coefficients for a transmission-mode NSMM is formidable. Possible ways forward include the multimode calibration procedure discussed in Chapter 2 or procedures for calibration of antenna-to-antenna transmission in the near field [56]. This is not a simple problem and the work in this area is ongoing.

References

[1] M. Fee, S. Chu, and T. W. Hänsch, "Scanning Electromagnetic Transmission Line Microscope with Subwavelength Resolution," *Optics Communications* 69 (1989) pp. 219–224.

[2] D. E. Steinhauer, C. P. Vlahacos, C. Canedy, A. Stanishevsky, R. Melngailis, R. Ramesh, F. C. Wellstood, and S. M. Anlage, "Imaging of Microwave Permittivity, Tunability, and Damage Recovery in (Ba,Sr)TiO3 Thin Films," *Applied Physics Letters* 75 (1999) pp. 3180–3182.

[3] M. Tabib-Azar, N. Shoemaker, and S. Harris, "Nondestructive Characterization of Materials by Evanescent Microwaves," *Measurement Science and Technology* 4 (1993) pp. 583–590.

[4] M. M. Wang, K. Haddadi, D. Glay, and T. Lasri, "Compact Near Field Microwave Microscope Based on the Multi-port Technique," *Proceedings of the 40th European Microwave Conference* (Paris, 2010) pp. 771–774.

[5] M. Golosovsky and D. Davidov, "Novel mm-wave Near-Field Resistivity Microscope," *Applied Physics Letters* 68 (1996) pp. 1579–1581.

[6] V. V. Talanov and A. R. Schwartz, "Near-Field Scanning Microwave Microscope for Interline Capacitance Characterization of Nanoelectronics Interconnect," *IEEE Transactions on Microwave Theory and Techniques* 57 (2009) pp. 1224–1229.

[7] C. Gao and X.-D. Xiang, "Quantitative Microwave Near-Field Microscopy of Dielectric Properties," *Review of Scientific Instruments* 69 (1998) pp. 3846–3851.

[8] S. J. Stranick and P. S. Weiss, "A Versatile Microwave Frequency-Compatible Scanning Tunneling Microscope," *Review of Scientific Instruments* 64 (1993) pp. 1232–1234.

[9] M. Tabib-Azar and Y. Wang, "Design and Fabrication of Scanning Near-Field Microwave Probes Compatible with Atomic Force Microscopy to Image Embedded Microstructures," *IEEE Transactions on Microwave Theory and Techniques* 52 (2004) pp. 971–979.

[10] K. Lai, W. Kundhikanjana, M. Kelly, and Z. X. Shen, "Modeling and Characterization of Cantilever-Based Near-Field Scanning Microwave Impedance Microscope," *Review of Scientific Instruments* 79 (2008) art. no. 063703.

[11] B. T. Rosner and D. W. van der Weide, "High-Frequency Near-Field Microscopy," *Review of Scientific Instruments* 73 (2002) pp. 2505–2525.

[12] S. M. Anlage, V. V. Talanov, and A. R. Schwartz, "Principles of Near-Field Microscope." In *Scanning Probe Microscopy* (S. Kalinin and A. Gruverman, eds.), (Springer, 2007) pp. 215–253.

[13] M. J. Yoo, T. A. Fulton, H. F. Hess, R. L. Willet, L. N. Dunkleberger, R. J. Chichester, L. N. Pfeiffer, and K. W. West, "Scanning Single-Electron Transistor Microscopy: Imaging Individual Charges," *Science* 276 (1997) pp. 579–582.

[14] G. J. Sloggett, N. G. Barton, and S. J. Spencer, "Fringing Fields in Disc Capacitors," *Journal of Physics A: Mathematical and General* 19 (1986) pp. 2725–2736.

[15] W. R. Smythe, *Static and Dynamic Electricity* (McGraw-Hill, 1950).

[16] H. P. Kleinknecht, J. R. Sandercock, and H. Meier, "An Experimental Scanning Capacitance Microscope," *Scanning Microscopy* 2 (1988) pp. 1839–1844.

[17] C. Gao, F. Duewer, and X.-D. Xiang, "Quantitative Microwave Evanescent Microscopy," *Applied Physics Letters* 75 (1999) pp. 3005–3007.

[18] S. V. Kalinin, E. Karapetian, and M. Kachanov, "Nanoelectromechanics of Piezoresponse Force Microscopy," *Physical Review B* 70 (2004) art. no. 184101.

[19] E. A. Eliseev, S. V. Kalinin, S. Jesse, S. L. Bravina, and A. N. Morozovska, "Electromechanical Detection in Scanning Probe Microscopy: Tip Models and Materials Contrast," *Journal of Applied Physics* 102 (2007) art. no. 014109.

[20] B. M. Law and F. Rieutord, "Electrostatic Forces in Atomic Force Microscopy," *Physical Review B* 66 (2002) art. no. 035402.

[21] G. M. Sacha, E. Sahagun, and J. J. Saenz, "A Method for Calculating Capacitances and Electrostatic Forces in Atomic Force Microscopy," *Journal of Applied Physics* 101 (2007) art. no. 024310.

[22] G. M. Sacha, "Method to Calculate Electric Fields at Very Small Tip-Sample Distances in Atomic Force Microscopy," *Applied Physics Letters* 97 (2010) art. no. 033115.

[23] S. Gomez-Monivas, J. J. Saenz, R. Carminati, and J. J. Greffet, "Theory of Electrostatic Probe Microscopy: A Simple Perturbative Approach," *Applied Physics Letters* 76 (2000) pp. 2955–2957.

[24] S. Hudlet, M. Saint Jeana, C. Guthmann, and J. Berger, "Evaluation of the Capacitive Force between an Atomic Force Microscopy Tip and a Metallic Surface," *European Physical Journal B* 2 (1998) pp. 5–10.

[25] G. Gomila, J. Toset, and L. Fumagalli, "Nanoscale Capacitance Microscopy of Thin Dielectric Films," *Journal of Applied Physics* 104 (2008) art. no. 024315.

[26] Š. Lanyi, "Effect of Tip Shape on Capacitance Determination Accuracy in Scanning Capacitance Microscopy," *Ultramicroscopy* 103 (2005) pp. 221–228.

[27] Š. Lanyi, "Shape Dependence of the Capacitance of Scanning Capacitance Microscope Probes," *Ultramicroscopy* 108 (2008) pp. 712–717.

[28] G. Gomila, G. Gramse, and L. Fumagalli, "Finite Size Effects and Analytical Modeling of Electrostatic Force Microscopy Applied to Dielectric Films," *Nanotechnology* 25 (2014) art. no. 255702.

[29] G. Gramse, M. Kasper, L. Fumagalli, G. Gomila, P. Hinterdorfer, and F. Kienberger, "Calibrated Complex Impedance and Permittivity Measurements with Scanning Microwave Microscopy," *Nanotechnology* 26 (2015) art. no. 149501.

[30] Y. Naitou, A. Yasaka, and N. Ookubo, "An Analytical Model for Capacitance between Probe Tip and Dielectric Film Deduced by High Frequency Electromagnetic Field Simulations," *Journal of Applied Physics* 105 (2009) art. no. 044311.

[31] A. Sommerfeld, "Über die Ausbreitung der Wellen in der drahtlosen Telegraphie," *Annalen der Physik* 28 (1909) pp. 665–736.

[32] A. Sommerfeld, "Über die Ausbreitung der Wellen in der drahtlosen Telegraphie," *Annalen der Physik* 81 (1926) pp. 1135–1153.

[33] E. F. Kuester and D. C. Chang, "Evaluation of Sommerfeld Integrals Associated with Dipole Sources above Earth," *Electromagnetics Laboratory/The MIMICAD Research Center* (1979) Paper 65.

[34] R. W. P. King, *Theory of Linear Antennas* (Harvard University Press, 1956).

[35] L. B. Felsen and N. Marcuvitz, *Radiation and Scattering of Waves* (IEEE Press, 1994).

[36] J. R. Wait, *Electromagnetic Waves in Stratified Media* (IEEE Press, 1996).

[37] I. V. Lindell and E. Alanen, "Exact Image Theory for the Sommerfeld Half-Space Problem, Part I: Vertical Magnetic Dipole," *IEEE Transactions on Antennas and Propagation* 32 (1984) pp. 126–133.

[38] I. V. Lindell and E. Alanen, "Exact Image Theory for Sommerfeld Half-Space Problem, Part II: Vertical Electric Dipole," *IEEE Transactions on Antennas and Propagation* 32 (1984) pp. 841–847.

[39] J. R. Wait, "Possible Influence of the Ionosphere on the Impedance of a Ground-Based Antenna," *Journal of Research of the National Bureau of Standards-D* 66 (1962) pp. 563–569.

[40] J. R. Wait, "Impedance Characteristics of Electric Dipoles over a Conducting Half-Space," *Radio Science* 4 (1969) p. 971.

[41] E. Alanen and I. Lindell, "Impedance of Vertical Electric and Magnetic Dipole above a Dissipative Ground," *Radio Science* 19 (1984) pp. 1469–1474.

[42] L. E. Vogler and J. L. Noble, "Curves of Input Impedance Change due to Ground for Dipole Antennas," *National Bureau of Standards* (1964) Monograph 72.

[43] A. Karbassi, D. Ruf, A. D. Bettermann, C. A. Paulson, D. W. van der Weide, H. Tanbakuchi, and R. Stancliff, "Quantitative Scanning Near Field Microwave Microscopy for Thin Film Dielectric Constant Measurement," *Review of Scientific Instruments* 79 (2008) art. no. 094706.

[44] M. Tabib-Azar, P. S. Pathak, G. Ponchak, and S. LeClair, "Nondestructive Superresolution Imaging of Defects and Nonuniformities in Metals, Semiconductors, Dielectrics, Composites, and Plants Using Evanescent Microwaves," *Review of Scientific Instruments* 70 (1999) pp. 2783–2792.

[45] H. P. Huber, M. Moertelmaier, T. M. Wallis, C. J. Chiang, M. Hochleitner, A. Imtiaz, Y. J. Oh, K. Schilcher, M. Dieudonne, J. Smoliner, P. Hinterdorfer, S. J. Rosner, H. Tanbakuchi, P. Kabos, and F. Kienberger, "Calibrated Nanoscale Capacitance

Measurements Using a Scanning Microwave Microscope," *Review of Scientific Instruments* 81 (2010) art. no. 113701.

[46] J. Hoffmann, M. Wollensack, M. Zeier, J. Niegemann, H-P. Huber, and F. Kienberger, "A Calibration Algorithm for Near Field, Scanning Microwave Microscopes," *Proceedings of the 2012 12th IEEE International Conference on Nanotechnology* (Birmingham, 2012).

[47] M. Farina, D. Mencarelli, A. Di Donato, G. Venanzoni, and A. Morini, "Calibration Protocol for Broadband Near-Field Microwave Microscopy," *IEEE Transactions on Microwave Theory and Techniques* 59 (2011) pp. 2769–2776.

[48] J. D. Chisum and Z. Popović, "Performance Limitations and Measurement Analysis of a Near-Field Microwave Microscope for Nondestructive and Subsurface Detection," *IEEE Transactions on Microwave Theory and Techniques* 60 (2012) pp. 2605–2615.

[49] T. Dargent, K. Haddadi, T. Lasri, N. Clement, D. Ducatteau, B. Legrand, H. Tanbakuchi, and D. Theron, "An Interferometric Scanning Microwave Microscope and Calibration Method for Sub-fF Microwave Measurements," *Review of Scientific Instruments* 84 (2013) art. no. 123705.

[50] M. Farina, A. Lucesoli, T. Pietrangelo, A. di Donato, S. Fabiani, G. Venanzoni, D. Mencarelli, T. Rozzia, and A. Morinia, "Disentangling Time in a Near-Field Approach to Scanning Probe Microscopy," *Nanoscale* 3 (2011) pp. 3589–3593.

[51] Z. Frait, V. Kambersky, Z. Malek, and M. Ondris, "Local Variations of Uniaxial Anisotropy in Thin Films," *Czechoslovak Journal of Physics* 10 (1960) p. 616.

[52] R. F. Soohoo, *Microwave Magnetics* (Harper and Row, 1985).

[53] R. Wang, F. Li, and M. Tabib-Azar, "Calibration Methods of a 2 GHz Evanescent Microwave Magnetic Probe for Noncontact and Nondestructive Metal Characterization for Corrosion, Defects, Conductivity, and Thickness Nonuniformities," *Review of Scientific Instruments* 76 (2005) art. no. 054701.

[54] T. An, N. Ohnishi, T. Eguchi, Y. Hasegawa, and P, Kabos, "Local Excitation of Ferromagnetic Resonance and Its Spatially Resolved Detection with an Open-Ended Radio-Frequency Probe," *IEEE Magnetics Letters* 1 (2010) art. no. 3500104.

[55] A. O. Oladipo, A. Lucibello, M. Kasper, S. Lavdas, G. M. Sardi, E. Proietti, F. Kienberger, R. Marcelli, and N. C. Panoiu, "Analysis of a Transmission Mode Scanning Microwave Microscope for Subsurface Imaging at the Nanoscale," *Applied Physics Letters* 105 (2014) art. no. 133112.

[56] D. M. Kerns, "Plane-Wave Scattering Matrix Theory of Antennas and Antenna-Antenna Interactions," *National Bureau of Standards* (1981) Monograph 162.

9 Radio Frequency Scanning Probe Measurements of Materials

9.1 Electromagnetic Characterization of Materials: Fundamental Concepts

Preceding chapters have described the near-field scanning microwave microscope (NSMM), while discussing both the underlying theory of operation and practical considerations for instrumentation. A primary application area for NSMMs and related microscopes is broadband, local metrology of materials with nanometer-scale spatial resolution. Several methods for spatially resolved characterization of materials are reviewed in this chapter. We begin by reviewing the fundamental concepts of electromagnetic materials metrology. Furthermore, since the extraction of quantitative information from NSMM measurements requires appropriate modeling of the measurement system, we will describe strategies for modeling probe–material interactions. Once the fundamental concepts and models have been established, we will review several applications of scanning-probe-based metrology to local characterization of electromagnetic materials.

Just as Maxwell's equations are fundamental to the characterization of electromagnetic devices, they are also fundamental to electromagnetic characterization of materials. In Chapter 2, we introduced the Maxwell-Lorentz microscopic equations. To proceed from the microscopic equations to a description of electromagnetic fields in materials, it is reasonable to start with the introduction of the constitutive relations for polarization and magnetization. In addition to the time-dependent charge and current distributions, the Maxwell-Lorentz equations also require specification of the relationship between the polarization P, magnetization M, and the electromagnetic fields E and B, as well as their space and time derivatives. These relations are called constitutive relations. Note that the magnetization and polarization may each depend on both the electric and magnetic fields. While these concepts are especially useful for emerging materials like ferromagnetic-ferroelectric composites, photonic bandgap crystals, or metamaterials, each class of materials may require a customized approach. Therefore, it is important to have a fundamental understanding of the constitutive relations in Maxwell's equations. In general, the primary objective is to describe the general electromagnetic response of a material, starting from the microscopic quantities and subsequently averaging to get the macroscopic quantities. This chapter is primarily concerned with the local, spatially resolved measurement techniques of spatially varying quantities at small length scales. Techniques for macroscopic microwave measurements of dielectric properties and techniques for electromagnetic characterization of bulk materials are addressed in numerous texts [1]–[4].

The constitutive relations can be expressed in the form

$$P \Leftrightarrow \{E, B\}, \quad M \Leftrightarrow \{\mathbf{B}, \mathbf{E}\}, \tag{9.1}$$

where $\Leftrightarrow$ indicates that this relationship can be either local or nonlocal in both time and space. Further, these relations can be either linear or nonlinear functions of the driving fields [1]. Note that the dielectric polarization is odd under parity transformation and even under time reversal while the magnetization is even under parity transformation and odd under time reversal. Although E and B are widely accepted as fundamental fields, here we will use E and H as driving fields that may be time dependent.

The introduction of constitutive relations for polarization and magnetization gives rise to complications due to the fact that they depend not only on the applied fields but also on the internal energy of the investigated material in its given state. In the case of magnetization, the response originates from moving charge and spin, as described by Bloch or Landau-Lifshitz equations [3]. These phenomenological equations relate the time-dependent magnetization to both the external magnetic field and the internal energy. The behavior of dielectric materials originates from charge as well electric multipole moments. The description of the time-dependent polarization response is more complex because the motion of charges and multipoles induces magnetic fields in the material in addition to electric fields. Furthermore, under ambient conditions only a small percentage of electric dipoles are fully aligned with an applied electric field due to thermal effects. Naturally, the situation becomes even more complicated in composite materials. For example, in magnetoelectric materials, complex coupling mechanisms include the electric-field-induced shift of the Landé g factor, spin–orbit interactions, exchange energies and the electric-field-induced shift of single ion anisotropy energies [5], [6]. The general approach presented here follows References [7] and [8] and includes the possibility of magnetoelectric coupling within the material, but the results can be extended to materials without such coupling. For micro- or nanoscale materials, the microscopic approach is necessary; however, some problems in RF nanoelectronics can be treated by averaging the microscopic quantities to find macroscopic quantities that can be used instead.

The standard materials relations in Maxwell's equations are often expressed in terms of the effective macroscopic polarization and magnetization. The magnetic induction is related to the magnetic field H and the magnetization M by

$$B = \mu_0 (H + M), \tag{9.2}$$

while the electric displacement is related to the electric field polarization and the macroscopic quadrupole density $\tilde{Q}$ by

$$D = \varepsilon_0 E + \tilde{P} - \nabla \cdot \left[\tilde{Q} \right] = \varepsilon_0 E + P. \tag{9.3}$$

$\tilde{P}$ is the macroscopic dipole density, while ε_0 and μ_0 are the permittivity and permeability of free space, respectively. In order to move from the macroscopic variables to the microscopic variables, we define the macroscopic variables in terms of microscopic densities. The microscopic dipole density with the quadrupole-moment distribution for N particles is

$$p = \sum_{j=1}^{N} r_j e_j \delta(r - r_j) - \frac{1}{2} \sum_{j=1}^{N} e_j \left(r_j r_j \right) \nabla \delta(r - r_j) = \sum_j \mathbf{p}_j. \tag{9.4}$$

The index $j = 1, 2\dots, N$. The position of the j-th particle is r_j. The charge on the j-th particle e_j can be positive or negative. The summation is over the bound charge, which is electrically neutral as a whole. The same condition also applies to the free charge. The motion of the free charge gives rise to kinetic energy that has to be included in the internal energy of the system. Similarly, the microscopic magnetic dipole density is

$$m = \sum_i \left\{ \gamma_i s_i \left(r_i \right) \delta(r - r_i) - \frac{\gamma_{ei}}{e_i} \pi_i \times \mathbf{p}_i \right\} \equiv \sum_i m_i + m_O. \tag{9.5}$$

Equation (9.5) clearly separates the magnetic dipole density into contributions from spin and from charge motion. s is the intrinsic spin.

The gyromagnetic ratio is

$$\gamma_i = g_i \mu_0 e_i / 2m_{si}, \tag{9.6}$$

where m_{si} is the mass of the electron, g_i is the spectroscopic Landé factor. In addition,

$$\gamma_{ei} = \mu_0 e_i / 2m_{si} \tag{9.7}$$

and the canonical momentum is given by

$$\pi_i = m_{si} \frac{dr_i}{dt} + e_i A_i, \tag{9.8}$$

where A_i is the vector potential. The expected values of macroscopic polarization and magnetization are found by averaging:

$$P(r,t) = <\mathbf{p}> = Tr \left[\mathbf{p} \rho(t) \right] \tag{9.9a}$$

$$M(r,t) = <m> = Tr \left[m \rho(t) \right] = \sum_i < m_i > (r,t) + < m_O > (r,t), \tag{9.9b}$$

where $\rho(t)$ is the statistical density operator satisfying Liouville's equations and $Tr(.)$ is the trace operator defined in Reference [8]. Note that both p and m

implicitly depend on functions of electric and magnetic field. By use of statistical mechanics, it is possible to derive the evolution equations for equilibrium and non-equilibrium conditions and the corresponding equations of motion for the magnetization and polarization, including the cross-coupling terms. Such a comprehensive derivation is beyond the scope of this chapter, but leads to the classical Landau-Lifshitz equation of motion for magnetization and the equation of motion for polarization. For the interested reader, the details of the derivation can be found in Reference [7].

More complex interactions such as magnetoelectric interactions may be incorporated into the equations of motion by including all possible relationships between the polarization and magnetization with the magnetic and electric fields. The macroscopic averages of the microscopic variables that are useful for measurement applications are [8]

$$P \approx \alpha_{ee} E_p + \alpha_{em} H_m + \chi_{es} \cdot [\Sigma_\varepsilon] + \chi_{eu}(T - T_0),\qquad(9.10\mathrm{a})$$

$$M \approx \alpha_{me} E_p + \alpha_{mm} H_m + \chi_{ms} \cdot [\Sigma_\varepsilon] + \chi_{mu}(T - T_0),\qquad(9.10\mathrm{b})$$

with

$$E_p(r,t) \approx [\chi_{ee}]^{-1} \cdot P(r,t) + [N_p] \cdot P(r,t),\qquad(9.11\mathrm{a})$$

$$H_m(r,t) \approx [\chi_{mm}]^{-1} \cdot M(r,t) + [N_m] \cdot M(r,t).\qquad(9.11\mathrm{b})$$

The tensors $[\chi_{ii}]$ are the electric and magnetic susceptibility tensors while the tensors $[N_i]$ are depolarization and demagnetization tensors. The coefficients α_{ij} are the corresponding magnetoelectric coupling coefficients. The tensor $[\Sigma_\varepsilon]$ represents the influence of strain. Finally, within a material the macroscopic fields differ from applied fields due to induced surface depolarization charges. For example, the total electric field can be expressed as

$$E_p(r,t) = E_0(r,t) - [N_p] \cdot P(r,t),\qquad(9.12)$$

where E_0 is the applied field.

This brief introduction of the microscopic electromagnetic field variables provides a foundation for describing complex materials at submicrometer length scales. The application of this analysis to specific cases can be quite challenging, requiring detailed knowledge of the relevant geometrical configurations and boundary conditions. The definitions of the electromagnetic field variables for microscopic characterization of materials that we have introduced here provide but a starting point for extending the theory to each particular case. Some insightful examples are described in Reference [3].

9.2 Impedance Circuit Models of Probe–Sample Interactions

9.2.1 Toward Materials Characterization with Near-Field Scanning Microwave Microscopy

We now focus on models that enable quantitative characterization of device and material properties by use of an NSMM. In Chapter 8, we discussed the interaction of the tip with the measured SUT or DUT when the NSMM probe tip was at a certain distance above the surface. We introduced the coupling capacitance and approaches to extract quantitative, calibrated data from the measurements. Here, we extend the models to incorporate material properties. Materials measurements performed by use of an NSMM usually come in one of two forms: height-dependent measurements with the lateral position of the tip fixed or spatially resolved NSMM images. For the latter case, we will focus primarily on images obtained in "contact mode," i.e., with the probe in direct mechanical and electrical contact with the SUT.

Recall that for NSMMs that are configured as a one-port system, the measured parameter is the reflection coefficient. The transmission coefficient is an additional measured parameter for two-port NSMM systems. Also recall that in an NSMM, the tip–sample interaction is in the near field. In most cases the wavelength of the applied field is much larger than the interaction dimensions and the sample dimensions are large with respect to the tip dimensions. Thus, the modeling of the probe's interaction with a material or device is often done using lumped-element impedance circuits. If it is necessary to consider propagation between two ports in a measurement system, then a transmission line model, or a hybrid combination of lumped element and transmission line models, is likely to prove useful. If the tip and the sample dimensions are comparable, full electromagnetic field calculations must be implemented. Returning to the case of lumped-element impedance modeling, once the impedance is known from the measurement, it is possible to use physical models to extract the quantitative parameters of the SUT or DUT. Where possible, we present models generalizable to any experimental case readers may encounter. In this chapter, we will demonstrate application of these models to some specific cases for different material parameter categories and show how the parameters of interest can be extracted from the measurements.

9.2.2 Near-Field, Lumped-Element Models

One modeling approach is to represent the tip–sample interaction by a complex, lumped-element impedance Z_S. In the near- or evanescent-field regime, if the tip–sample interaction is lossless, then no net power is transmitted into the sample. The stored reactive energy of the tip–sample interaction may be represented by a reactance X_S in the lumped-element impedance. On the other hand, if the sample or tip is lossy, energy is dissipated. Then the lumped-element impedance must include a resistive part, R_S. The combined lumped-element impedance is

$$Z_s = R_s + jX_s, \tag{9.13}$$

where j is the imaginary unit. (As discussed in Chapter 8, Z_s is in series with the coupling capacitance.) Z_S depends on the tip geometry, the sample properties, and the tip-sample distance. Figure 9.1 shows an elementary, lumped-element circuit representing the tip coupled to the sample: C_{Cpl} is the coupling capacitance, C_{Str} is the stray or parasitic capacitance, and R_{tip} is the probe resistance. Having previously discussed the coupling and the stray capacitance in Chapter 8, here we turn to models of the sample impedance. Several reasonable and practically useful models of the sample impedance were introduced for bulk materials in Reference [9]. X_S and R_S can be obtained from the integration of the Poynting vector over the whole sample volume. They can be expressed as [9], [10]:

$$X_s = \frac{4\omega}{|I|^2}\int(w_m - w_e)dV \tag{9.14}$$

and

$$R_s = \frac{\omega}{I^2}\int\left(\frac{\sigma}{\omega}|E|^2 + \varepsilon_0\varepsilon''|E|^2 + \mu_0\mu''|H|^2\right)dV, \tag{9.15}$$

where I is the amplitude of the input current at the tip terminal. The scalar permittivity $\varepsilon_s = \varepsilon_0(\varepsilon' - j\varepsilon'')$ and the permeability $\mu = \mu_0(\mu' - j\mu'')$. The variables $w_m = \dfrac{B \cdot H^*}{4}$ and $w_e = \dfrac{E \cdot D^*}{4}$ in Equation (9.14) are magnetic and electric field densities (the asterisk represents complex conjugation of the variable). The three terms in the integrand of Equation (9.15) represent the conductive, magnetic, and dielectric losses in the investigated material, respectively.

Examination of the total circuit impedance in Fig. 9.1(a) leads to the conclusion that maximum sensitivity to Z_s can be achieved when the coupling impedance is small with respect to Z_s and the parasitic impedance is negligible. This can usually be achieved by making the distance between the tip and the sample much smaller than the diameter of the probe tip (e.g., when the tip is in contact with the sample). A more rigorous statement of the near-field condition is:

$$|k|d_{tip} \ll 1; \quad k = \omega(\varepsilon_0\varepsilon_s\mu_0\mu_s)^{1/2}, \tag{9.16}$$

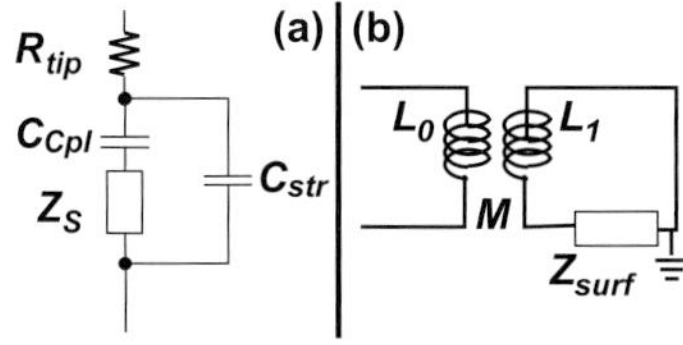

Figure 9.1.   General circuit models of tip-sample system.
(a) Lumped-element circuit model of probe tip coupled to sample. (b) Lumped-element circuit model for a magnetic loop probe coupled to sample.
Adapted from S.-C. Lee, C. P. Vlahacos, B. J. Feenstra, A. Schwartz, D. E. Steinhauer, F. C. Wellstood, and S. M. Anlage, *Applied Physics Letters* 77 (2000) pp. 4404–4406, with permission from AIP Publishing.

where d_{tip} is the effective tip diameter that defines the tip–sample interaction area and ω is the radial operation frequency of the microscope. Thus, in the near field [9]

$$Z_s \approx \frac{1}{j\omega\varepsilon_0\varepsilon_s d_{tip}} \qquad (9.17)$$

and the lumped-element model for a lossy bulk material becomes:

$$Z_s = \frac{tan\delta}{\omega\varepsilon_0\varepsilon' d_{tip}} - j\frac{1}{\omega\varepsilon_0\varepsilon' d_{tip}}. \qquad (9.18)$$

Equation (9.18) can be modified for different classes of materials.

For example, in a low loss dielectric material with loss tangent $tan\delta$ 1, Equation (9.18) can be further simplified by neglecting the first term. For a normal metal, Equation (9.18) simplifies to

$$Z_s = \frac{1}{d_{tip}\sigma}, \qquad (9.19)$$

which is simply the DC resistance of the interaction volume. For bulk semiconductors free of native oxides, R_S and X_S are of the same order and the sample impedance can be expressed as

$$Z_s = \frac{1}{d_{tip}\sigma + j\omega\varepsilon_0\varepsilon' d_{tip}}, \qquad (9.20)$$

where σ is the conductivity of the semiconductor. Alternatively, the dielectric properties of semiconductors and other materials can be represented by a plasma-like Lorentz-Drude permittivity [4]

$$\varepsilon_p = \varepsilon' - \frac{\omega_p^2}{\omega(\omega + jv)}, \qquad (9.21)$$

where $\omega_p^2 = \dfrac{e^2 N v}{\varepsilon_0 m_{eff}}$ is the plasma frequency, v is the collision frequency, e is the charge, m_{eff} the effective mass of the carrier, and N is the concentration. Note that in these expressions, a harmonic $\exp(j\omega t)$ time dependence is assumed. The impedance is then obtained by inserting this permittivity into Equation (9.17) in place of ε_S.

Inductive loop probes provide an effective approach for imaging magnetic materials, albeit with poorer resolution than sharp tip probes. A circuit model of a loop probe is illustrated in Fig. 9.1(b). In this case, the tip-sample impedance can be expressed as [11]

$$Z_s = \frac{\omega^2 M^2}{G Z_{surf} + j\omega L_1}, \qquad (9.22)$$

where G is the geometric factor, Z_{surf} the complex surface impedance, M is the mutual tip sample inductance, and L_1 is the effective inductance of the sample.

For a scanning-tunneling-microscope-based, resonant NSMM, a hybrid approach that combines lumped elements and transmission lines is used. Specifically, a transmission line is used to model the transmission resonator while a lumped-element model is used to characterize the tip–sample interaction [12]. The influence of the sample is modeled as a parallel combination of the tunneling impedance Z_t and the probe impedance Z_p

$$Z_S = (Z_t \, Z_p)/\left(Z_t + Z_p\right). \tag{9.23}$$

Recall that in an STM, an electrically biased tip is positioned about 1 nm from a conducting sample, giving rise to a tunneling current and corresponding tunneling impedance. If the tunneling impedance is mostly due to dissipative losses in the tunnel junction, this impedance is effectively just a tunneling resistance and $Z_t \approx Re\left(Z_t\right) = R_t$. In a typical tunneling microscope, when the tip height h is on the order of a nanometer, the tunneling resistance can be approximated by

$$R_t\left(h\right) = R_0 (1+\left(h/h_0\right)^{\alpha}). \tag{9.24}$$

The constants R_0, h_0, and α are best obtained from a comparison with measurements. When the probe is raised outside of the tunneling range, $R_t \rightarrow \infty$ and does not contribute. There can be a significant difference in values of R_0 for AC and DC operation. In the cases when the simple resistance approximation is not valid, the tunneling impedance can be expressed as $Z_t = R_t(1+ j\omega\tau)$. The tunneling impedance should have an inductive character and therefore a corresponding characteristic time constant τ proportional to L_t / R_t, where L_t is the tip inductance.

9.2.3   Transmission Line Models

In some cases, a transmission line model provides a more suitable approach to calculation of Z_S. Multilayer thin film samples, in which the incident signal may penetrate into subsurface layers, are particularly amenable to this approach. The multilayer material is modeled as a series of cascaded transmission line segments with each segment representing a layer of the material. For example, a transmission line can be used to calculate Z_S for the specific case of a multilayered solar cell [13], as illustrated in Fig. 9.2. The complex impedance of the multilayer solar cell can be expressed as

$$Z_S \approx Z_n \frac{Z_{sub} + jZ_n \gamma_n t_n}{Z_n + jZ_{sub} \gamma_n t_n}, \tag{9.25}$$

where Z_n, γ_n, and t_n are the impedance, electromagnetic wave propagation constant, and thickness of the n-type silicon layer of the stack, respectively. If the impedances of the deepest two layers in Figures 9.2(a) and 9.2(b) are combined into a single parameter Z_{sub}, it can be expressed as

$$Z_{sub} \approx jZ_0 \gamma_0 t_{sub}. \tag{9.26}$$

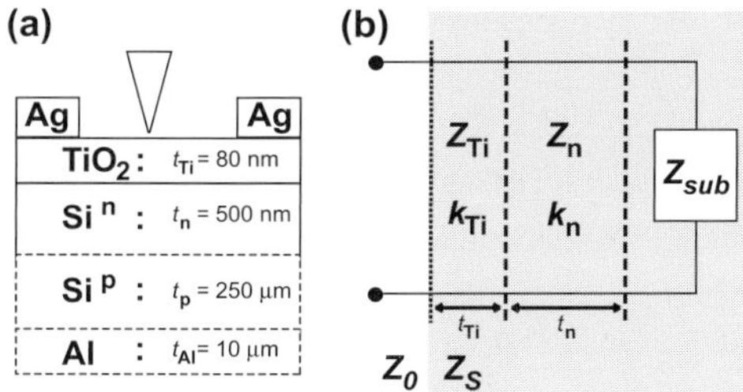

Figure 9.2. Model of multilayer photovoltaic system.
(a) Schematic and (b) equivalent circuit model of a probe tip positioned above a multilayer, photovoltaic structure.
Adapted from A. Hovsepyan, A. Babajanyan, T. Sargsyan, H. Melikyan, S. Kim, J. Kim, K. Lee, and B. Friedman, *Journal of Applied Physics* 106 (2009) art. no. 114901, with permission from AIP Publishing.

Z_0 is the impedance of the air, γ_0 the propagation constant of the air, and t_{sub} is the thickness of the substrate. Note that Equations (9.25) and (9.26) are derived under the assumptions that $\gamma t \ll 1$ and that the effects of the TiO$_2$ layer may be neglected.

A more general approach that applies to NSMM measurements of layered media accounts for the wavelike nature of the excitation source. In References [14] and [15], the surface impedance of a multilayer planar structure was calculated for the case of vertical and horizontal dipoles. Because most NSMM probes act as vertical electric dipoles, we will summarize that approach. The reader can find the impedance due to horizontal electric dipole and magnetic dipoles in the referenced literature. Consider a sample with M layers indexed by $m = 0, 1..., M$ (layer $m = 0$ corresponds to the free space above the layered half space). The m-th layer has conductivity σ_m, permittivity ε_m, and permeability μ_m. The algebraic form of the NSMM reflection coefficient for the vertical electrical dipole over such a multilayer structure is identical to that of the reflection coefficient calculated for a plane wave incident perpendicularly on a planar layered material. This is due to the fact that the field of a dipole over a stratified half space can be regarded as a spectrum of plane waves whose angle of incidence and reflection is determined by a parameter $\zeta = k_0 \sin(\theta)$, where k_0 is the free space wave vector and θ is the angle of incidence.

The input impedance of the layered material is Z_1, which can be expressed through recursive equations [14], [15]:

$$Z_1 = K_1(\zeta) \frac{Z_2(\zeta) + K_1(\zeta)\tanh(\beta_1 t_1)}{K_1(\zeta) + Z_2(\zeta)\tanh(\beta_1 t_1)}, \tag{9.27a}$$

$$Z_2 = K_2(\zeta) \frac{Z_3(\zeta) + K_2(\zeta)\tanh(\beta_2 t_2)}{K_2(\zeta) + Z_3(\zeta)\tanh(\beta_2 t_2)}, \tag{9.27b}$$

and

$$Z_m = K_m(\zeta) \frac{Z_{m+1}(\zeta) + K_m(\zeta)\tanh(\beta_m t_m)}{K_m(\zeta) + Z_{m+1}(\zeta)\tanh(\beta_m t_m)}, \tag{9.27c}$$

with

$$Z_M = K_M(\zeta), \tag{9.28}$$

where t_m is the thickness of the m-th layer. The very last layer is assumed to be a half space. The parameters β_m, γ_m, and K_m are defined as follows:

$$\beta_m(\zeta) = \sqrt{\zeta^2 + \gamma_m^2}, \tag{9.29a}$$

$$\gamma_m^2 = (j\sigma_m \mu_m \omega - \varepsilon_m \mu_m \omega^2), \tag{9.29b}$$

$$K_m(\zeta) = \frac{\beta_m(\zeta)}{\sigma_m + j\omega\varepsilon_m}. \tag{9.29c}$$

The generalized reflection coefficient is

$$\Gamma(\zeta) = \frac{K_0(\zeta) - Z_1(\zeta)}{K_0(\zeta) + Z_1(\zeta)}. \tag{9.30}$$

This solution is reminiscent of the one-port reflection coefficient derived for a transmission line [4], [14]. When the incident plane wave is perpendicular to the surface, $\theta = 0$, $\beta_m = \gamma_m$, and $K_m = Z_m$. In the limit where the multilayer structure consists of very thin layers, the equations for the impedance and reflection coefficient of the layered medium are reduced to ordinary differential equations of the so-called Ricatti form that can be solved as an initial value problem [16].

9.3   Resonant Cavity Models and Methods

9.3.1   Resonant-Cavity-Based, Near-Field Scanning Microwave Microscopy

Up to now, we have described the interaction of the tip and the sample by use of a sample impedance. Alternatively, we can treat the whole NSMM system as a specialized transmission cavity [17]. Such a treatment of NSMM was first introduced in Reference [18] and many NSMM instruments intended for material characterization rely on an engineered resonant system inserted between a network analyzer and the tip-sample junction [19]–[21]. Some resonant NSMM implementations operate in a linear regime [17], while others operate in a nonlinear regime [22], [23].

In order to understand resonant NSMM measurements, it is necessary to understand how the tip-sample system can be represented by a load on a resonant circuit. There are several lumped-element circuit models that represent a resonant circuit connected to the NSMM probe tip. For example, in Reference [24], the coaxial resonator circuit is modeled as a sixth order LCR notch filter, while in Reference [18] the resonator circuit is modeled by a simple series or parallel RLC circuit. Here, we follow the approach of Reference [17], as shown in Fig. 9.3. The equivalent lumped-element circuit of a resonator is coupled to a source as well as the tip-sample junction. The junction is represented by a lumped-element circuit, as previously introduced in Fig. 9.1. The application of this model depends upon which method is used to measure the reflection coefficient Γ from the NSMM. (In some places, the reflection coefficient is alternately referred to as S_{11} and we will use both notations.)

There are two common measurement techniques. In the first technique, the source frequency is swept in a narrow range around the resonance frequency and Γ is measured as a function of frequency for each probe position. In the second technique, at each probe position, Γ is measured at a single, constant frequency chosen to be near, but not equal to the resonant frequency. Each of these techniques has advantages and disadvantages. The former technique provides a richer set of measurement data, but the latter is simpler and faster. Thus, the constant frequency technique is the more common approach, particularly in commercial NSMMs.

9.3.2   Resonant Cavity Measurements with Swept Frequency

In resonator-based NSMM measurements, the majority of the microwave energy is concentrated in the resonant structure or cavity. The probe tip is a weakly coupled extension of this cavity and therefore the tip–sample interaction acts as a perturbation of the cavity response. Thus, perturbation theory, which is well established for microwave resonators, can be used. The influence of the perturbation upon the resonator is characterized by a shift of the resonance frequency f_r and a change in the loss of the cavity that is represented by the quality factor Q. From perturbation theory, the relative shift in the resonant frequency can be expressed as [4], [25]:

$$\frac{\Delta f_r}{f_r} = -\frac{\int\left(\left(E_1 \cdot D_0 - E_0 \cdot D_1\right) - \left(H_1 \cdot B_0 - H_0 \cdot B_1\right)\right)dV}{\int(E_0 \cdot D_0 - H_0 \cdot B_0)dV}, \tag{9.31}$$

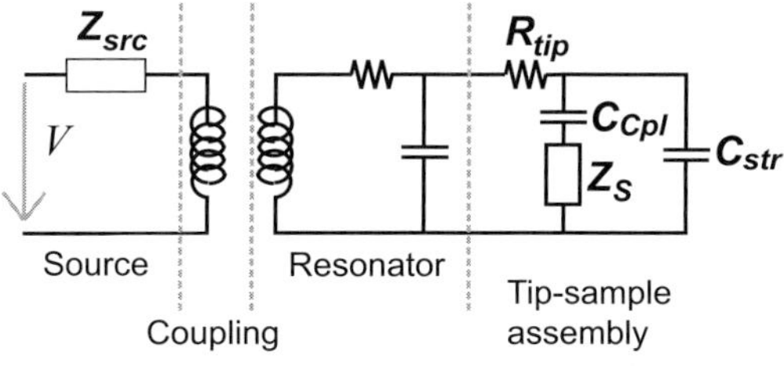

Figure 9.3.   Equivalent lumped-element circuit model of a resonant NSMM. The circuit model comprises, from left to right in the figure, the source, resonator, and the tip-sample system.
Adapted from C. Gao and X.-D. Xiang, *Review of Scientific Instruments* 69 (1998) pp. 3846–3851, with permission from AIP Publishing.

where $\boldsymbol{D}_i, \boldsymbol{E}_i$, and $\boldsymbol{B}_i, \boldsymbol{H}_i$ are perturbed ($i = 1$) and unperturbed ($i = 0$) electric and magnetic fields and the integration must be done over the whole volume of the sample. In terms of the scalar permittivity and permeability, this equation can be expressed as

$$\frac{\Delta f_r}{f_r} = \frac{\int (\Delta\varepsilon' \boldsymbol{E}_1 \cdot \boldsymbol{E}_0 + \Delta\mu' \boldsymbol{H}_1 \cdot \boldsymbol{H}_0) dV}{\int (\varepsilon_0 \boldsymbol{E}_0^2 + \mu_0 \boldsymbol{H}_0^2) dV}. \tag{9.32}$$

The resonator quality factor is given by

$$Q = \frac{\omega W_{st}}{P_{loss}}, \tag{9.33}$$

where W_{st} is the stored energy in the cavity and P_{loss} is the loss of the cavity. The change in the quality factor is

$$\Delta\left(\frac{1}{Q}\right) = \frac{\int (\Delta\varepsilon'' \boldsymbol{E}_1 \cdot \boldsymbol{E}_0 + \Delta\mu'' \boldsymbol{H}_1 \cdot \boldsymbol{H}_0) dV}{\int (\varepsilon_0 \boldsymbol{E}_0^2 + \mu_0 \boldsymbol{H}_0^2) dV} = -\frac{\Delta f_r}{f_r} \tan\delta. \tag{9.34}$$

For some specific cases, it is possible to use a simple, transmission line approach to obtain the frequency and quality factor shifts from Γ [27]. For a resonator of length L, it is possible to write Γ as [27]

$$\Gamma = \frac{\Gamma_c + \Gamma_t e^{-2\gamma L} - 2\Gamma_c \Gamma_t e^{-2\gamma L}}{1 - \Gamma_c \Gamma_t e^{-2\gamma L}}, \tag{9.35}$$

where Γ_c is the cavity coupling reflection coefficient, Γ_t is the reflection coefficient at the tip, and γ is the propagation constant of the transmission line. From Equation (9.35), it follows that if the reflection coefficient is only weakly dependent on frequency, i.e., if $f_0 \partial |\Gamma_{c,t}|/\partial\omega \quad 1$ and $f_0 \partial(\arg\Gamma_{c,t})/\partial\omega \quad 1$, then [26]

$$\Delta f_r \approx \frac{1}{2\pi} \frac{\sum_i \partial\varphi/\partial c_i}{\partial\varphi/\partial\omega} = \frac{1}{2\pi} f_0 \sum_x \frac{\partial \arg(\Gamma_t)}{\partial c_x} dc_x, \tag{9.36}$$

where $\varphi = \arg(\Gamma_c \Gamma_t e^{-2\gamma L})$ and f_0 is the unperturbed, fundamental resonance frequency of the resonator. The circuit variable set $\{c_x\} = \{C_{str}, C_{cpl}, Z_s(C_s, R_s)\}$ is drawn from the equivalent circuit in Fig. 9.3 or other equivalent circuits. The circuit variables are directly related to the change of the resonance frequency in the experiment. If one can neglect the tip resistance R_t, Equation (9.36) can be simplified to

$$\frac{\Delta f_r}{f_r} \approx -2Z_0 f_0 \sum_i \frac{\partial C_t}{\partial c_i} dc_i. \tag{9.37}$$

C_t is the effective tip capacitance derived from the given lumped-element circuit. Under the same assumption, the change of the quality factor can be expressed as [26]

$$\Delta\left(\frac{1}{Q}\right) = \frac{4}{\omega} f_0 Z_0 \Delta G. \tag{9.38}$$

ΔG is the change in the parallel conductance of the sample (defined by the sample impedance $Z_s(R_s, C_s)$) from Fig. 9.3 and represents the changes in the loss within the sample.

The exact forms of the solutions to Equations (9.31), (9.32), and (9.34) depend on the geometry of the system, material properties, and boundary conditions. Therefore, we will not include here a derivation of general analytic solutions to those equations in terms of a Green's function or other formulation. Experimentally, the local permittivity and permeability may be determined by use of Equations (9.31), (9.32), and (9.34), as follows. The frequency shift, i.e., the left-hand side of Equations (9.31) and (9.32), is obtained from measurement of the reflection coefficient during a frequency sweep around the resonance frequency. To obtain the shift, this measurement must be carried out both when the resonator is unperturbed (far from the sample) and when the resonator is loaded (near the sample). The measurement of the frequency shift of the cavity with respect to the unperturbed resonance frequency can be obtained directly.

Obtaining the shift in the cavity quality factor ΔQ from the measurements can be confusing. Intuitively, one might assume the quality factor of the resonant cavity is obtained directly from the fitting the observed frequency dependence in the vicinity of the resonant frequency. This is incorrect and therefore we will spend some time to define the correct approach following the procedure described in Reference [28]. Considering the circuit model of the resonant cavity in Fig. 9.3, it is possible to define an external quality factor Q_E associated with the power dissipation due to coupling of the cavity to the external circuitry (e.g., the iris in waveguide cavities). It is also possible to define a quality factor Q_V associated with the dissipation within the cavity volume (e.g., the combined cavity and sample dissipation volumes). The overall Q_L factor of the cavity then can be expressed as

$$\frac{1}{Q_L} = \frac{1}{Q_E} + \frac{1}{Q_V}. \tag{9.39}$$

It is useful to introduce reduced parameters $\delta_E = Q_E \Delta_\omega$, $\delta_V = Q_V \Delta_\omega$, and $\delta_L = Q_L \Delta_\omega$, with $\Delta_\omega = \dfrac{\omega}{\omega_r} - \dfrac{\omega_r}{\omega}$. In addition, by introducing $\beta = Q_V / Q_E = \left(\dfrac{1 \mp S_{11r}}{1 \pm S_{11r}}\right)$, where S_{11r} is the reflection coefficient at resonance frequency, as a coupling coefficient of the cavity, it is possible to relate the reduced parameters to the measured reflection coefficient $S_{11}(\omega)$ as [28]

$$\delta_V = \pm \sqrt{\frac{(1+\beta)^2 S_{11}^2 - (1-\beta)^2}{1 - S_{11}^2}}. \tag{9.40}$$

The $\pm$ sign in Equation (9.40) is determined by the sign of Δ_ω, which is negative for $\omega_r < \omega$ and positive for $\omega_r > \omega$. The other reduced parameters can be determined from δ_V by $\delta_E = \delta_V / \beta$ and $\delta_L = \delta_V / (1+\beta)$, plotting the reduced parameters as a function of Δ_ω and obtaining a linear fit yields Q_E, Q_V, and Q_L. An example of the measured frequency dependence of the reflection coefficient and the deduction of the quality factors is shown in Fig. 9.4. Specifically, Fig. 9.4(a) shows the frequency dependence of the reflection coefficient and Fig. 9.4(b) shows the parameter δ_V as a function of Δ_ω. It is possible to use a similar procedure to obtain the quality factor for a transmission cavity in which S_{21} is measured.

Measurements of the complex permittivity of bulk materials based on NSMM provide a practical example of material characterization through measurement of the frequency shift and change in the quality factor [29]. We consider specific examples of frequency-swept measurements that reveal a number of issues that are important for quantitative characterization of material properties. A typical experimental configuration of an NSMM resonator is described in Reference [17]. The bulk SUTs studied therein include dielectrics, semiconductors, and metals. The experimental data confirmed that the shift of the resonance frequency is most sensitive to the tip-sample coupling capacitance C_{Cpl} and the capacitive contribution to the sample impedance Z_s (see Fig. 9.3), while the change of the quality factor is most sensitive to the resistive contribution to Z_s (although, in general, both shifts are functions of C_{Cpl} and the total sample impedance Z_s).

In the specific examples of fused silica and silicon, the shift in the resonance frequency and the quality factor were measured by use of frequency-swept measurements at each height in the range from 0.1 to 1000 μm above the sample. The experimental results are shown in Fig. 9.5 where $\Delta f / f_0$ is plotted against Q / Q_0. The unperturbed values of f_0 and Q_0 of the unloaded resonator are measured at the height of 1000 μm. As seen in Fig. 9.5, there is a significant, height-dependent frequency shift but little change in quality factor for the fused silica, as expected for a low loss material. For the silicon, in which the loss is significantly higher, there is a significant change in both the quality factor and the resonant frequency. To extract the material properties from the experimentally observed dependences for dielectric and semiconducting materials, the coupling capacitance can be successfully modeled by the quasi-static image charge method discussed in Chapter 8. The sample properties can be modeled by use of the lumped-element approach, as described in Equations (9.16) through (9.19). In this example, when the sample is modeled as a series RLC circuit the input impedance of the resonator is [27]

$$Z_{in} = R_{res}(1 + 2j(Q\Delta f / f_0)), \tag{9.41}$$

where R_{res} is the series resistance of the resonator equivalent circuit.

There are several different ways to visualize height-dependent NSMM measurements, each of which brings out different features. In general, different materials with different complex permittivity values will present different loads to the resonator. Data plotted with Δf as a function of Q, as in Fig. 9.5 visually distinguish different classes of materials – dielectrics, semiconductors, and metals – based on differences in the loss. By contrast, plotting Q as a function of height offers little distinction between the respective responses of metal and dielectrics. Finally, a plot

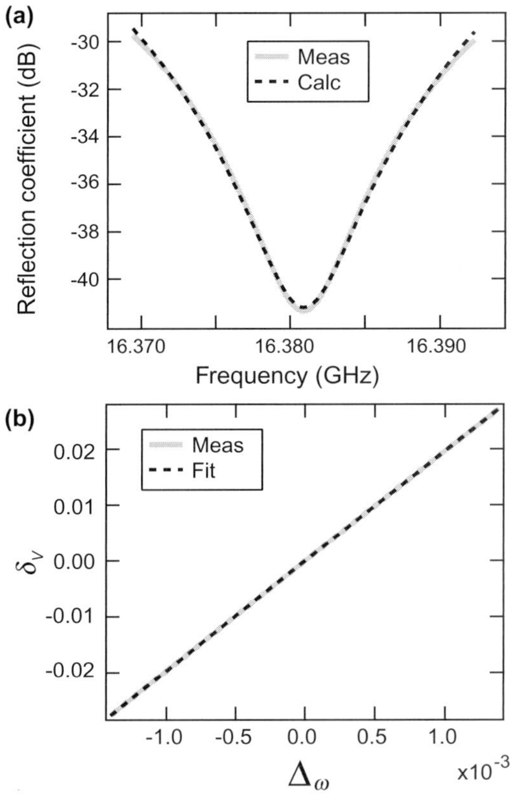

Figure 9.4. Obtaining the cavity quality factor from resonant, NSMM measurements. (a) A comparison of measured (solid gray line) and calculated (dashed black line) reflection coefficients as a function of frequency. (b) The experimental reduced parameter δ_V (solid gray line) is determined from the measured reflection coefficient. A linear fit (dashed black line) to the experimental data provides the quality factor Q_V associated with the dissipation within the cavity volume. Further, the fitted values of δ_V are used to find the calculated reflection coefficient in (a).

of the product $(\Delta f / f_0)(Q / Q_0)$ versus height highlights the capacitive coupling to the sample load.

A pure quasi-static model of the tip–sample interaction fails for metals. In particular, it is necessary to include a calculation of the near-field antenna impedance as it is brought close to a metal surface, as discussed in Chapter 8. The reason for this is that there is non-negligible resistive loss in the antenna (probe tip) as it is brought close to the metal surface.

9.3.3 Calibration, Uncertainty, and Sensitivity

To model and extract the correct material properties, an unknown critical parameter – the effective tip diameter d_{tip} – must be obtained. One technique to find d_{tip}

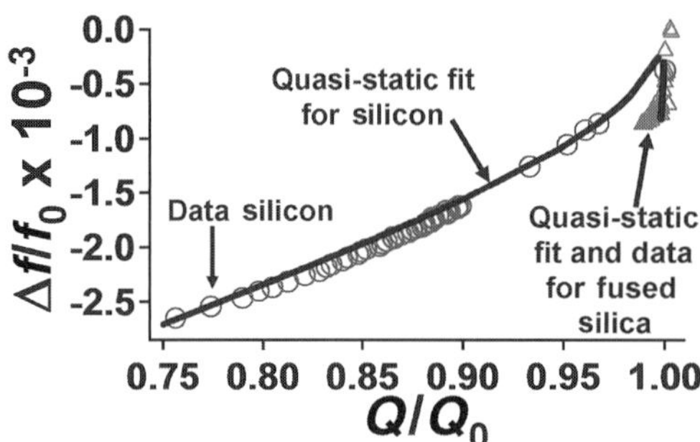

Figure 9.5.   Measurements of silicon and fused silica using NSMM. The normalized shift of the resonance frequency is shown as a function of the change in the resonator quality factor for bulk silicon (open circles) and bulk fused silica (open triangles). Fitted curves for a quasi-static model are shown as solid lines for both samples. Reprinted from A. Imtiaz, T. Baldwin, H. T. Nembach, T. M. Wallis, and P. Kabos, *Applied Physics Letters* 90 (2007) art. no. 243105, with permission from AIP Publishing.

is via the calibration procedures discussed in Chapter 8. Another possibility is to use a low loss dielectric with known permittivity as a calibration sample. This is essentially a simplified calibration procedure for height-dependent NSMM of bulk materials. To obtain the effective tip diameter from the measured dependence of Δf versus height for the known, low-loss calibration material, one need only fit that data with the single fitting parameter d_{tip}.

Another calibration approach is presented in Reference [30], in which the permittivity of a thin film was obtained. The experimental system comprised a loop-coupled cavity operating as a transmission resonator and a probe tip protruding from the bottom of that cavity. The system was calibrated by measuring three reference samples: magnesium oxide, neodymium gallate, and yttrium-stabilized zirconia. The permittivity of the reference samples was known from measurements made by use of a split post resonator technique. The transmission coefficient of the cavity resonator S_{21} was measured both for standard samples as well as the thin film at two frequencies: 1.8 GHz and its third harmonic 4.4 GHz. To reduce statistical error, measurements were made at different positions on the film. After these measurements the film was patterned with a CPW to measure its permittivity through a complementary method. The measured transmission coefficients were fitted to the equation [30]

$$S_{21}(f) = S_{21L} + \frac{S_{21}(f_0)}{1 + jQ_L\Delta\omega} e^{j\phi}; \quad \Delta\omega = \frac{f^2 - f_0^2}{f^2}, \tag{9.42}$$

where S_{21L} is the nonresonant leakage between the ports. The phase shift ϕ is introduced because the measurement ports are physically separated from the coupling loops. $S_{21}(f_0)$ is the transmission coefficient at resonance. The data can be subsequently processed following the perturbation approach introduced in Equations (9.31) through (9.34).

An analysis of the measurement uncertainty is critical for any quantitative measurements of materials with NSMM. In this case, the total uncertainty was estimated from statistical uncertainties in the measurement of tip diameter d_{tip}, film thickness t, calibration constant A, and normalized frequency shift as

$$\sigma(\varepsilon_f) = \left[\left(\frac{\delta\varepsilon_f}{\delta(d_{tip}/2)} \right)^2 \sigma(d_{tip}/2)^2 + \left(\frac{\delta\varepsilon_f}{\delta t} \right)^2 \sigma(t)^2 + \left(\frac{\delta\varepsilon_f}{\delta A} \right)^2 \sigma(A)^2 + \left(\frac{\delta\varepsilon_f}{\delta(\Delta f/f')} \right)^2 \sigma(\Delta f/f')^2 \right]^{\frac{1}{2}}$$

(9.43)

where ε_f is the permittivity of the film. The differential terms can be obtained from analytical expressions of the frequency shift derived from models of the tip–sample interaction, such as the charge image model [17].

The sensitivity of NSMM for measurement impedances is [23]

$$S_f = \frac{g_s \omega_0}{4\pi} \frac{C_c^2}{C_0(C_c + C_s)^2},$$

(9.44)

where $g_s = A_{eff}/(\xi_s)$, A_{eff} corresponds to the effective tip area, ξ_s is the decay length of the evanescent wave, C_c is the coupling capacitance, C_0 is the capacitance of the equivalent circuit of the resonator, and ω_0 is the unloaded, radial resonance frequency. To distinguish between the regions of different permittivity it is necessary to fulfill the condition

$$\frac{\Delta\varepsilon}{\varepsilon} = \left(\frac{V_{n(rms)}}{V_{in}} \right)/(S_f S_r \varepsilon)$$

(9.45)

with S_r defined as [22]

$$S_r = \frac{dS_{11}}{d\omega} \approx \frac{Q'}{\omega_0'}\left(1 - \frac{\Delta\omega}{\omega_0'} \right); \; \Delta\omega = \omega - \omega_0'.$$

(9.46)

Q' and ω_0' are the perturbed quality factor and resonance frequency of the resonator due to tip–sample interaction, $V_{n(rms)} = \sqrt{(4k_B TBR)}$ is the rms value of the noise voltage generated by resistance R at temperature T, k_B is Boltzmann constant, B is the bandwidth of the system, and V_{in} is the probe input voltage.

9.3.4 Resonant Cavity Measurements of Semiconductors

Semiconductors are foundational for nanoelectronic applications. As the dimensions of semiconductors are scaled to the nanometer range, important questions arise about the applicability of standard semiconductor device theories at these scales. Moreover, such scaled systems are inherently more sensitive to the effects of interfaces and defects. Thus, scanning probe techniques are vital experimental tools for spatially resolved characterization of these materials. As we will discuss further in Chapter 11, NSMM is but one of many different scanning probe techniques used for the local, nanoscale characterization of semiconducting materials [32].

When a metal NSMM probe tip is in contact with a native or deposited oxide layer on a semiconductor SUT, the tip-sample interface may be treated as a metal-oxide-semiconductor (MOS) system. In order to model the MOS system, a number of modifications must be made to the model shown in Fig. 9.3. For the tip sample assembly, a nanoscale, parallel plate MOS capacitor with effective tip diameter can be used in most cases as a first approximation. As the tip is in contact with the surface of the sample, there is no need to model the coupling capacitance separately from the sample response. In contact mode, the field is concentrated under the tip and the stray capacitance can be incorporated into the effective tip diameter. This simplifies the sample model, e.g., to a series connection of the oxide and sample capacitances [24]. To further simplify the model, charges at the oxide-semiconductor interface may be neglected. It is important to remember that any 50 Ω shunt resistor that is sometimes present in commercial system configurations has to be included in the model.

Experimentally, a DC bias V_{tip} is often applied to the probe tip and varied systematically throughout the measurements. In the case of silicon samples, it has been shown that the sample response under such a varying tip bias can be effectively modeled by use of standard MOS theories [33]. As V_{tip} is varied, the resonant circuit is perturbed due to changes in the depletion within the semiconductor, resulting in a change in the reflection coefficient measured by the NSMM. If a small, low-frequency voltage is added to V_{tip}, then the derivative dS_{11}/dV_{tip} may be measured simultaneously by use of a lock-in technique. Just as macroscopic capacitance-voltage curves provide an avenue to measure dopant concentration in semiconductors, local, tip-bias-dependent NSMM measurements provide a way to calibrate the local dopant concentration and quantify their distribution throughout the sample [32].

This requires a calibration sample with regions of known dopant concentration N_A. An ideal calibration sample has a flat surface with no significant topographic variation such that any changes in the measured reflection coefficient are due to materials contrast. One particular implementation, to which we will refer in this chapter, comprises a series of regions of variably doped silicon, each with a different dopant concentration. Each region is a 1.5 μm-wide strip, such that the materials contrast appears as a series of parallel stripes.

Tip-bias-dependent measurements of such a calibration sample must first be converted from an S_{11} measurement into a measurement of the sample capacitance C_{sample} (or dS_{11}/dV_{tip} into dC_{sample}/dV_{tip}). The resulting capacitance-voltage measurement curves may then be compared to corresponding simulated curves for selected values of N_A, which provides an approach for converting S_{11} measurements into dopant concentrations measurements. Examples of tip-bias-dependent S_{11} measurements and C_{sample} calculations for various dopant concentrations in

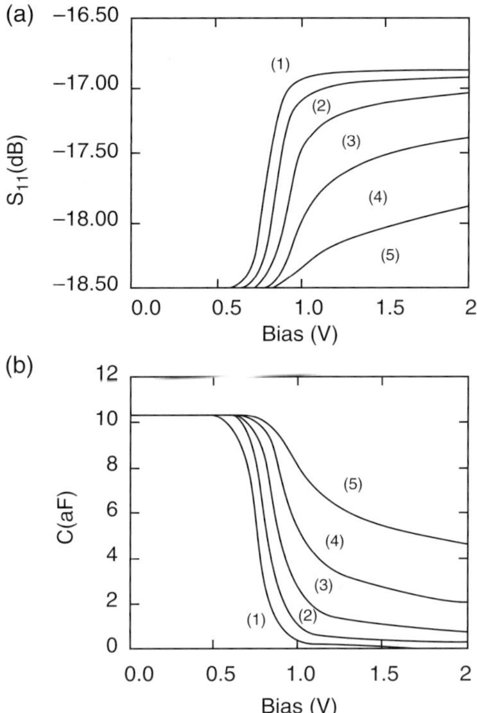

Figure 9.6. Dependence of NSMM reflection coefficient and sample capacitance as a function of tip bias voltage.
Calculated (a) S_{11} and (b) sample capacitance as a function of tip bias at 18 GHz for p-doped silicon at five doping levels: (1) $N_A = 1 \times 10^{15}$ cm^{-3}, (2) $N_A = 1 \times 10^{16}$ cm^{-3}, (3) $N_A = 1 \times 10^{17}$ cm^{-3} (4) $N_A = 1 \times 10^{18}$ cm^{-3}, (5) $N_A = 1 \times 10^{19}$ cm^{-3}. An effective tip area of 30×30 nm and 1 nm native oxide thickness are assumed. Reprinted from J. Smoliner, H. P. Huber, M. Hochleitner, M. Moertelmaier, and F. Kienberger, *Journal of Applied Physics* 108 (2010) art. no. 064315, with permission from AIP Publishing.

p-doped silicon are shown in Fig. 9.6. In order to accurately simulate the MOS system and extract accurate dopant concentration from the measurements, additional parameters such as oxide thickness and the effective tip diameter have to be known [32]. The simulation package FASTC2D [34], [35] incorporates the relevant parameters into a quasi-static, three-dimensional model of an MOS capacitor formed between tip and the oxidized semiconductor surface, leading to rapid conversion of dC_{sample}/dV_{tip} measurements to dopant profiles. An example of calibrated dopant profiling is shown in Fig. 9.7 [36]. Further details of dopant profiling in semiconductors, including calibration samples, measurement techniques, and modeling, are discussed in Chapter 11.

Another critical parameter that influences NSMM results is the operating frequency. Naively, one might assume there is no significant difference to be expected

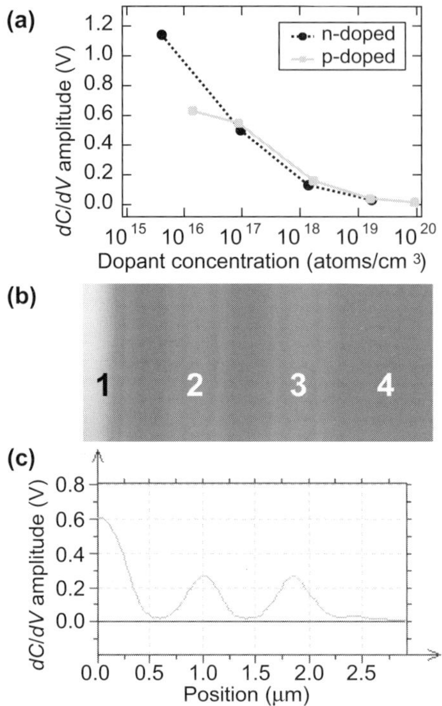

Figure 9.7. Dopant profiling with NSMM.
(a) Dopant calibration curves showing the dC/dV amplitude with respect to the dopant concentration for n-doped (black circles) and p-doped (gray squares) reference samples. Lines have been added to guide the eye. The data were obtained from dC/dV images by averaging the dC/dV amplitude at various regions of known doping concentrations. (b) dC/dV image of a p-doped test sample with four different doping densities: 6×10^{15} (region 1), 8×10^{16} (region 2), 1×10^{17} (region 3), and 1×10^{19} atoms/cm³ (region 4). Regions 2 and 3 are 300 nm wide. (c) Average cross-section of the image shown in (b). Adapted from H. P. Huber et al., *Journal of Applied Physics* 111 (2012) art. no. 014301, with permission from AIP Publishing.

in the frequency range commonly used by NSMMs (1 GHz to 50 GHz). In practice, the selection of the operating frequency may have a significant impact on the experimental results. This is because the measured response in many semiconducting materials is frequency dependent. For example, frequency-dependent materials contrast due to change of sheet resistance was observed in boron-doped silicon [37]. A more systematic investigation of the influence of the frequency was carried out on the "striped" calibration sample described earlier [32], [38]. In order to compare results at different operating frequencies, a phase shifter is inserted into the microwave signal path between the VNA and the NSMM resonator, as illustrated in Fig. 9.8(a). With V_{tip} equal to zero, at each operating frequency, the phase shifter is used to tune the reactance of the input impedance such that in contact with the sample the reflection coefficient has a magnitude

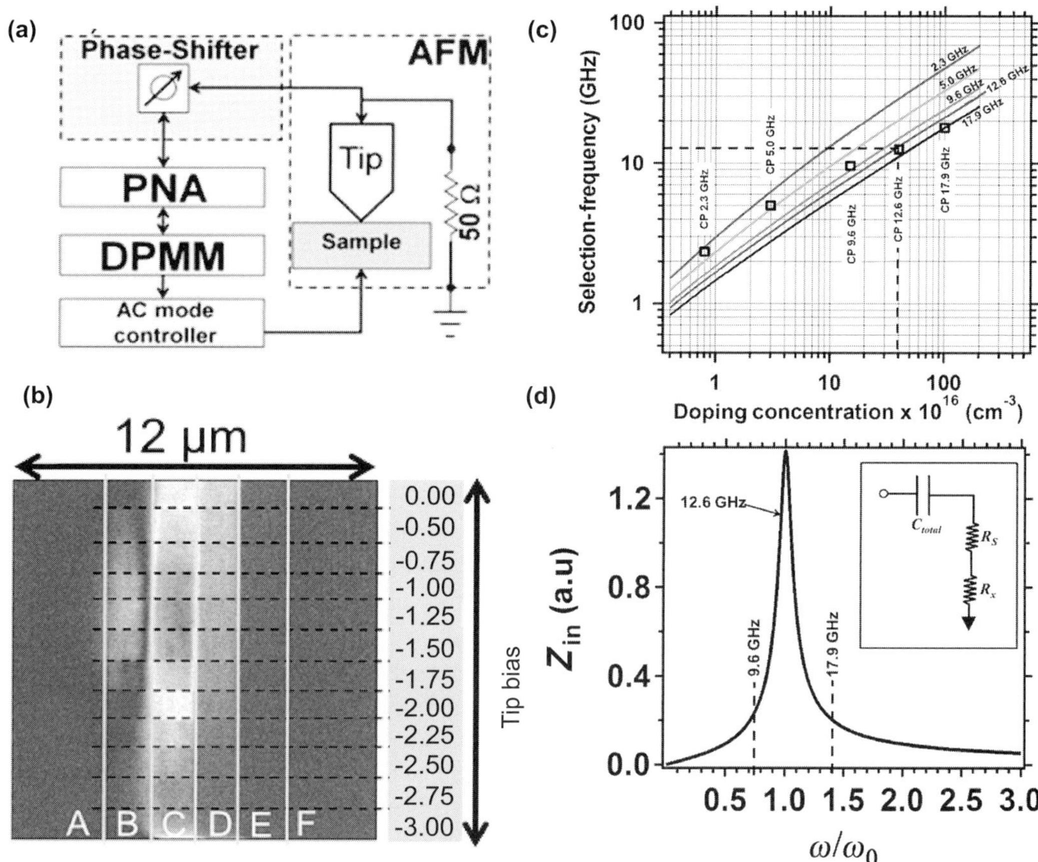

Figure 9.8. Frequency-dependent sensitivity of NSMM.
(a) Schematic of the AFM-based NSMM with the phase shifter. (b) An image of $d(S_{11})/dV$, acquired at 5 GHz, illustrating the structure of the sample. Solid white vertical lines delineate the different doped regions: A = bulk silicon, B = 10^{16} cm^{-3}, C = 10^{17} cm^{-3}, D = 10^{18} cm^{-3}, E = 10^{19} cm^{-3}, and F = 10^{20} cm^{-3}. The dashed black lines demarcate regions of various applied tip bias voltages. (c) Calculated selection frequency vs. doping concentration (solid lines) for 2.3 GHz, 5.0 GHz, 9.6 GHz, 12.6 GHz, and 17.9 GHz. The rectangles represent the measured concentrations from the highlighted areas at a given frequency. (d) Calculated input impedance vs. frequency. The operating frequency 12.6 GHz is most sensitive to concentrations near 10^{17} cm^{-3}, while 9.6 GHz and 17.9 GHz are most sensitive to 10^{16} cm^{-3} and 10^{18} cm^{-3}, respectively. Reprinted from A. Imtiaz, T. M. Wallis, S.-H. Lim, H. Tanbakuchi, H.-P. Huber, A. Hornung, P. Hinterdorfer, J. Smoliner, F. Kienberger, and P. Kabos, *Journal of Applied Physics* 111 (2012) art. no. 093727, with permission from AIP Publishing.

of about −50 dB. In Fig. 9.8(b), the vertical lines in the NSMM image delineate areas of different p-type dopant concentrations. The horizontal, black, dashed lines represent changes in the DC tip bias. The series of DC tip biases is denoted at the right-hand edge of the image. For this image acquired at 5 GHz, there is increased

NSMM sensitivity in the stripe with a doping concentration of 10^{17} cm^{-3} while the NSMM contrast for the other doping concentrations is suppressed. As the operating frequency is switched to other values – 2.3 GHz, 9.6 GHz, 12.6 GHz, and 17.9 GHz – increased sensitivity is observed in different ranges of dopant concentrations. Figure 9.8(d) shows a calculation of the frequency-dependent input impedance for a given dopant concentration. The maximum in the input impedance represents an optimal operating frequency or "selection frequency." Fig. 9.8(c) shows calculated selection frequencies as a function of the doping concentration. The experimental results are in reasonable agreement with these theoretical predictions and generally support the application of standard MOS theory to such measurements. These measurements underscore the significant potential of NSMM as a broadband tool that reveals physical behavior that can't be observed via single-frequency measurements, albeit with restrictions that arise from the fact that the experimentally utilized frequencies are constrained by the intrinsic response of the NSMM resonator.

The observed, frequency-dependent response of NSMM is attributed to the underlying physics within the sample as described by standard MOS diode theory [33], [39]. From MOS diode theory the space charge density may be calculated as a function of the surface potential defined with respect to intrinsic Fermi level. Knowing the space charge density, it is possible to solve for the surface charge density $Q_s(\psi_s)$, where ψ_s is the surface potential. Next it is necessary to calculate the depletion layer capacitance $C_{depl} = \partial Q_s / \partial \psi_s$, which is a function of the tip-bias voltage. In the NSMM measurements, one has to consider only the high-frequency branch of this capacitance. The microscope is sensitive to changes in the series combination of the oxide and C_{depl} capacitances. In addition, incident microwave power is dissipated in the sample through the surface resistance R_{surf}, which is given by

$$R_{surf} = \rho_s / \delta_{sd}, \qquad (9.47)$$

where ρ_s is the resistivity of the locally doped region and $\delta_{sd} = \sqrt{\rho_s / \mu_0 \pi f}$ is the skin depth of the penetrating microwave field. This resistance is in series with the sheet resistance of the semiconductor

$$R_{sh} = \rho_s / d_{eff}, \qquad (9.48)$$

where d_{eff} is an effective length scale corresponding roughly to the distance from the top of the sample to the nearest grounding conductor. This optimal NSMM operating frequency, or "selection frequency," is inversely proportional to the selection time constant

$$\tau_{sel} = \xi \frac{(R_{surf} + R_{sh})}{C_{tot}}, \qquad (9.49)$$

where

$$C_{tot} = \frac{C_{depl} C_{ox}}{C_{depl} + C_{ox}}. \qquad (9.50)$$

The parameter ξ is dimensionless and is proportional to $\left(\dfrac{4\delta_{sd}}{d_{tip}}\right)^{2}$. Note that NSMM measurements in the frequency range between 1 GHz and 50 GHz of n-doped samples show no frequency dependence. This is attributed to the fact that the mobility of the holes in a p-type sample is about ten times smaller than the mobility of electrons in n-type samples. Thus, p-type samples will display a higher resistivity and in turn a higher τ_{sel} relative to n-type samples. In order to observe frequency-dependent contrast in an n-type sample, the NSMM operating frequency must be significantly higher than 50 GHz.

9.3.5 Nonlinear Dielectric Microscopy of Materials

Another application area for resonant NSMM is the imaging of spontaneous polarization of ferroelectric materials based on measurement of the spatial variation of the nonlinear dielectric constant [40]–[44]. The design, shown in Fig. 9.9, is capable of nanometer resolution. The microscope's signal path consists of an LC resonator in series with a needle probe. The operating frequency of the probe lies in the range between 1 GHz and 6 GHz. An oscillating electric field

$$v(t) = V cos(\omega_p t) \tag{9.51}$$

is applied in addition to the applied field between the tip and the back electrode under the sample, resulting in a frequency-modulated signal. The nonlinear

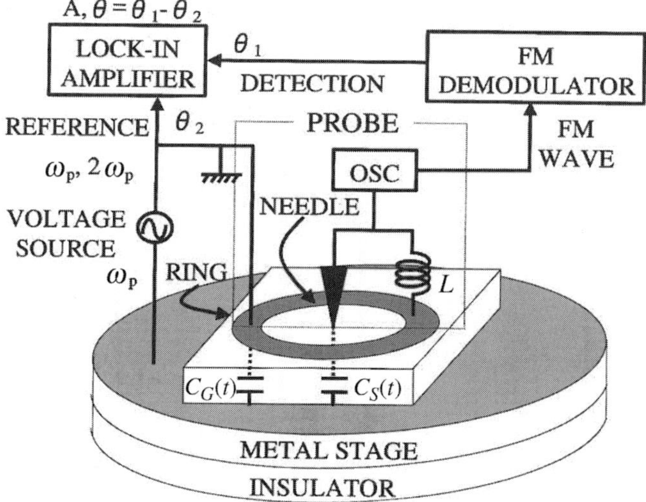

Figure 9.9. Schematic diagram of scanning nonlinear dielectric microscope. Reprinted from Y. Cho, S. Kazuta, and K. Matsuura, *Applied Physics Letters* 75 (1999) pp. 2833–2835, with permission from AIP Publishing.

response of the sample volume beneath the tip to this field leads to a modulation of the sample capacitance $\Delta C_s(t)$ that in turn leads to a modulation of the probe oscillating frequency. After demodulation, a lock-in amplifier is used to record the capacitance variation.

The voltage-dependent capacitance variation can be quantitatively related to the local, nonlinear dielectric constant as follows. The tip-sample capacitance as a function of the voltage v can be expanded in a Taylor series

$$C_s(t) = C_{s0} + \left(\frac{dC_s}{dv}\right)v + \frac{1}{2}\left(\frac{d^2C_s}{dv^2}\right)v^2 + \frac{1}{6}\left(\frac{d^3C_s}{dv^3}\right)v^3 + \ldots \tag{9.52}$$

The displacement can be expressed as a function of the electric field [44]

$$\boldsymbol{D} = \boldsymbol{P}_s + \left[\varepsilon(2)\right]\cdot\boldsymbol{E} + \frac{1}{2}\left[\varepsilon(3)\right]:\boldsymbol{E}^2 + \frac{1}{6}\left[\varepsilon(4)\right]:\boldsymbol{E}^3 + \frac{1}{24}\left[\varepsilon(5)\right]:\boldsymbol{E}^4 + \ldots, \tag{9.53}$$

where $\boldsymbol{P}_s$ is a spontaneous polarization. The variables $\left[\varepsilon(i)\right]$ $i = 2, 3,\ldots$ are permittivity tensors. For $i = 2$, this represents the linear permittivity and is a second rank tensor. For higher values of the index i, the corresponding tensors of increasing rank represent the nonlinear permittivity. The odd, nonlinear terms in Equation (9.53) are sensitive to polarization direction. The ratio of the change in the capacitance $\Delta C_s(t) = C_s(t) - C_{s0}$ to the static value of the capacitance can be expressed (for isotropic media) as [44]

$$\frac{\Delta C_s(t)}{C_{s0}} \approx \frac{\varepsilon(3)}{\varepsilon(2)}E_p\cos\left(\omega_p t\right) + \frac{1}{4}\frac{\varepsilon(4)}{\varepsilon(2)}E_p^{\,2}\cos\left(2\omega_p t\right) + \frac{1}{24}\frac{\varepsilon(5)}{\varepsilon(2)}E_p^{\,3}\cos\left(3\omega_p t\right) + \ldots. \tag{9.54}$$

where ω_p is the radial lock-in reference frequency.

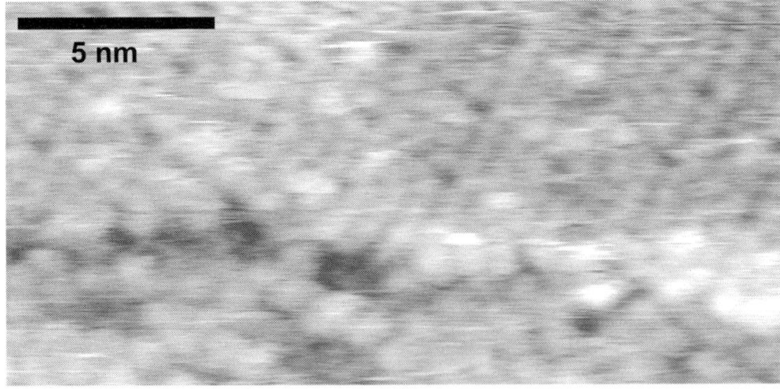

Figure 9.10. Atomic resolution image obtained by use of nonlinear dielectric microscopy [42]. Topography of a reconstructed Si(111)-7×7 surface obtained from the $\varepsilon(4)$ signal in scanning nonlinear dielectric microscope [42]. © IOP Publishing. Adapted with permission. All rights reserved.

An atomic-scale image acquired from the $\varepsilon(4)$ nonlinear dielectric signal is shown in Fig. 9.10. The well-known arrangement of silicon adatoms in a reconstructed Si(111)-7×7 surface is clearly discernable.

9.4　Measurements of Thin Films and Low-Dimensional Materials

9.4.1　Materials for RF Nanoelectronics

Bulk samples provide a useful testbed for establishing the capabilities of NSMM to qualitatively and quantitatively characterize materials, but the ultimate objective is to characterize technologically relevant systems, such as thin films and low-dimensional systems. The spatially resolved measurement of nanomaterials is critical for enabling next generation electronic devices. Materials of interest include thin films with thicknesses on the same order as other critical dimensions, such as grain size or domain size in ferroelectric or magnetoelectric samples. This category also extends to two-dimensional materials where the thickness of the material is a single atomic layer, such as graphene and the TMDs.

While our focus here is on broadband scanning probe microscopy as a tool for quantitative imaging and local spectroscopy of materials, NSMM is but one of many complementary techniques for electromagnetic imaging at the nanoscale [45]. These include free space antenna measurements, resonant cavity approaches, and guided-wave methods, including waveguides, coaxial systems, and microstrip lines. Inevitably, these methods require a clear understanding of the interaction of electromagnetic waves with dielectric materials across macroscopic, mesoscopic, and microscopic length scales [46] as well as an understanding of how materials behave when integrated into devices [47].

9.4.2　Dielectric Film Characterization

The measurement of thin dielectric films deposited on a dielectric substrate represents an interesting example and a particular challenge, especially if the permittivity of the film is not much different from the permittivity of the substrate [48]. One effective strategy is to measure the ratio of the normalized frequency shift of NSMM resonator loaded with a nonmagnetic, dielectric thin film to the frequency shift with the resonator loaded by the bare substrate alone. This ratio does not depend on the calibration constant of the microscope and is equal to the ratio of the perturbed energies of the resonator. In the context of an imaging system, this strategy is particularly appealing if an image can be obtained that includes regions of the film as well as regions of bare substrate. To obtain the permittivity of the film, it is necessary to determine the relationship between the permittivity and the perturbed energy of the resonator either from an appropriate analytical model or by use of numerical techniques. For example, the numerical value of the permittivity may be obtained by comparing measured value of the normalized frequency shift to the results from simulation, as in [49], by fitting the numerical dependences with functions of the type

$$-\frac{\Delta f_r}{f_0} = A\left(\frac{p_1}{p_2(2g/d_{tip})^{p3} + p_4\left(2g/d_{tip}\right)+1}\right) + F, \qquad (9.55)$$

where parameters p_i are extracted from the numerical simulations for a range of permittivity, d_{tip} was introduced earlier, while A and F are fitting constants [49]. The value of this approach is limited because the fitting functions are not unique and may have different forms. In addition, the number of fitting parameters is large. A similar method may be used to extract the material losses by measuring the ratio of changes in the quality factor instead of the ratio of changes in the frequency.

9.4.3　Measurements of Graphene

Atomically thin structures are among the most important materials for nanoelectronics. The explosion of research of these materials was triggered by the work of Nobel Prize winners Geim and Novoselov on graphene. Graphene has attracted attention due to its outstanding transport properties as well as new fundamental physics [50]. Since the initial work by Geim and Novoselov, a broader range of atomically thin materials has been explored. Heterogeneous stacks of atomically thin materials, sometimes referred to as van der Waals heterostructures, are of particular interest for nanoelectronics applications [51]. By optimizing the ordering and composition of these layered structures, the dynamical response of these systems may be tuned over a broad frequency range, opening new possibilities for both applications and fundamental study. Due to its broadband capability, subsurface measurement capability and sensitivity to doping levels, NSMM is a potentially useful tool for investigation of the properties of graphene. Early applications of NSMM to graphene [52] focused on mapping local conductivity, but in order to realize the full potential of NSMM for atomically thin materials, it will be necessary to determine if the sensitivity of NSMM is sufficient to characterize an atomically thin layer at the surface of a heterostructure and, ideally, additional layers buried beneath the surface layer.

In principle, the tip–sample interaction for any two-dimensional material should be modeled as a distributed circuit. However, in many cases the standard simplified lumped-element model introduced earlier is appropriate (see Fig. 9.3). It is convenient to visualize the complex admittance Y of the tip-sample system in a Re(Y) vs. Im(Y) coordinate system, as is commonly done for spreading resistance data [52]. This representation is similar to well-known Cole-Cole diagrams that consist of a semicircle with the center on the imaginary axis (for admittance). An example is shown in the left panel of Fig. 9.11. The radius of the semicircle is

$$r_{sc} = \frac{C_{cpl}}{2\omega(1+C_{ox}/C_{cpl})} \qquad (9.56)$$

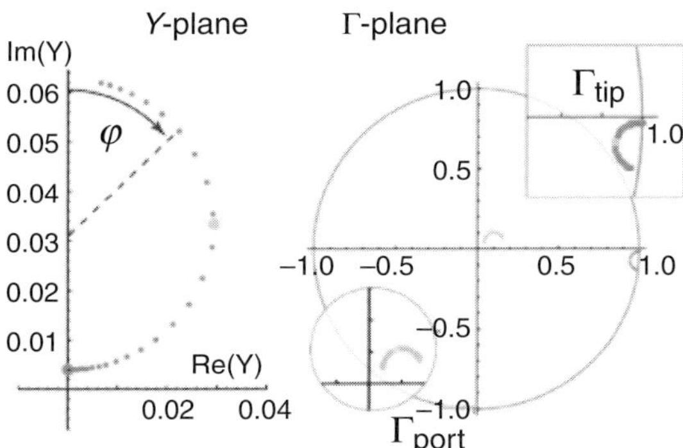

Figure 9.11. Visualization of the complex admittance. (Left panel) Y-plane spreading resistance plot. The dots represent a series of logarithmically scaled spreading resistances. (Right panel) The corresponding spreading resistances plotted in the complex reflection coefficient (Γ) plane [52]. © IOP Publishing. Reprinted with permission. All rights reserved.

and the angular distance

$$\varphi = 2\arctan(\omega\left(C_{cpl} + C_{ox}\right)R_s) \tag{9.57}$$

where C_{ox} is the capacitance of the oxide layer and C_{cpl} is the tip-sample coupling capacitance. For comparison, a visualization of the complex admittance in the complex reflection coefficient plane is shown in the right panel of Fig. 9.11. The reflection coefficient plane representation is a conformal mapping of the admittance plane representation.

Many early NSMM measurements of low-dimensional materials focused on graphene and an abundance of work has been done to develop corresponding sample models. The relationship between the spreading resistance R_s of the graphene and the graphene sheet resistance R is given by [52]

$$R_s = \frac{R}{2\pi}ln\frac{r_g}{r_t}, \tag{9.58}$$

where r_g is the inner radius of the ground electrode and r_t is the radius of the contact area to the graphene sample. In the experimental configuration describe in Reference [52], r_g is usually large and the exact shape of the graphene patch does not significantly influence the result of Equation (9.58). Investigation of the sheet resistance of graphene layers with different thicknesses led to a simplified resistive model tied to the measurement of the reflection coefficient [53]. An alternative characterization method measures the microwave conductivity of graphene by use of a high-Q resonator [54]. Like many implementations of NSSM, the resonator method measures the shifts in the resonant frequency and the quality factor, therefore the analysis is also applicable to NSMM measurements. First, resonator measurements are made on both the bare substrate and the substrate with graphene layer. The change of the resonator loss due to presence of the graphene can be obtained by taking the difference of the measured quality factors. In turn, this difference can be related to the shift of the resonance frequency due to presence of graphene sheet, the material dimensions, and material properties. Application of Equation (9.34) results in [54]

$$\Delta\left(\frac{1}{Q}\right) = \Delta\left(\frac{1}{Q_g}\right) - \Delta\left(\frac{1}{Q_s}\right) = \varepsilon_g'' \frac{\Delta f_s t_g}{(\varepsilon_s' - 1)t_s}, \tag{9.59}$$

where t_g is the thickness of the graphene layer and t_s is the thickness of the substrate. Similar notation distinguishes the permittivity of the graphene ε_g from that of the substrate ε_s. The sheet resistance is related to graphene conductivity by

$$R_s = \frac{1}{\sigma t_g}. \tag{9.60}$$

Given the dependence between the conductivity and imaginary component of the dielectric constant, $\sigma = 2\pi f_0 \varepsilon_0 \varepsilon_g''$, the sheet resistance is [54]

$$R_s = \frac{\Delta f_s}{2\pi f_0 \varepsilon_0 \Delta\left(\frac{1}{Q}\right)(\varepsilon_s' - 1)t_s}. \tag{9.61}$$

Returning to the particular case of NSMM, it is useful to introduce a simple model of the impedance of a graphene sheet as well as other, similar two-dimensional materials. In one example [55], the tip–graphene interaction in an NSMM was described by a lumped-element model consisting of series connection of oxide capacitance, spreading resistance of graphene, and the capacitance between the substrate and the shield of the probe. Here, we develop a general approach based on a two-dimensional electron gas (2DEG) [56], [57]. The 2DEG impedance can be expressed as [57]

$$Z_{2DEG} = R_{2DEG} + j\omega L_k; \; R_{2DEG} = \frac{m_{eff}}{Ne^2 v}; \; L_k = \frac{m_{eff}}{Ne^2} \tag{9.62}$$

where R_{2DEG} is the 2DEG resistance per square and L_k is the kinetic inductance. More generally, the impedance for a sample of any thickness is determined by use of the Drude conductivity as [56]

$$Z_{gr} = \frac{g_F + 2a}{N} \frac{1 + j\omega\tau}{\sigma} \tag{9.63}$$

where g_F and a are form factors depending on the geometry of the probe, N is number of layers in the graphene stack, τ is scattering time and $\sigma = \mu n_d e$ is the low-frequency Drude two-dimensional conductivity.

In order to develop a general model there is an additional parameter that has to be considered, namely the quantum capacitance. Quantum capacitance was introduced in Reference [58] in connection with two-dimensional electron gases and discussed in Reference [59] as an approach for modeling one-dimensional nano-scale devices. The quantum capacitance is defined as

$$C_Q = \frac{\partial Q}{\partial V_l}, \tag{9.64}$$

where V_l is a local electrostatic potential. The quantum capacitance is conventionally defined per unit length for one-dimensional systems and per unit area for two-dimensional systems. The quantum capacitance depends critically on the form of the DOS. We now give a few examples of the DOS and corresponding quantum capacitances.

In general, the DOS is given by

$$g(E) = \frac{m_{eff}}{\pi\hbar^2} v(E), \tag{9.65}$$

where $v(E)$ is the number of contributing bands at a given energy. Assuming that $v(E)$ may be approximated by a quadratic dependence on energy E, the quantum capacitance for a two-dimensional system is [59]

$$C_Q = \frac{v m_{eff} e^2}{2\pi\hbar^2}\left[2 - \frac{\sinh(E_G/2kT)}{\cosh\left(\dfrac{E_G/2 - eV_l}{2kT}\right)\cosh\left(\dfrac{E_G/2 + eV_l}{2kT}\right)}\right] \tag{9.66}$$

where E_G is the band gap. For $E_G = 0$ this reduces to [58]

$$C_Q = \frac{v m_{eff} e^2}{2\pi\hbar^2} \tag{9.67}$$

Alternatively, for undoped, single-layer graphene, a linear DOS may be used. If the DOS has the form [60, 61]

$$g(E) = \frac{g_s g_v}{2\pi(\hbar v_F)^2} |E| \tag{9.68}$$

where g_s is the spin degeneration and g_v is the valley parameter, then the quantum capacitance is given by

$$C_Q = \frac{2e^2 kT}{\pi(\hbar v_F)^2} \ln\left[2\left(1 + \cosh(\frac{eV_l}{k_B T})\right)\right], \tag{9.69}$$

where k_B is the Boltzmann constant. Under the condition $eV_l = k_B T$, Equation (9.69) simplifies to

$$C_Q \approx e^2 \frac{2eV_l}{\pi(\hbar v_F)^2} = \frac{2e^2}{\hbar v_F \sqrt{\pi}}\sqrt{n}, \tag{9.70}$$

where n is the carrier concentration and e is the electron charge.

In the one-dimensional case

$$g(E) = \frac{1}{\pi \hbar^2} v(E) \sqrt{\frac{2m_{eff}}{E}} \qquad (9.71)$$

and the calculation of the quantum capacitance is more complicated. Under special conditions, if $v(E) = v$ can be considered constant and the DOS can be approximated as

$$g(E) = \frac{v}{hv_F}, \qquad (9.72)$$

where v_F is the Fermi velocity, then the one-dimensional quantum capacitance simplifies to

$$C_Q = \frac{2ve^2}{hv_F}. \qquad (9.73)$$

Experimentally, the local quantum capacitance of graphene or other low-dimensional system can be obtained by use of scanning probe microscopy and spectroscopy [62]. Although such measurements have historically been performed with scanning capacitance microscopes, they can also be performed with NSMM. The NSMM tip is positioned near the sample and the reflection coefficient is measured as a function of the tip bias voltage. As needed, a small AC signal may be added to the DC bias to facilitate a more sensitive spectroscopic measurement by use of a lock-in technique.

Consider such a spectroscopic measurement of a mechanically exfoliated graphene flake supported by a SiO_2 layer grown on a Si substrate. Spectroscopic measurements are taken at tip positions above the graphene flake as well as above the bare SiO_2. The ratio of the reflection coefficient measured with the tip positioned over the graphene to that measured with the tip positioned over the oxide is calculated. Once again, the system consisting of the probe tip over the oxide and Si substrate can be reasonably modeled as an MOS structure, but the model must be modified such that the quantum capacitance is in series with the MOS impedance. The quantum capacitance per unit area C_Q may then be related to the oxide capacitance per unit area C_{ox} by

$$C_Q = C_{ox} \frac{V_b}{\Delta V_{gr}} \qquad (9.74)$$

where V_b is the bias voltage and ΔV_{gr} is the potential drop across the thickness of the graphene within the effective probe area. The local potential in graphene is dependent on the charge density around the contact and can be expressed as [62]

$$V_{gr}(r) = \frac{\sqrt{\pi}\hbar v_F}{e} \sqrt{|n(r)|}. \qquad (9.75)$$

The potential drop within the effective probe area is then obtained by integrating the potential over the effective tip area. The charge density distribution $n(r)$ depends on the geometry and structure of the measured system and must be calculated either numerically or analytically.

9.4.4 Measurements of Transition Metal Dichalcogenides

Let's turn now from the case of exfoliated graphene to the case of two-dimensional semiconductors, including TMDs. Electrical properties of several TMDs, including metallic and semiconducting systems, are shown in Table 9.1 [63]. Other two-dimensional semiconductor materials include silicine, phosphorene, and hexagonal boron nitride. Unlike graphene, semiconducting TMDs have a band gap and are thus suitable for transistor applications.

Representative TMD materials MoS_2 and WSe_2 have been investigated by NSMM [64], revealing that the sensitivity of NSMM is sufficient to image monolayer semiconducting materials and perform quantitative spectroscopy on them. The signal-to-noise ratio was found to be dependent on the polarity of the tip bias. Figure 9.12(a) shows NSMM images of a mechanically exfoliated, few-layer $W_{1-x}Nb_xSe_2$ patch on a SiO_2/Si substrate while Fig. 9.12(b) shows NSMM images

Table 9.1 Properties of Transition Metal Dichalcogenides. Single-layer (1L) and bulk band gaps are listed for the semiconducting materials.

Reprinted by permission from Macmillan Publishers Ltd: Nature Nanotechnology. Q. Hua Wang, K. Kalantar-Zadeh, A. Kis, J. N. Coleman, and M. S. Strano, *Nature Nanotechnology* 7 (2012) pp. 699–712. © 2012.

	$-S_2$		$-Se_2$		$-Te_2$	
	Electronic Characteristics	Ref.	Electronic Characteristics	Ref.	Electronic Characteristics	Ref.
Nb	Metal; Superconductor; Charge Density Wave	[67]	Metal; Superconductor; Charge Density Wave	[67],[68]	Metal	[72]
Ta	Metal; Superconducting; Charge Density Wave	[67],[68]	Metal; Superconducting; Charge Density Wave	[67],[68]	Metal	[72]
Mo	Semiconductor 1L: 1.8 eV Bulk: 1.2 eV	[69],[70]	Semiconductor 1L: 1.5 eV Bulk: 1.1 eV	[71],[70]	Semiconductor 1L: 1.1 eV Bulk: 1.0 eV	[71],[73]
W	Semiconductor 1L: 1.9 eV - 2.1 eV Bulk: 1.2 eV	[66],[71], [70]	Semiconductor 1L: 1.7 eV Bulk: 1.2 eV	[72],[70]	Semiconductor 1L: 1.1 eV	[72]

of a single layer MoS_2. To acquire these images a lock-in technique was used to obtain the real and imaginary parts of the derivative of the reflection coefficient ($S_R' = \Re(dS_{11}/dV_{tip})$, $S_C' = \Im(dS_{11}/dV_{tip})$). The images demonstrate that changes in the DC tip bias induce different levels of contrast in the S_R' image. Further investigation revealed that for a given material, contrast was observed for only one polarity of the tip bias, which was directly correlated with the polarity of the charge carriers in the material. There is little contrast in the S_C' images, regardless of tip bias polarity or magnitude.

Further insight into the interpretation of the NSMM measurements was gained through finite-element, multiphysics simulation of the NSMM-TMD system. The objective of the simulation process is to connect the simulated parameters with the materials properties of the sample by relating them to the measured, bias-dependent data. While this approach is phenomenological, it provides insight into the underlying physics with the NSMM probe-TMD sample system. In this case, it is useful to understand the relation between the charge carrier density and the bias voltage V_{tip}. Specifically, the real part of the measured admittance of the TMD materials was modeled as a function of conductivity. The capacitive behavior of the TMDs was captured by modeling the material as a dielectric sheet with the relative permittivity $\varepsilon_r = 7$. If the band edge DOS is approximated by a quadratic function, the relation between the carrier density n_q and the conductivity of the given material can be expressed as [65], [66]

$$n_q \propto \mp\left(aV_{tip} - E_0\right)^3 = \sigma_i \tag{9.76}$$

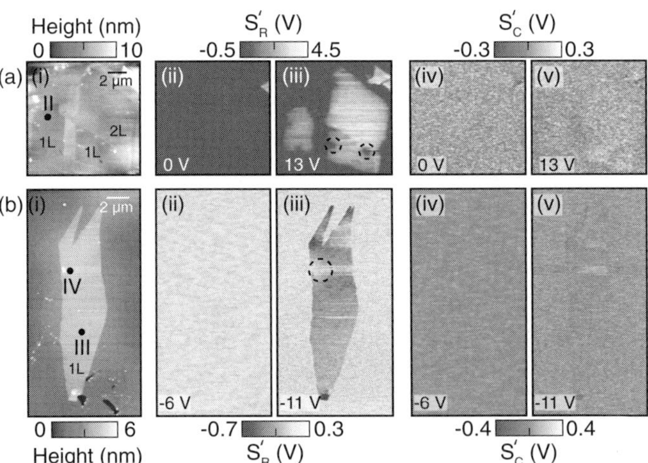

Figure 9.12.   Images of TMDs using NSMM.
AFM topography (i) and NSMM images (ii-v) of (a) an exfoliated few-layer $W_{1-x}Nb_xSe_2$ sample and (b) a single layer MoS_2 sample with tip bias as indicated. Reprinted with permission from S. Berweger et al., *Nano Letters* 15 (2015) pp. 1122–1127. © 2015, American Chemical Society.

where the minus sign corresponds to the n-type semiconductor (as in MoS_2) and the plus sign to the p-doped semiconductor (as in WSe_2). The constant a reflects the scaling between E and V_{tip} and E_0 accounts for doping at zero bias i.e., the Fermi level shift. It is important to note that a reasonable guess from existing published data or independent measurements must be used to make a reasonable estimate of the mobility of the carriers, as this parameter also enters the relation between the charge carriers and the conductivity of the material. Based on the comparison of the model with the experimental data, the tip-bias dependent contrast is attributed to changes of the TMD conductivity. Measurements and modeling agree that both MoS_2 and WSe_2 show strong unipolar charge transport that correlates with the polarity-dependent contrast in NSMM images.

In addition to NSMM imaging, local spectroscopy can be performed at any tip position above the investigated sample. Of especial importance is spectroscopy on single-layer regions, which provides local measurements of electronic structure as well as sensitivity to local defects. Examples of local spectroscopic measurements of TMD samples are shown in Fig. 9.13. The predicted spectra of the phenomenological, numerical modeling are in good agreement with the experimental data. In order to fit the capacitive part of the spectra, it was necessary to include the quantum capacitance in series with the geometric, electrostatic capacitance. Interpretation of NSMM measurements of semiconducting, two-dimensional materials is more challenging than the interpretation of NSMM measurements of graphene, requiring not only consideration of the effects of quantum capacitance, but also accurate modeling of the sample conductivity, as well as a comprehensive model of the probe, two-dimensional material, and supporting substrate.

The successful characterization of thin films and low-dimensional materials by use of NSMM suggests that the technique can have a broader impact for characterization not only of any two-dimensional system, but also more complicated,

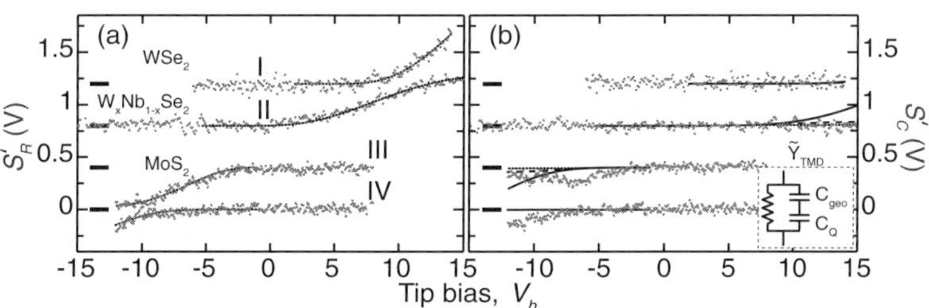

Figure 9.13. Local NSMM spectroscopy of TMDs.
Resistive (a) and capacitive (b) components of the derivative of the reflection coefficient $S'_{R,C}$-voltage sweeps taken from single layer locations in Fig. 9.12. The solid lines in (a) are finite-element simulations fit to the resistive part of the measured signal. Solid and dashed lines in (b) are fits to the capacitive part of the signal with and without taking into account the quantum capacitance in series with geometric capacitance (see insert). Reprinted with permission from S. Berweger et al., *Nano Letters* 15 (2015) pp. 1122–1127. © 2015, American Chemical Society.

multilayer heterostructures. With the advantage of relatively high penetration depth at microwave frequencies, this technique is of particular importance for investigation of subsurface interfaces and defects on the properties of heterostructures.

References

[1] S. R. de Groot and L. G. Suttorp, *Foundations of Electrodynamics* (Elsevier, 1972).

[2] U. Kaatze, "Techniques for Measuring the Microwave Dielectric Properties of Materials," *Metrologia* 47 (2010) pp. S91–S113.

[3] L. D. Landau and E. M. Lifshitz, *Electrodynamics of Continuous Media* (Addison-Wesley, 1960).

[4] J. D. Jackson, *Classical Electrodynamics*, 3rd edn (Wiley, 1999).

[5] R. Hornreich and S. Shtrikman, "Statistical Mechanics and Origin of the Electromagnetic Effect in Cr_2O_3," *Physical Review* 161 (1967) pp. 506–512.

[6] G. T. Rado, "Statistical Theory of Magnetoelectric Effects in Antiferromagnetics," *Physical Review* 128 (1962) pp. 2546–2556.

[7] J. Baker-Jarvis and P. Kabos, "Dynamic Constitutive Relations for Polarization and Magnetization," *Physical Review E* 64 (2001) art. no. 056127.

[8] J. Baker-Jarvis, P. Kabos, and C. Holloway, "Nonequilibrium Electromagnetics: Local and Macroscopic Fields and Constitutive Relationships," *Physical Review E* 70 (2004) art. no. 036615.

[9] S. M. Anlage, V. V. Talanov, and A. R. Schwartz, "Principles of Near Field Microscopy." In *Scanning Probe Microscopy,* vol. 1, (S. Kalinin and A. Gruverman, eds.) (Springer, 2007).

[10] A. N. Reznik and N. V. Yurasova, "Electrodynamics of Microwave Near Field Probing: Application to Medical Diagnostics," *Journal of Applied Physics* 98 (2005) art. no. 114701.

[11] S.-C Lee, C. P. Vlahacos, B. J. Feenstra, A. Schwartz, D. E. Steinhauer, F. C. Wellstood, and S. M. Anlage, "Magnetic Permeability Imaging of Metals with Scanning Near-Field Microwave Microscope," *Applied Physics Letters* 77 (2000) pp. 4404–4406.

[12] A. N. Reznik, "Electromagnetic Model for Near Field Microwave Microscope with Atomic Resolution: Determination of Tunnel Junction Impedance," *Applied Physics Letters* 105 (2014) art. no. 083512.

[13] A. Hovsepyan, A. Babajanyan, T. Sargsyan, H. Melikyan, S. Kim, J. Kim, K Lee, and B. Friedman, "Direct Imaging of Photoconductivity of Solar Cells by Using Near-Field Scanning Microwave Probe," *Journal of Applied Physics* 106 (2009) art. no. 114901.

[14] J. R. Wait, *Electromagnetic Waves in Stratified Media* (Pergamon Press, 1962).

[15] J. R. Wait, *Wave Propagation Theory* (Pergamon Press, 1981).

[16] W. Ch. Chew, *Waves and Fields in Inhomogeneous Media* (Van Nostrand Reinhold, 1990).

[17] C. Gao and X.-D. Xiang, "Quantitative microwave near field microscopy of dielectric properties," *Review of Scientific Instruments* 69 (1998) pp. 3846–3851.

[18] G. P. Kochanski, "Nonlinear Alternating-Current Tunneling Microscopy," *Physical Review Letters* 62 (1989) pp. 2285–2288.

[19] M. Tabib-Azar, D. -P. Su, A. Pohar, S. R. LeClair, and G. Pochak, "0.4 µm Spatial Resolution with 1 GHz (λ=30cm) Evanescent Microwave Probe," *Review of Scientific Instruments* 70 (1999) pp. 1725–1729.

[20] M. Tabib-Azar and S. R. LeClair, "Novel Hydrogen Sensors Using Evanescent Microwave Probes," *Review of Scientific Instruments* 70 (1999) pp. 3707–3713.

[21] R. A. Kleinsmit, M. K. Kazimierczuk, and G. Kozlowski, "Sensitivity and Resolution of Evanescent Microwave Microscope," *IEEE Transactions on Microwave Theory and Techniques* 54 (2006) pp. 639–647.

[22] S. J. Stranick and P. S. Weiss, "A Versatile Microwave-Frequency-Compatible Scanning Tunneling Microscope," *Review of Scientific Instruments* 64 (1993) pp. 1232–1234.

[23] Y. Cho, "Scanning Nonlinear Dielectric Microscope with Super High Resolution," *Japanese Journal of Applied Physics* 46 (2007) pp. 4428–4434.

[24] J. Smoliner, H. P. Huber, M. Hochleitner, M. Moertelmaier, and F. Kienberger, "Scanning Microwave Microscopy/Spectroscopy on Metal Oxide Semiconductor Systems," *Journal of Applied Physics* 108 (2010) art. no. 064315.

[25] D. A. Hill, *Electromagnetic Field in Cavities*, IEEE Press Series on Electromagnetic wave theory (John Wiley and Sons, 2009).

[26] A. Tselev, S. M. Anlage, Z. Ma, and J. Melngailis, "Broadband Dielectric Microwave Microscopy on Micron Length Scales," *Review of Scientific Instruments* 78 (2007) art. no. 044701.

[27] D. M. Pozar, *Microwave Engineering*, 2nd edn. (Wiley, 1998).

[28] K. D. McKinstry and C. E. Patton, "Methods for Determination of Microwave Cavity Quality Factors from Equivalent Circuit Models," *Review of Scientific Instruments* 60 (1989) pp. 439–443.

[29] A. Imtiaz, T. Baldwin, H. T. Nembach, T. M. Wallis, and P. Kabos, "Near Field Microscope Measurements to Characterize Bulk Material Properties," *Applied Physics Letters* 90 (2007) art. no. 243105.

[30] D. J. Barker, T. J. Jackson, P. M. Suherman, M. S. Gashinova, and M. J. Lancaster, "Uncertainties in the Permittivity of Thin Films Extracted from the Measurements with Near Field Microwave Microscopy Calibrated by Image Charge Model," *Measurement Science and Technology* 25 (2014) art. no. 105601.

[31] R. A. Oliver, "Advances in AFM for the Electrical Characterization of Semiconductors," *Reports on Progress in Physics* 71 (2008) art. no. 076501.

[32] H. P. Huber, I. Humer, M. Hochleitner, M. Fenner, M. Moertelmaier, C. Rankl, A. Imtiaz, T. M. Wallis, H. Tanbakuchi, P. Hinterdorfer, P. Kabos, J. Smoliner, J. J. Kopanski, and F. Kienberger, "Calibrated Nanoscale Dopant Profiling Using Scanning Microwave Microscope," *Journal of Applied Physics* 111 (2012) art. no. 014301.

[33] S. M. Sze, *The Physics of Semiconductor Devices* (Wiley, 1981).

[34] J. J. Kopanski, J. F. Marchiando, and J. R. Lowney, "Scanning Capacitance Microscopy Applied to Two-Dimensional Dopant Profiling of Semiconductors," *Materials Science and Engineering B: Solid-State Materials for Advance Technology* 44 (1997) pp. 46–51.

[35] J. F. Marchiando, J. J. Kopanski, and J. R. Lowney, "Model Database for Determining Dopant Profiles from Scanning Capacitance Microscope Measurements," *Journal of Vacuum Science and Technology B* 16 (1998) pp. 463–470.

[36] J. Smoliner, B. Basnar, S. Golka, E. Gornik, B. Lo"ffler, M. Schatzmayr, and H. Enichlmair, "Mechanism of Bias-Dependent Contrast in Scanning-Capacitance Microscopy Images," *Applied Physics Letters* 79 (2001) pp. 3182–3184.

[37] A. Imtiaz, S. M. Anlage, J. D. Barry, and J. Melngailis, "Nanometer-Scale Material Contrast Imaging with a Near-Field Microwave Microscope," *Applied Physics Letters* 90 (2007) art. no. 143106.

[38] A. Imtiaz, T. M. Wallis, S.-H. Lim, H. Tanbakuchi, H.-P. Huber, A. Hornung, P. Hinterdorfer, J. Smoliner, F. Kienberger, and P. Kabos, "Frequency Selective Contrast on Variable Doped p-type Silicon with a Scanning Microwave Microscope," *Journal of Applied Physics* 111 (2012) art. no. 093727.

[39] E. H. Nicollian and J. R. Brews, *MOS (Metal Oxide Semiconductor) Physics and Technology* (Wiley, 1982).

[40] Y. Cho, A. Kirihara, and T. Saeki, "Scanning Nonlinear Dielectric Microscope," *Review of Scientific Instruments* 67 (1996) pp. 2297–2303.

[41] Y. Cho, S. Kazuta, and K. Matsuura, "Scanning Nonlinear Dielectric Microscopy with Nanometer Resolution," *Applied Physics Letters* 75 (1999) pp. 2833–2835.

[42] R. Hirose, K. Ohara, and Y. Cho, "Observation of the Si(111) 7x7 Atomic Structure Using Noncontact Scanning Nonlinear Dielectric Microscopy," *Nanotechnology* 18 (2007) art. no. 084014.

[43] K. Tanaka, Y. Kurihashi, T. Uda, Y. Daimon, N. Odagawa, R. Hirose, Y. Hiranaga, and Y. Cho, "Scanning Nonlinear Dielectric Microscopy Nano-science and Technology for Next Generation High Density Ferroelectric Data Storage," *Japanese Journal of Applied Physics* 47 (2008) pp. 3311–3325.

[44] Y. Cho, "Scanning Nonlinear Dielectric Microscopy," *Journal of Materials Research* 26 (2011) pp. 2007–2016.

[45] D. A. Bonnel, D. N. Basov, M. Bode, U. Diebold, S. V. Kalinin, V. Madhavan, L. Novotny, M. Salmeron, U. D. Schwartz, and P. S. Weiss, "Imaging Physical Phenomena with Local Probes: From Electrons to Photons," *Reviews of Modern Physics* 84 (2012) pp. 1343–1381.

[46] J. Baker-Jarvis and S. Kim, "The Interaction of Radio-Frequency Fields with Dielectric Materials at Macroscopic to Mesoscopic Scales," *Journal of Research of the National Institute of Standards and Technology* 117 (2012) pp. 1–60.

[47] P. Queffelec, M. Le Floc'h, and P. Gelin, "Broad-Band Characterization of Magnetic and Dielectric Thin Films Using a Microstrip Line," *IEEE Transactions on Instrumentation and Measurement* 47 (1998) pp. 956–963.

[48] S. Huang, H. M. Christen, and M. E. Reeves, "Parameter-Free Extraction of Thin-Film Dielectric Constants from Scanning Near Field Microwave Microscope Measurements," Naval Research Laboratory, Washington DC, Technical Report ADA524131, (2010). (Also in arXiv:00909.3579v1 [cond-mat.sci] September 19, 2009).

[49] J. H. Lee, S. Hyun, and K. Char, "Quantitative Analysis of Scanning Microwave Microscopy on Dielectric Thin Film by Finite Element Calculation," *Review of Scientific Instruments* 72 (2001) pp. 1425–1434.

[50] G. Fiori, F. Bonaccorso, G. Iannaccone, T. Palacios, D. Neumaier, A. Seabaugh, S. K. Banerjee, and L. Colombo, "Electronics Based on Two-Dimensional Materials," *Nature Nanotechnology* 9 (2014) pp. 768–779.

[51] A. K. Geim and I. V. Grigorieva, "Van der Waals heterostructures," *Nature* 499 (2013) pp. 419–425.

[52] A. Tselev, N. V. Lavrik, I. Vlassiouk, D. P. Briggs, M. Rutgers, R. Proksch, and S. V. Kalinin, "Near-Field Microwave Scanning Probe Imaging of Conductivity Inhomogeneities in CVD Graphene," *Nanotechnology* 23 (2012) art. no. 385706.

[53] T. Monti, A. Di Donato, D. Mencarelli, G. Venanzoni, A. Morini, and M. Farina, "Near-Field Microwave Investigation of Electrical Properties of Graphene-ITO Electrodes for LED Applications," *Journal of Display Technology* 9 (2013) pp. 504–510.

[54] L. Hao, J. Gallop, S. Goniszewski, O. Shaforost, N. Klein, and R. Yakimova, "Non-contact Method for Measurement of Microwave Conductivity of Graphene," *Applied Physics Letters* 103 (2013) art. no. 123103.

[55] A. Tselev, V. K. Sangwan, D. Jariwala, T. Marks, L. J. Lauhon, M. C. Hersam, and S. V. Kalinin, "Near Field Microwave Microscopy of High-k Oxides Grown on Graphene with an Organic Seed Layer," *Applied Physics Letters* 103 (2013) art. no. 243105.

[56] V. V. Talanov, C. Del Barga, L. Wickey, I. Kalichava, E. Gonzales, E. A. Shaner, A. V. Gin, and N. G. Kalugin, "Few-Layer Graphene Characterization by Near-Field Scanning Microwave Microscopy," *ACS Nano* 4 (2010) pp. 3831–3838.

[57] S. Kang, P. J. Burke, L. N. Pfeiffer, K. W. West, "AC Ballistic Transport in a Two-Dimensional Electron Gas Measured in GaAs/AlGaAs Heterostructures," *Physical Review B* 72 (2005) art. no. 165312.

[58] S. Luryi, "Quantum Capacitance Devices," *Applied Physics Letters* 52 (1988) pp. 501–503.

[59] D. L. John, L. C. Castro, and D. L. Pulfrey, "Quantum Capacitance in Nanoscale Device Modeling," *Journal of Applied Physics* 96 (2004) pp. 5180–5184.

[60] T. Fang, A. Konar, H. Xing, and D. Jena, "Carrier Statistics and Quantum Capacitance of Graphene Sheets and Ribbons," *Applied Physics Letters* 91 (2007) art. no. 092109.

[61] J. Xia, F. Chen, J. Li, and N. Tao, "Measurement of the Quantum Capacitance of Graphene," *Nature Nanotechnology* 4 (2009) pp. 505–509.

[62] F. Giannazzo, S. Sonde, V. Raineri, and E. Rimini, "Screening Length and Quantum Capacitance in Graphene by Scanning Probe Microscopy," *Nano Letters* 9 (2009) pp. 23–29.

[63] Q. Hua Wang, K. Kalantar-Zadeh, A. Kis, J. N. Coleman, and M. S. Strano, "Electronics and Optoelectronics of Two-Dimensional Transition Metal Dichalcogenides," *Nature Nanotechnology* 7 (2012) pp. 699–712.

[64] S. Berweger, J. C. Weber, J. John, J. M. Velazquez, A. Pieterick, N. A. Sanford, A. V. Davydov, B. Brunschwig, N. S. Lewis, T. M. Wallis, and P. Kabos, "Microwave Near-Field Imaging of Two-Dimensional Semiconductors," *Nano Letters* 15 (2015) pp. 1122–1127.

[65] E. S. Kadantsev and P. Hawrylak, "Electronic Structure of Single MoS_2 Monolayer," *Solid State Communications* 152 (2012) pp. 909–913.

[66] A. Kuc, N. Zibouche, and T. Heine, "Influence of Quantum Confinement on the Electronic Structure of the Transition Metal Sulfide TS_2," *Physical Review B* 83 (2011) art. no. 245213.

[67] A. R. Beal, H. P. Hughes, and W. Y. Liang, "The Reflectivity Spectra of Some Group VA Transition Metal Dichalcogenides," *Journal of Physics C* 8 (1975) pp. 4236–4248.

[68] J. A. Wilson, F. J. Di Salvo, and S. Mahajan, "Charge-Density Waves and Superlattices in the Metallic Layered Transition Metal Dichalcogenides," *Advances in Physic.* 24 (1975) pp. 117–201.

[69] K. F. Mak, C. Lee, J. Hone, J. Shan, and T. F. Heinz, "Atomically Thin MoS2: A New Direct-Gap Semiconductor," *Physical Review Letters* 105 (2010) art. no. 136805.

[70] K. K. Kam and B. A. Parkinson, "Detailed Photocurrent Spectroscopy of the Semiconducting Group-VI Transition-Metal Dichalcogenides," *Journal of Physical Chemistry* 86 (1982) pp. 463–467.

[71] L. Liu, S. B. Kumar, Y. Ouyang, and J. Guo, "Performance Limits of Monolayer Transition Metal Dichalcogenide Transistors," *IEEE Transactions on Electron Devices* 58 (2011) pp. 3042–3047.

[72] Y. Ding, Y. Wang, J. Ni, L. Shi, S. Shi, and W. Tang, "First Principles Study of Structural, Vibrational and Electronic Properties of Graphene-Like MX2 (M=Mo, Nb, W, Ta; X=S, Se, Te) Monolayers," *Physica B: Condensed Matter* 406 (2011) pp. 2254–2260.

[73] *Gmelin Handbook of Inorganic and Organometallic Chemistry*, 8th edn, vol. B7 (Springer, 1995).

10   Measurement of Active Nanoelectronic Devices

10.1   Applications of RF Nanoelectronics

The preceding chapters have described a variety of measurement methods for the characterization of RF nanoelectronics. For example, we have described how on-wafer measurement and de-embedding techniques can be extended to devices and circuits that incorporate nanoscale building blocks. We went on to discuss how local microwave microscopy techniques enable us to characterize defects, interfaces, and the constituent materials inside a device. Taken together, this suite of techniques provides a comprehensive picture that relates overall device performance to intra-device circuit and material parameters. In a nanoelectronic world in which the placement of each individual atom can impact device performance, this capability has potentially powerful implications for RF nanoelectronic design, fabrication, and operation. However, the full impact of these measurement techniques will only be realized when they are applied to state-of-the-art technological challenges, both within the field of nanoelectronics and beyond. To that end, the final five chapters of this book are dedicated to application of the measurement techniques that we have introduced. These chapters not only demonstrate the relevance of these methods to contemporary research and engineering, but also serve as practical examples of how the general RF and microwave measurement techniques can be adapted and extended to specific measurement problems.

Many of the most promising applications of nanomaterials involve active, nanoelectronic devices. For example, the intrinsic transport properties of single-walled CNTs enable field effect transistors (FETs) that allow higher current densities and lower distortion than their conventional counterparts [1]. In addition, CNTFETs offer the tantalizing possibility of FETs with cutoff frequencies in the terahertz range, at least in theory. Given that a typical, 100 nm-long CNT has a capacitance on the order of $C = 4$ aF and a quantized resistance as small as $R = 6.25$ kΩ, a rough calculation of the RC time constant of such a nanotube is on the order of 160 fs, corresponding to a frequency of about 6 THz [2]. Early measurements of the transconductance in CNT transistors also suggested potential cutoff frequencies well into the terahertz regime [3]. However, in practice, the extrinsic cutoff frequencies of realizable CNTFETs have been on the order of tens of gigahertz [4]–[6].

The realization of CNTFETs with terahertz-scale, extrinsic cutoff frequencies faces significant, potentially insurmountable challenges. For instance, materials-science challenges include the isolation of large numbers of uniform, semiconducting CNTs. Also, while the intrinsic properties of CNTs favor high cutoff frequencies, optimization of the extrinsic cutoff frequency requires minimization of contact reactance and other parasitics. These fabrication challenges are also intimately tied to the need for impedance matching. Moreover, quantum mechanical effects such as kinetic inductance and quantum capacitance can play significant roles in device performance, depending upon the intended application and desired operating frequency [2], [7]. Ultimately, the cutoff frequencies of CNTFETs appear unlikely to surpass those of less expensive, conventional FETs. However, the unique properties of CNTs may yet be leveraged for specialized applications such as high-linearity amplifiers and mixers [1].

Building upon efforts to produce CNT-based, RF nanotransistors, the field of carbon-based nanoelectronics has naturally extended to include graphene-based, RF nanotransistors. In fact, following the experimental demonstration of ambipolar field effects and high electron mobility in graphene [8], [9], significant progress has been made in the fabrication and optimization of graphene-based FETs. Notably, graphene is not a natural choice for transistor applications as it lacks a band gap, though a band gap may be opened by patterning a graphene sheet into nanoribbons. The lack of a band gap makes it difficult to achieve a high on/off ratio with graphene nanotransistors, thus limiting their effectiveness as switches. As in the case of CNTFETs, extrinsic cutoff frequencies of graphene FETs have reached, but not exceeded, tens of gigahertz [10], [11]. Recent advances in graphene transistor technology include the engineering of FETs-based heterostructures such as graphene-boron nitride [12] and graphene-fluorographene [13]. An example of a graphene-fluorographene heterostructure FET is shown in Fig. 10.1. Like graphene, TMDs are two-dimensional materials, but unlike graphene, TMDs have an intrinsic band gap, making them more suitable for transistor-based applications [14]. In particular, TMD-based transistors display high on/off ratios on the order of 1 to 10^8 [15].

In Chapters 4, 5, and 6, we introduced RF measurement techniques for simplified, passive nanoelectronic devices. While these structures serve as useful, introductory examples, a more comprehensive approach is required for quantitative, broadband characterization of nanotransistors and other active devices. Here, building upon established measurement methods and equivalent circuit models for traditional FETs, we review systematic techniques for characterization of RF nanotransistors that are extendable to other active nanoelectronic devices. In this chapter, after a brief review of modeling and measurement of conventional transistors, we illustrate the measurement methodology and device modeling through a specific example of a GaN nanowire-based FET. The methods are enabled by on-wafer calibration procedures, accurate equivalent circuit models, calibration structures for extraction of parasitic reactance, and numerical optimization processes for parameter extraction.

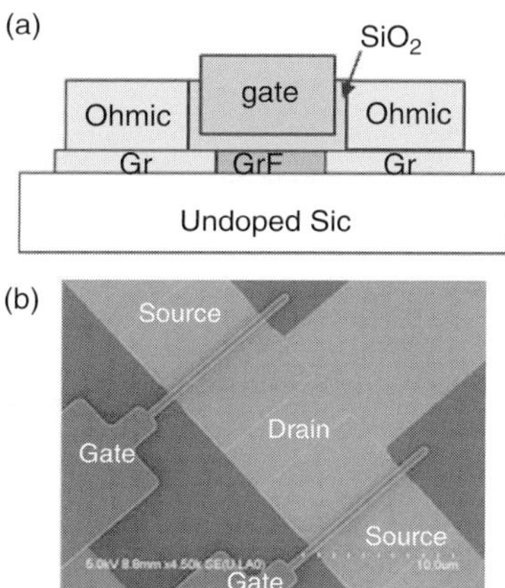

Figure 10.1.   A graphene-fluorographene heterostructure field effect transistor.
(a) A schematic (side view) of a field effect transistor based on a lateral graphene (Gr) /
fluorographene (GrF) / graphene (Gr) heterostructure. (b) An SEM image (top view) of
the fabricated transistor. © 2013 IEEE. Reprinted, with permission from J. S. Moon et al.,
IEEE Electron Device Letters 34 (2013) pp. 1190–1192.

10.2   Modeling and Measurement of Active Devices

10.2.1   Small-Signal Models of Conventional Transistors

Throughout this book, we have introduced a large number of models of RF and
microwave systems, including broadband, two-port nanoscale devices and a variety
of different local microwave probe-sample configurations. By and large, our motivation for introducing these models has been to extract estimates of physical and
material parameters from broadband measurements. In commercial applications
such as wireless communications, circuit designers develop and rely upon active
device models to maintain efficient design cycles and minimize time to market [16].
These models may take on a variety of forms and often take advantage of the computational capabilities of commercial finite-element solvers, computer-aided design
software, circuit simulators, and multi-physics packages. Physical models attempt
to describe not only the electromagnetic properties of constituent materials and
circuit elements, but also thermal and mechanical effects within a device. Given the
complexity of state-of-the-art FETs and other active devices, comprehensive physical models can be both elaborate and computationally intensive. By contrast, compact circuit models strive to capture the fundamental behavior of a device, often
sacrificing some measure of detailed understanding and insight for computational
efficiency. Compact circuit models may be based on the designer's knowledge of the
operational principles or they may be purely phenomenological. Though a great
deal of work has been done in the field of FET modeling, the ongoing push to
incorporate new materials and to operate at higher frequencies implies that work

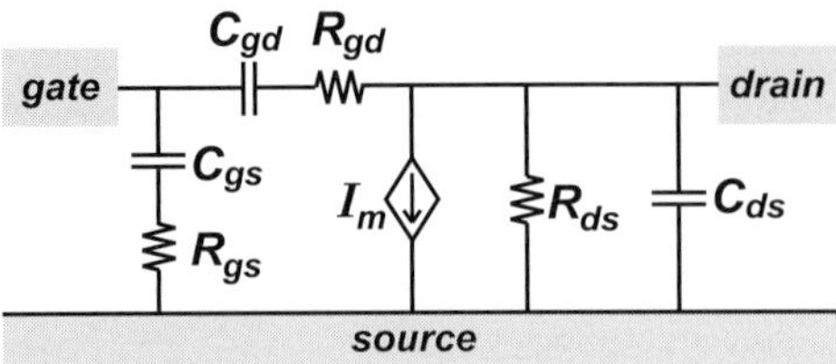

Figure 10.2. Equivalent circuit model for an intrinsic FET.
I_m is the modulated current. R_{gs} (C_{gs}), R_{gd} (C_{gd}), and R_{ds} (C_{ds}) are the gate-source, gate-drain, and drain-source resistances (capacitances), respectively.

in this area is far from complete. Here, we will focus on compact, equivalent circuit models of an FET operating in the small-signal regime. We will assume that the reader is familiar with the core concepts of transistors and their operation.

Several important assumptions underlie equivalent circuit models and the measurement techniques that rely upon them. For instance, though the values of some equivalent circuit parameters may be determined from DC or low-frequency measurements, use of these values in high-frequency models implicitly assumes that these circuit parameters are frequency independent. As we will illustrate below, parasitic elements are often obtained from "cold" state measurements in which the device is not biased. If these "cold" state parameters are subsequently carried over to models of active states, then the underlying assumption is that the parasitic elements do not depend on the application of a bias to the device. Note that while the parasitic elements may not explicitly depend on the bias, in certain cases they may indirectly be affected by the bias through effects such as device heating. Finally, keep in mind that though individual circuit elements are often represented as lumped elements, they may represent effective behaviors that incorporate an aggregation of distributed, nonlocal effects. The relationship between equivalent circuit model elements and physical parameters of interest is likely to be complex. Thus, the most accurate determination of physical and material parameters from active device measurements will be determined by use of detailed, physical models.

Figure 10.2 shows a typical small-signal equivalent circuit model. Figure 10.2 shows only the intrinsic transistor, excluding parasitic elements, such as those related to contact impedance, the host structure, or measurement fixtures. The modulated current I_m flows in the active region of the device. The coupling between the drain, source, and gate are described by three resistive elements (R_{gs}, R_{gd}, and R_{ds}) and three capacitive elements (C_{gs}, C_{gd}, and C_{ds}). In this chapter, we will use this model of the intrinsic transistor as the foundation for a specialized model of a nanowire FET.

10.2.2   Microwave Measurements of Conventional Transistors

There are long-standing methods for measurement and characterization of conventional transistors. The field of RF transistor measurements is vast, incorporating a wide variety of characterization techniques, including noise measurements, pulsed scattering parameter measurements, and load-pull techniques. Here, we focus on methods that are based on calibrated, small-signal scattering parameter measurements, as such methods will provide a foundation for our discussion of RF characterization of nanotransistors. However, a more comprehensive approach complements scattering parameter measurements with DC current and voltage measurements. While microwave measurements are closer to the central topic of this book, DC electrical measurements are widely used to characterize the intrinsic and extrinsic properties of transistors. Further, DC measurements can also be relevant to characterization of microwave performance, providing methods to predict RF output power and noise. Moreover, full modeling of transistor behavior, including nonlinearities and large signal behavior, requires accurate measurements of the DC voltage-current relationships. Note that DC measurements can also be used to establish equivalent circuit parameters such as the gate, source, and drain resistances. Detailed descriptions of DC characterization of conventional transistors are available in the literature. For example, Reference [17] provides a systematic approach to determination of the basic properties of a GaAs FET, including the effective gate length, channel thickness, and gate length, from DC measurements.

Microwave measurements of transistors and other active devices require reliable calibration and de-embedding techniques. As we noted in the early chapters of this book, there are a number of reliable calibration techniques that enable the de-embedding of complex scattering parameter measurements in guided-wave, fixtured, and on-wafer environments. As in the case of passive devices, extension of these methods to nanoelectronic systems requires careful consideration of the underlying assumptions about the device and the propagating modes, in particular. In Chapter 5, we showed through measurements, numerical simulation, and validation that the on-wafer, multiline TRL calibration was extendable to coplanar-waveguide-based nanoelectronic devices. We will take advantage of the portability of multiline TRL in an example measurement in this chapter. In general, a variety of calibration techniques are applicable to microwave and RF characterization of transistors and other active devices. These include SOLT and line-reflect-reflect-match (LRRM). Calibrated measurements of RF transistors often face an additional challenge in that the devices are often embedded in coaxial or other guided-wave fixtures that facilitate the measurement. The effects of the fixtures must be removed from the measurements. This is done by measurements where possible, and simulations as necessary.

Even after the effects of the test platform, on-wafer probes, and fixtures have been removed through calibration, the challenge of separating parasitic effects from intrinsic properties remains. Unfortunately, even relatively simple, compact circuit models often incorporate ten or more unknown circuit elements. The determination of a large number of circuit elements from a single measurement of two-port, complex scattering parameters presents a complicated, potentially insurmountable optimization problem. Broadly speaking, the optimization problem can be simplified by performing additional measurements that ideally correspond to a subset of the unknown circuit elements. For example, measurement of a simpler device that excludes the active element(s) can allow independent determination of some parasitic impedances. However, this strategy requires the fabrication of another device (in addition to any calibration devices required for the preceding de-embedding step). Another strategy is to determine parasitic elements from scattering parameters that are measured under so-called cold bias conditions in which the drain is unbiased [18]. Effectively, measurement of both "cold bias" and active, biased scattering parameters provides more inputs into the optimization process. Even with these strategies in place, determination of the unknown circuit parameters requires nontrivial numerical optimization. Where possible, estimated circuit values should be validated by alternate approaches, including complementary DC measurements and finite-element modeling.

10.3   Determination of Equivalent Circuit Parameters for a Nanotransistor

The ongoing challenge of de-embedding of intrinsic device characteristics from scattering parameters is heightened for active, RF nanoelectronic devices. As with the case for passive devices, significant measurement challenges arise in large part from the inherent impedance mismatch between commercial test equipment and many nanoelectronic devices. Furthermore, given the unique properties of nanoscale materials, the development of accurate physical and circuit models of devices remains a work in progress. Naturally, the models for active devices are more complex than the models presented earlier in this book for passive devices. Fortunately, in the case of nanotransistors, we can build upon an extensive, preexisting body of knowledge, as described earlier. In particular, we will use traditional, semiconductor FET equivalent circuits as a starting point and introduce appropriate corrections as needed for the specialized case of nanoelectronic transistors.

In order to describe the de-embedding strategy, we will follow References [19] and [20]. As an illustrative example, we will apply the technique to a metal semiconductor field effect transistor (MESFET) that incorporates an individual GaN nanowire. Figure 10.3(a) shows an illustration of a planar, nanowire MESFET with a double-finger gate incorporated into a CPW. The CPW host structure was fabricated by use of standard photolithography techniques and the individual nanowires were aligned within the device by use of dielectrophoresis [21], [22]. The photolithographic fabrication process was refined such that the drain and source contacts to the nanowire are ohmic while the gate contacts are Schottky contacts. The device wafer also included a set of calibration devices for a multiline TRL calibration [23] and several empty, nanowire-free devices. Thus, the wafer layout followed the schematic first introduced in Fig. 6.3, but with the passive, nanowire bridge devices replaced by active, nanowire MESFETs.

An equivalent circuit model of the nanowire MESFET is shown in Fig. 10.4. Note that in the example presented here, parasitic elements are determined from scattering parameters under "cold bias" conditions [18], but the model in Fig. 10.4

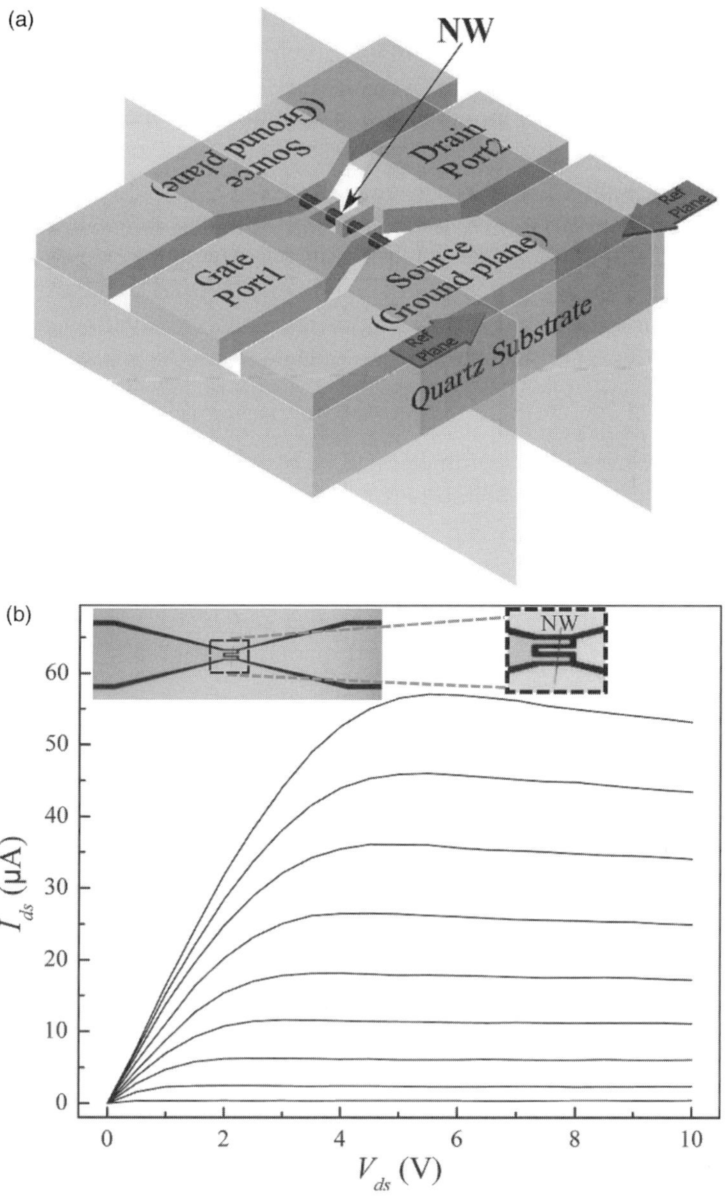

Figure 10.3. Planar, dual-gate nanowire MESFET.
(a) A schematic of a planar nanowire MESFET. The dual-gate structure, including the active GaN nanowire (NW) element, is incorporated into a CPW. (b) The DC current (I_{ds}) vs voltage (V_{ds}) characteristics of a typical GaN nanowire MESFET. From bottom to top, the curves correspond to gate biases ranging from $V_{gs} = -4$ V to $V_{gs} = 0$ V, in 0.5 V steps. The insets show top-view, optical images of the device [20]. © IOP Publishing. Reproduced with permission. All rights reserved.

corresponds to the MESFET in an active, biased state. The circuit parameters in this model can be conceptually separated into three groups, corresponding to the intrinsic transistor, the contacts, and the host structure. The first group corresponds to the intrinsic transistor, specifically the active nanowire element in the MESFET. The part of the circuit model that represents the intrinsic transistor is comparable to the more general case shown in Fig. 10.2. Here, the modulated current I_m flows in the active, nanowire element. The capacitive couplings between the drain, source, and gain (C_{gs}, C_{gd}, and C_{ds}) also remain. The gate-drain and gate-source resistances (R_{gs} and R_{gd}) are negligible in the GaN nanowire MESFET and have been removed. The drain-source resistance R_{ds} has been re-cast as the channel resistance R_c to highlight the fact that this resistance is associated with the active conduction channel within the nanowire element. The next group of parameters is related to the properties of the leads and Schottky contact. The lead and contact parasitics are represented by five circuit parameters, three of which encapsulate the inactive and dynamic properties of the Schottky gate: the gate resistance R_g, the Schottky resistance R_{sch}, and the Schottky capacitance C_{sch}. R_s and R_d are the contact resistances for the source and drain, respectively. The final group of circuit elements is associated with the host structure and includes three parasitic capacitances. C_{id} is the parasitic capacitance associated with the interdigitated fingers that provide contacts to the nanowire. C^R_{tp} and C^L_{tp} correspond to the tapered CPW segments on the right and left sides of the device, respectively. In order to determine all of the circuit parameters, our strategy will be to sequentially consider the host structure parasitics, followed by the lead and contact parasitics, and lastly the active nanowire element. Finally, we note that based upon inspection of the device symmetry, C_{gs} and C_{gd} are expected to be of the same order. Likewise, the values of C^R_{tp} and C^L_{tp} should be close.

As a first step, calibrated, on-wafer measurements are made in order to extract the complex scattering parameters of the calibration structures, nanowire transistors, and empty transistors. In the particular case of GaN nanowire MESFETs, multiline TRL was chosen as the calibration method as it enables calculation of the CPW propagation constant, and in turn, the translation of the reference planes to positions near to the device, as illustrated in Fig. 10.1. However, alternate on-wafer approaches may be used in place of TRL, with the choice depending upon the device layout as well as the availability of calibration structures, among other factors. Initially, measurements of the empty devices and MESFETs are carried out in the cold state with the gate and drain unbiased. As GaN is photoconductive, the devices were measured only in a dark state after prolonged isolation from any optical illumination. The two-port, calibrated scattering parameters of the empty device in the cold state are $S_{ij\,meas}^{empty}$, where i and j indicate the port indices ($i = 1,2$; $j = 1,2$). Similarly, the two-port, calibrated scattering parameters of the transistor device in the cold state are $S_{ij\,meas}^{cold}$. Subsequently, the MESFETs are measured in the active state of normal operation with the devices biased accordingly. The two-port, calibrated scattering parameters of the transistor device in the active state are $S_{ij\,meas}^{active}$. In the analysis below, the scattering parameters $S_{ij\,meas}^{empty}$, $S_{ij\,meas}^{cold}$, and $S_{ij\,meas}^{active}$ are

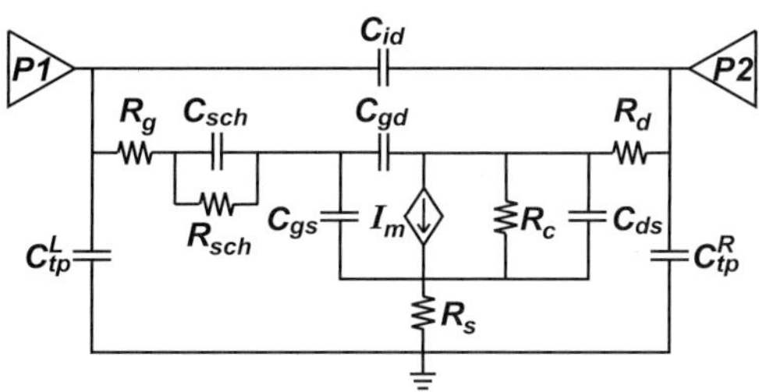

Figure 10.4.　Equivalent circuit model for a nanowire MESFET. The complete equivalent circuit model, including the host structure parasitic elements, is shown. *P1* and *P2* correspond to Ports 1 and 2, respectively. The symbols for the individual circuit elements are defined in the text [20].

transformed to admittance matrix elements ($Y_{ij\,meas}^{empty}$, $Y_{ij\,meas}^{cold}$, and $Y_{ij\,meas}^{active}$) and impedance matrix elements ($Z_{ij\,meas}^{empty}$, $Z_{ij\,meas}^{cold}$, and $Z_{ij\,meas}^{active}$), as needed.

With the calibrated, on-wafer measurements complete, the next step is the determination of the parasitic capacitances associated with the host structure. These parameters can be determined from the scattering parameters $S_{ij\,meas}^{empty}$. In Chapters 4 and 6, we introduced an analogous "empty device" method for characterization of passive, two-port devices. Here, the method is extended to include three parasitic capacitive terms: C_{id}, C_{tp}^{R}, and C_{tp}^{L}. The equivalent circuit shown Fig. 10.4 can be modified to represent an empty device by removing all circuit elements related to the active nanowire, the leads, and the contacts. This leaves the three parasitic capacitances – C_{id}, C_{tp}^{R}, and C_{tp}^{L} – configured in a pi-network. Thus, C_{id}, C_{tp}^{R}, and C_{tp}^{L} are given by [20]

$$C_{id} = -\frac{1}{2\omega}\left[\Im\left(Y_{12\,meas}^{empty}\right) + \Im\left(Y_{21\,meas}^{empty}\right) \right], \tag{10.1a}$$

$$C_{tp}^{R} = \frac{1}{\omega}\Im\left(Y_{22\,meas}^{empty}\right) - C_{id}, \tag{10.1b}$$

$$C_{tp}^{L} = \frac{1}{\omega}\Im\left(Y_{11\,meas}^{empty}\right) - C_{id}, \tag{10.1c}$$

where ω is the radial frequency. We use the notations $\Re(X)$ and $\Im(X)$ to represent the real and imaginary parts of a complex variable X, respectively. The pi-model of the empty device can then be expressed as an admittance matrix Y_{model}^{empty} in the form

$$Y_{model}^{empty} = \begin{bmatrix} j\omega\left(C_{tp}^{L} + C_{id}\right) & -j\omega C_{id} \\ -j\omega C_{id} & j\omega\left(C_{tp}^{R} + C_{id}\right) \end{bmatrix}. \tag{10.2}$$

The calibrated admittance parameters of the transistor $Y_{ij\,meas}^{cold}$ and $Y_{ij\,meas}^{active}$ can now be corrected to remove the host structure parasitic elements by

$$Y_{meas}^{cold-corr} = Y_{meas}^{cold} - Y_{model}^{empty} \qquad (10.3a)$$

and

$$Y_{meas}^{act-corr} = Y_{meas}^{act} - Y_{model}^{empty}. \qquad (10.3b)$$

The simple algebraic form of Equations (10.3a) and (10.3b) relies on the assumption that the network of host structure parasitic elements is in parallel with the rest of the device. Note that this assumption may not be valid for all device configurations and ideally should be verified by both measurements and numerical modeling.

Having corrected the calibrated measurements by removing the parasitic effects of the host structure, we turn now to the isolation of the circuit elements associated with the active nanowire element. This requires removing the parasitic effects of the leads and contacts. Earlier, the measurement of the empty transistor device enabled the isolation of the host structure's parasitic circuit elements from the rest of the circuit model. Unfortunately, there is no analogous reference device or measurement strategy for isolating the parasitic effects of the leads and contacts.

As a first step, consider the group of circuit elements associated with the leads and contacts. Since these elements form a tee-network, the equivalent circuit can be naturally expressed as an impedance matrix:

$$\mathbf{Z}_{model}^{lead} = \begin{bmatrix} R_g + R_S + \dfrac{R_{sch}}{1 + j\omega R_{sch} C_{sch}} & R_S \\[2ex] R_S & R_d + R_S \end{bmatrix}. \qquad (10.4)$$

Next, consider the group of circuit elements associated with the internal nanowire element. Since this group of elements forms a pi-network, an admittance matrix representation is most suitable:

$$\mathbf{Y}_{model}^{nw} = \begin{bmatrix} j\omega\left(C_{gs} + C_{gd}\right) & -j\omega C_{gd} \\[2ex] -j\omega C_{gd} & \dfrac{1}{R_{ds}} + j\omega\left(C_{ds} + C_{gd}\right) \end{bmatrix}. \qquad (10.5)$$

It is vital to note that while the contributions of C_{gs}, C_{gd}, and C_{ds} are often negligible for thin-film-based MESFETs, in the case of the nanowire MESFET example, they can't be neglected. The admittance matrices $\mathbf{Z}_{model}^{lead}$ and $\mathbf{Z}_{model}^{nw}$ can be combined to model the combination of the leads and the nanowire element: $\mathbf{Z}_{model}^{cold-corr}$ ($\mathbf{Z}_{model}^{nw}$ is the impedance matrix that corresponds to $\mathbf{Y}_{model}^{nw}$). Initial estimates of R_g, R_s, and R_d are determined from the high-frequency behavior of the corrected cold device measurements $\mathbf{Z}_{meas}^{cold-corr}$ [20]:

$$R_S \approx \dfrac{\Re\left(Z_{12\,meas}^{cold-corr}\right) + \Re\left(Z_{21\,meas}^{cold-corr}\right)}{2}, \qquad (10.6a)$$

$$R_g + R_s \approx \Re\left(Z_{11meas}^{cold-corr}\right), \tag{10.6b}$$

$$R_d + R_s \approx \Re\left(Z_{22meas}^{cold-corr}\right). \tag{10.6c}$$

In a similar fashion, initial estimates of R_{sch}, R_c, C_{gd}, C_{gs}, and C_{gd} are determined from the low-frequency behavior of $\mathbf{Z}_{meas}^{cold-corr}$ [20]

$$R_g + R_s + R_{sch} + \left(\frac{C_{gd}}{C_{gd} + C_{gs}}\right)^2 R_c \approx \Re\left(Z_{11meas}^{cold-corr}\right), \tag{10.7a}$$

$$R_s + \frac{C_{gd}}{C_{gd} + C_{gs}} R_c \approx \frac{1}{2}\left(\Re\left(Z_{12meas}^{cold-corr}\right) + \Re\left(Z_{21meas}^{cold-corr}\right)\right), \tag{10.7b}$$

$$R_g + R_s + R_c \approx \Re\left(Z_{22meas}^{cold-corr}\right), \tag{10.7c}$$

in conjunction with an initial estimation that all of the capacitive elements are 1 fF. These initial estimates are subsequently used as seed values for a numerical fitting procedure that optimizes all of the parameters such that the differences between the model $\mathbf{Z}_{model}^{cold-corr}$ and the experimental data $\mathbf{Z}_{meas}^{cold-corr}$ are minimized. The numerical optimization routine interfaces with a commercial circuit simulator that calculates the scattering parameters from the circuit model. The resulting values of the parasitic circuit elements R_g, R_s, R_d, R_{sch}, and C_{sch} are now fixed and used to calculate the $\mathbf{Z}_{model}^{lead}$. However, the estimates of the nanowire circuit elements R_c, C_{gd}, C_{gs}, and C_{gd} are discarded.

The final values of R_c, C_{gd}, C_{gs}, and C_{gd} are determined from the active state measurements. The admittance matrix of the active nanowire element can be de-embedded from the original measurements by

$$Y_{meas}^{NW} = Y_{meas}^{act-corr} - Y_{model}^{lead} = Y_{meas}^{act} - Y_{model}^{empty} - Y_{model}^{lead}. \tag{10.8}$$

Comparison of Y_{meas}^{NW} to Equation (10.5) leads to simple expressions for the unknown nanowire parameters:

$$C_{gd} = \frac{-\Im\left(Y_{12meas}^{NW}\right)}{\omega}, \tag{10.9a}$$

$$C_{gs} = \frac{\Im\left(Y_{12meas}^{NW}\right) + \Im\left(Y_{11meas}^{NW}\right)}{\omega}, \tag{10.9b}$$

$$C_{ds} = \frac{\Im\left(Y_{12meas}^{NW}\right) + \Im\left(Y_{22meas}^{NW}\right)}{\omega}, \tag{10.9c}$$

and

$$R_c = \frac{1}{\Re\left(Y_{22\,meas}^{NW}\right)}. \tag{10.9d}$$

Typical equivalent circuit values for a GaN nanowire device such as the one shown in Fig. 10.1 are [19]: $C_{tp}^{L} = 11.6$ fF; $C_{tp}^{R} = 11.3$ fF; $C_{id} = 1.5$ fF; $C_{sch} = 3.5$ fF; $R_{sch} = 16000\ \Omega$; $C_{ds} = 0.0$ fF; $C_{gs} = 5.5$ fF; $C_{gd} = 2.2$ fF; $R_{g} = 3680\ \Omega$; $R_{d} = 108\ \Omega$; and $R_{s} = 270\ \Omega$.

Two additional transistor metrics can be directly obtained from Y_{meas}^{NW}: the transconductance g_m and the transit time τ. The transconductance is defined by

$$g_m \equiv \left.\frac{\delta I_{ds}}{\delta V_{gs}}\right|_{V_{ds}=constant}. \tag{10.10}$$

Thus, larger values of transconductance correspond to an increased sensitivity of the source-drain current I_{ds} to the gate-source voltage V_{ds}. The transconductance is related to the admittance parameters of the active nanowire element through [20]

$$g_m = \left\{ \Re\left(Y_{21\,meas}^{NW}\right)^2 + \left[\Im\left(Y_{21\,meas}^{NW}\right) - \Im\left(Y_{12\,meas}^{NW}\right)\right]^2 \right\}^{1/2}. \tag{10.11}$$

The modulated current is related to the transconductance as follows:

$$I_m = g_m V_i e^{-j\omega\tau}, \tag{10.12}$$

where V_i is the voltage across C_{gs}, ω is the operating frequency, and τ is the transit time. The transit time quantifies the inherent response delay of an FET. It is the governing factor in determination of the cutoff frequency and can be determined from [20]

$$\tau = \frac{\cos^{-1}\left[\Re\left(Y_{21\,meas}^{NW}\right)/g_m\right]}{\omega}. \tag{10.13}$$

Typical values for a GaN nanowire MESFET are $g_m = 16.6\ \mu S$ and $\tau = 9.8$ ps with a drain-source voltage V_{ds} of 5 V [20].

It is important to notice that we have made a broad, simplifying assumption that all of the equivalent circuit parameters are independent of bias voltage. While this may be a reasonable assumption for some of the parameters, such as the host structure parasitic capacitances, it is almost certainly untrue for the elements associated with the active nanowire element. A more complete characterization would necessarily require calibrated measurements of $S_{ij\,meas}^{active}$ under varying bias conditions and subsequent re-calculation of the voltage-dependent circuit elements for each bias condition.

The uncertainties in the calibrated, small-signal scattering parameters of the nanotransistor can be calculated following the analysis presented in Chapter 6. The uncertainty of the admittance parameters is calculated from

$$\Delta Y_{ij\,meas}^{cold-corr} = \sqrt{\left(\Delta Y_{ij\,meas}^{empty}\right)^2 + \left(\Delta Y_{ij\,meas}^{active}\right)^2 + \left(\Delta Y_{ij}^{M-M}\right)^2}, \qquad (10.14)$$

where $\Delta Y_{ij\,meas}^{empty}$ and $\Delta Y_{ij\,meas}^{active}$ are the uncertainties in the admittance matrices of the empty device and the nanowire device in the active state, respectively. Uncertainties in the admittance matrices Y_{ij} are determined from the uncertainties in the measured scattering parameters S_{kl} through

$$\Delta Y_{ij} = \sqrt{\sum_{k,l}\left(\frac{\delta Y_{ij}}{\delta S_{kl}} S_{kl}\right)^2}. \qquad (10.15)$$

ΔY_{ij}^{M-M} is the difference between the modeled and measured admittances for the empty device. Statistical (Type A) uncertainties are neglected in Equation (10.14) as they are significantly smaller than the systematic errors. In general, additional contributions to the uncertainty may arise from the device fabrication process, which may lead to variability in material properties and feature geometry from device to device. Here, these contributions were found to be negligible compared to the uncertainty terms included in Equation (10.14), which is not surprising given that all of the nanotransistors, empty devices, and calibration structures were fabricated on the same wafer. Measured and modeled scattering parameters for the MESFET device are shown in Fig. 10.5, along with calculated uncertainties. As we saw with a passive nanowire device in Chapter 6, the magnitude of the uncertainties approaches the magnitude of the measured transmission. For example, the values of $|S_{21}|$ and $|S_{12}|$ on a linear scale at 4 GHz, are 0.0017 (+0.0011, −0.0007) and 0.0011 (+0.0003, −0.0009). While this prototype device is suitable for demonstration of measurement methods, the low signal values preclude most practical applications. Increased transmission in nanowire- and nanotube-based FETs can be facilitated by use of multiple nanowires in parallel [24] or aligned arrays of nanotubes [25].

In summary, several broader lessons and strategies are revealed by the case study of the GaN nanowire MESFET. First, as was the case for passive nanoelectronic devices, calibration and de-embedding play a critical role in the characterization of active nanoelectronic devices. This is true both for on-wafer devices as well as devices embedded in fixtures. Calibration and de-embedding enable us to distinguish intrinsic behavior from the overall device response. Moreover, the accuracy and reproducibility of the calibrated measurements will depend not only on the chosen method, but also on practical issues such as the skill of the user, the quality of the test platform, and the stability of the measurement environment. Another lesson from this case study is that the choice of equivalent circuit model directs certain aspects of the measurement strategy. Here, we saw that the equivalent circuit model could be modularized into three groups of circuit elements. As a result, the measurement process sequentially addressed each circuit element group. We also observed that specialized calibration structures can play an important role in determining parasitic elements. Here, the empty device served as a calibration structure and enabled the determination of several parasitic capacitances, thus reducing the number of unknown elements in the circuit model. Though we did not take

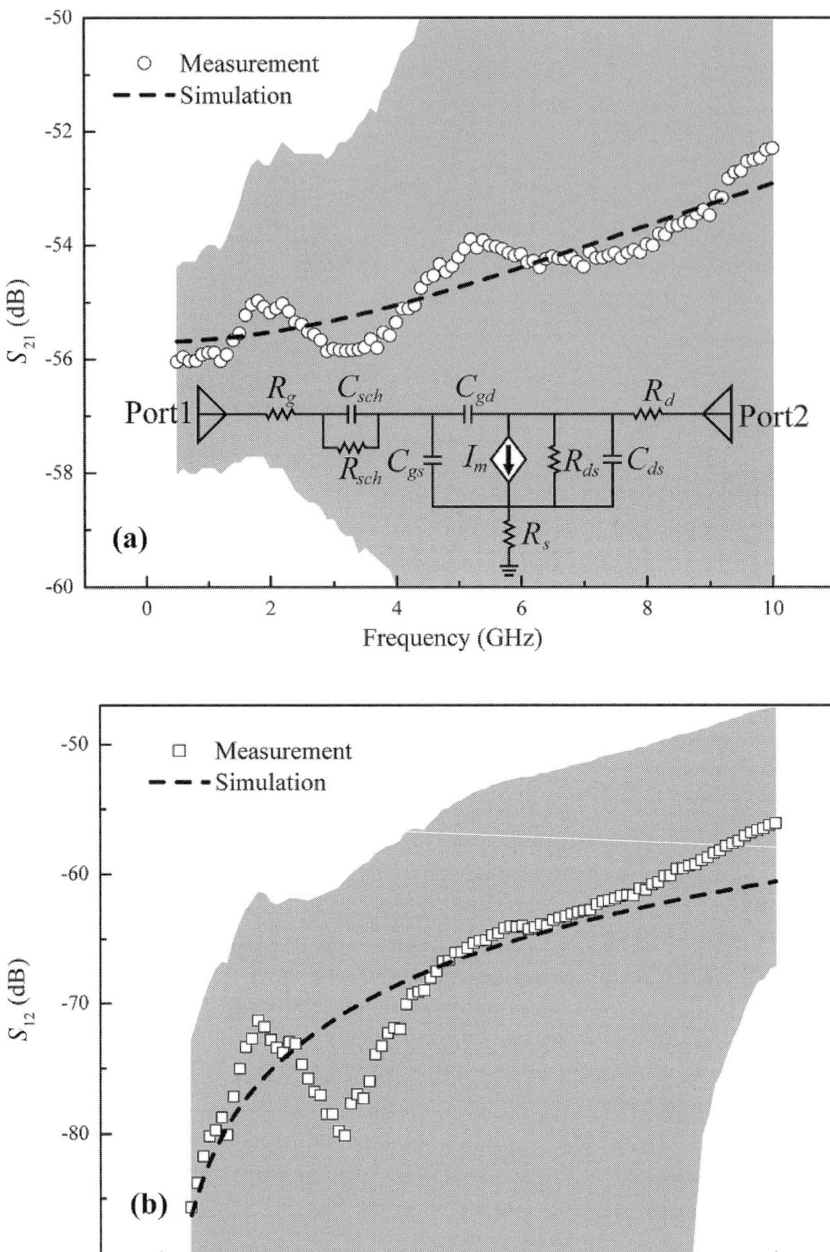

Figure 10.5. Scattering parameters of a GaN nanowire MESFET.
The measured (open shapes) and simulated (dashed lines) scattering parameters of the MESFET are shown for (a) forward transmission, S_{21} and (b) reverse transmission, S_{12}. The simulation is based on the equivalent circuit model shown as an inset in (a). Estimated uncertainties are shown in gray. Note that the scale has been adjusted to allow a detailed comparison on the simulated and measured values. As a result, the shaded uncertainty bands extend past the edge of the graphs [20].

advantage of DC measurements in this particular example, DC measurements can also play an important role in determining circuit models as well as large-signal and nonlinear effects. Finally, we saw that numerical simulations and optimization are often necessary for full characterization and validation of active device measurements.

References

[1] M. Schroter, M. Claus, P. Sakalas, M. Haferlach, and D. Wang, "Carbon Nanotube FET Technology for Radio-Frequency Electronics: State-of-the-Art Overview," *IEEE Journal of the Electron Devices Society* 1 (2013) pp. 9–20.

[2] P. J. Burke, "AC Performance of Nanoelectronics: Towards a Ballistic THz Nanotube Transistor," *Solid State Electronics* 48 (2004) pp. 1981–1986.

[3] T. Durkop, S. A. Getty, E. Cobas, and M. S. Fuhrer, "Extraordinary Mobility in Semiconducting Carbon Nanotubes," *Nano Letters* 4 (2004) pp. 35–39.

[4] A. Le Louarn, F. Kapche, J.-M. Bethoux, H. Happy, G. Dambrine, V. Derycke, P. Chenevier, N. Izard, M. F. Goffman, and J.-P. Bourgoin, "Intrinsic Current Gain Cutoff Frequency of 30 GHz with Carbon Nanotube Transistors," *Applied Physics Letters* 90 (2007) art. no. 233108.

[5] L. Nougaret, H. Happy, G. Dambrine, V. Derycke, J. P. Bourgoin, A. A. Green, and M. C. Hersam, "80 GHz Field-Effect Transistors Produced Using High Purity Semiconducting Single-Walled Carbon Nanotubes," *Applied Physics Letters* 94 (2009) art. no. 243505.

[6] M. Schroter, P. Lolev, D. Wang, S. Lin, N. Samarakone, M. Bronikowksi, Z. Yu, P. Sampat, P. Syams, and S. McKernan, "A 4" Wafer Photostepper-Based Carbon Nanotube FET Technology for RF Applications," *2011 IEEE MMT-S International Microwave Symposium Digest (MTT)* (2011) pp. 1–4.

[7] C. Rutherglen and P. Burke, "Nanoelectromagnetics: Circuit and Electromagnetic Properties of Carbon Nanotubes," *Small* 5 (2009) pp. 884–906.

[8] K. Novoselov, A. Geim, S. Morozov, D. Jiang, Y. Zhang, S. Dubonos, I. Grigorieva, and A. Firsov, "Electric Field Effect in Atomically Thin Carbon Films," *Science* 306 (2004) pp. 666–669.

[9] Y. Zhang, Y.-W. Tan, H. L. Stormer, and P. Kim, "Experimental Observation of the Quantum Hall Effect and Berry's Phase in Graphene," *Nature* 438 (2005) pp. 201–204.

[10] Y.-M. Lin, C. Dimitrakopoulos, K. A. Jenkins, D. B. Farmer, H. Y. Chiu, A. Grill, and P. Avouris, "100-GHz Transistors from Wafer-Scale Epitaxial Graphene," *Science* 327 (2010) p. 662.

[11] J. S. Moon, D. Curtis, D. Zehnder, S. Kim, D. K. Gaskill, G. G. Jernigan, R. L. Myers-Ward, C. R. Eddy, P. M. Campbell, K.-M. Lee, and P. Asbeck, "Low-Phase Noise Graphene FETs in Ambipolar RF Applications," *IEEE Electron Device Letters* 32 (2011) pp. 270–272.

[12] L. Britnell, R. V. Gorbachev, R. Jalil, B. D. Belle, F. Schedin, M. I. Katsnelson, L. Eaves, S. G. Morozov, N. M. R. Peres, J. Leist, A. K. Geim, K. S. Novoselov, and L. A. Ponomarenko, "Field-Effect Tunneling Transistor Based on Vertical Graphene Heterostructures," *Science* 335 (2012) pp. 947–950.

[13] J. S. Moon, H.-C. Seo, F. Stratan, M. Antcliffe, A. Schmitz, R. S. Ross, A. A. Kiselev, V. D. Wheeler, L. O. Nyakiti, D. K. Gaskill, K.-M. Lee, and P. M. Asbeck, "Lateral Graphene Heterostructure Field-Effect Transistor," *IEEE Electron Device Letters* 34 (2013) pp. 1190–1192.

[14] Q. Hua Wang, K. Kalantar-Zadeh, A. Kis, J. N. Coleman, and M. S. Strano, "Electronics and Optoelectronics of Two-Dimensional Transition Metal Dichalcogenides," *Nature Nanotechnology* 7 (2012) pp. 699–712.

[15] B. Radisavljevic, A. Radenovic, J. Brivio, V. Giacometti, and A. Kis, "Single Layer MoS_2 Transistors," *Nature Nanotechnology* 6 (2011) pp. 147–150.

[16] P. Aaen, J. A. Pla, and J. Wood, *Modeling and Characterization of RF and Microwave Power FETs* (Cambridge University Press, 2007).

[17] H. Fukai, "Determination of the Basic Device Parameters of a GaAs MESFET," *Bell System Technical Journal* 58 (1979) pp. 771–797.

[18] W. R. Curtice and R. L. Camisa, "Self-Consistent GaAs FET Models for Amplifier Design and Device Diagnostics," *IEEE Transactions on Microwave Theory and Techniques* 32 (1984) pp. 1573–1578.

[19] D. Gu, T. M. Wallis, P. Blanchard, S.-H. Lim, A. Imtiaz, K. A. Bertness, N. A. Sanford, and P. Kabos, "De-embedding Parasitic Elements of GaN Nanowire Metal Semiconductor Field Effect Transistors by Use of Microwave Measurements," *Applied Physics Letters* 98 (2011) art. no. 223109.

[20] D. Gu, T. M. Wallis, P. Kabos, P. Blanchard, K. A. Bertness, and N. A. Sanford, "Microwave Measurements and Systematic Circuit-Model Extraction of Nanowire Metal Semiconductor Field-Effect Transistors," *Measurement Science and Technology* 23 (2012) art. no. 105602.

[21] P. T. Blanchard, K. A. Bertness, T. E. Harvey, L. M. Mansfield, A. W. Sanders, and N. A. Sanford, "MESFETs Made from Individual Nanowires," *IEEE Transactions on Nanotechnology* 7 (2008) pp. 760–765.

[22] P. A. Smith, C. D. Norquist, T. N. Jackson, T. S. Mayer, B. R. Martin, J. Mbindyo, and T. E. Mallouk, "Electric-Field Assisted Assembly and Alignment of Metallic Nanowires," *Applied Physics Letters* 77 (2000) pp. 1399–1401.

[23] R. B. Marks, "A Multiline Method of Network Analyzer Calibration," *IEEE Transactions on Microwave Theory and Techniques* 39 (1991) pp. 1205–1215.

[24] S. W. Hong, T. Banks, and J. A. Rogers, "Improved Density in Aligned Arrays of Single-Walled Carbon Nanotubes by Sequential Chemical Vapor Deposition on Quartz," *Advanced Materials* 22 (2010) pp. 1826–1830.

[25] G. J. Brady, Y. Joo, S. S. Roy, P. Gopalan, and M. S. Arnold, "High Performance Transistors via Aligned Polyfluorene-Sorted Carbon Nanotubes," *Applied Physics Letters* 105 (2014) art. no. 083107.

11 Dopant Profiling in Semiconductor Nanoelectronics

11.1 Introduction

As nanoelectronic device dimensions are scaled down to atomic sizes, device performance becomes more and more sensitive to the exact arrangement of atoms, including individual dopants and defects, within the device. Thus, there is ongoing demand for spatially resolved measurements of dopant concentration. Due to the predominance of silicon-based micro- and nanoelectronics, the need for dopant profiling in semiconductors is particularly acute. This need has been underscored by the inclusion of dopant profiling in the *International Technology Roadmap for Semiconductors*: "Materials characterization and metrology methods are needed for control of interfacial layers, dopant positions, defects, and atomic concentrations relative to device dimensions. One example is three-dimensional dopant profiling" [1]. The ongoing development of alternative nanoelectronic devices based on emerging low-dimensional materials such as CNTs and graphene also will benefit from enhanced capabilities to identify and characterize dopants. Ideally, dopant profiling tools are nondestructive, exhibit nanometer-scale or better spatial resolution, and are sensitive to both surface and subsurface features.

A variety of scanning-probe-based, microwave techniques are excellent candidates for dopant profiling. The NSMM, which we have described in detail throughout this book, is one such technique. An NSMM's combined capabilities to perform nondestructive, contact-free electrical measurements and to measure subsurface defects make it particularly attractive for dopant profiling. Other instruments, such as the scanning capacitance microscope (SCM), the scanning kelvin probe microscope (SKPM), and the scanning spreading-resistance microscope (SSRM) are also useful for characterizing semiconductors. Taken together, this suite of complementary scanning-probe techniques is capable of local measurement of work functions, surface barrier heights, dopant profiles, and resistance. To date, most dopant profiles obtained with scanning probe microscopes have been two-dimensional techniques, measuring only at or near the sample surface. In order to obtain the depth-dependent dopant concentration with such a technique, special sample preparation is required. For example, a cross-sectional sample of a device may be prepared via focused ion beam milling. However, the subsurface sensitivity of near-field scanning microwave microscopy makes the NSMM a strong candidate to provide quantitative, three-dimensional, subsurface resolution, as will be discussed in more detail in Chapter 12.

In Chapter 9, we reviewed a number of different, application-specific models of the tip-sample system, highlighting how such models are required for extraction of characteristic material parameters such as conductivity and complex permittivity. Thus, our discussion of dopant profiling begins with an overview of tip-sample models for semiconductor samples, including a brief review of some basic aspects of semiconductor physics. From there, we will detail approaches for performing quantitative dopant profiling with two RF scanning probe systems: scanning capacitance microscopy and near-field scanning microwave microscopy. These approaches require the development of calibration techniques and the selection of suitable reference samples. The chapter will conclude with a review of other scanning probe techniques that are used for spatially resolved characterization of semiconductors. As these additional techniques are not RF- or microwave-based measurements, our discussion of them will be limited.

11.2   Tip-Sample Models for Semiconductor Samples

11.2.1   Capacitive Models

Consider a contact-mode, scanning probe measurement of a semiconductor sample. Often, we will assume that the semiconductor is coated by a thin dielectric film. This film may represent a native oxide layer or a thin film that is deliberately deposited or grown on the sample. When the metal probe tip is in contact with the sample, the tip-sample system forms an MOS junction. Given the abundant knowledge about the physics of MOS systems, this representation suggests a natural approach to dopant profiling with scanning probe systems. However, the complexity of MOS physics and the deviations of real-world systems from ideal models require careful consideration. Throughout the remainder of this chapter, we develop dopant profiling techniques that assume idealized MOS structures. Though this approach has been successfully and widely applied to silicon, extension to other semiconductors will require a more complex physical description. For example, in the case of compound semiconductors such as GaN and AlGaN, interface states, trapped charges, and Fermi level pinning must be taken into account [2].

In order to perform dopant profiling, we will naturally focus on tip-sample models applicable to the case of semiconducting samples. We begin with a highly simplified picture of the interaction, approximating the system as a parallel plate capacitor. Then, we will present a metal-semiconductor model and a determination of the depletion layer capacitance. From there, we proceed to a more detailed description of the tip-sample junction in terms of an MOS capacitor junction. The simplest representation of a metal scanned probe interacting with a dielectric-covered, semiconducting sample is a parallel plate model, in which the two plates represent the probe tip and the semiconductor substrate, respectively. The parallel plate capacitance is

$$C = \frac{\varepsilon_0 \varepsilon_r A}{d},\tag{11.1}$$

where ε_0 is the permittivity of free space, ε_r is the relative permittivity of the intermediate dielectric, A is the area of the capacitor plates, and d is the separation between the plates. This simple model assumes uniform permittivity and uniform thickness of the oxide. In addition, this model ignores the tip shape as well as the semiconducting nature of the substrate (unless the semiconductor is degenerately doped). Note that a complete, lumped-element tip-sample model likely will require inclusion of the stray capacitance, tip resistance, and semiconductor sheet resistance, in addition to the parallel plate capacitance.

A natural extension of the parallel plate model is to include a reasonable approximation of the tip shape. Consider a probe tip in the shape of a truncated cone of aperture angle θ that is terminated by a spherical surface of radius R, as illustrated in Fig. 11.1. If the tip is positioned a distance z above the oxide surface, then the capacitance is [3]:

$$C = 2\pi\varepsilon_0 R\, ln\left[1 + \frac{R(1 - \sin\theta)}{z + d/\varepsilon_r}\right].\tag{11.2}$$

Once again, oxide uniformity is assumed. Given the underlying assumptions in Equations (11.1) and (11.2), we may consider one or more of the geometric parameters to represent effective electromagnetic dimensions as opposed to mechanical dimensions. For example, in Reference [4], the effective electromagnetic probe radius of an NSMM was extracted by use of Equation (11.2).

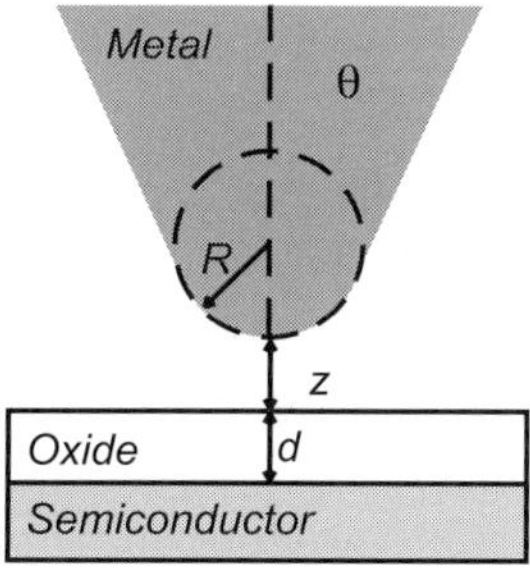

Figure 11.1. Metal-oxide-semiconductor junction formed by a probe tip and an oxidized semiconductor. The metal probe tip of a scanning probe microscope is characterized by geometric parameters: the effective tip radius R and the effective cone angle θ. A native or deposited oxide film of thickness d sits on the semiconductor SUT. In some systems, such as those based on contact-mode AFMs, the tip height z is zero and the metal probe is in direct mechanical contact with the oxide layer. Adapted from G. Gomila, J. Toset, and L. Fumagalli, *Journal of Applied Physics* 104 (2008) art. no. 024315, with permission from AIP Publishing.

11.2.2   Metal-Semiconductor Models

While elementary capacitive models provide an initial estimate of material parameters, advanced tip-sample models must accurately reflect the semiconducting nature of the sample. In this chapter, we will borrow heavily from established semiconductor physics models and related analyses [5]. In most cases, we will cite established results and provide only cursory discussion. Given the complexity of metal-semiconductor heterostructures, these results necessarily require a number of simplifying assumptions. For instance, many models assume idealized interfaces. Most models have been developed for multilayer, thin film geometries and thus do not account for the specialized tip shape found in scanning probe microscopes. As a result, one must proceed with care and attention to detail when selecting an appropriate tip-sample model. Ideally, any model will be complemented and validated by numerical simulations. The advanced models presented here are best thought of as a conceptual starting point that likely will require modification in order to suit specific applications.

Before discussing an ideal MOS system, we consider a metal-semiconductor system, such as a metal tip in a scanning probe microscope that is placed in direct mechanical contact with a semiconductor sample. Such a model would apply, for example, to an NSMM probe in contact with a clean silicon sample that is held in a vacuum environment to prevent the formation of a native oxide layer on the surface. As these experimental conditions are less common than ambient conditions, metal-semiconductor junctions are less common in scanning-probe-systems than MOS-like junctions. However, we briefly discuss metal-semiconductor systems here, introducing important concepts such as the depletion layer. At thermal equilibrium, charge in the vicinity of the interface is redistributed such that the Fermi levels of the two materials are coincident. As a result, a depletion layer will develop in the semiconductor at the material interface. The depletion layer is characterized by a depletion width W that is given by [5]

$$W = \sqrt{\frac{2\varepsilon_r \varepsilon_0}{q N_D}\left(V_{bi} - V - \frac{k_B T}{q}\right)}, \tag{11.3}$$

where ε_r is the relative permittivity of the semiconductor, q is the unit charge, N_D is the dopant density, V_{bi} is the built-in potential, V is the tip bias voltage, k_B is the Boltzmann constant, and T is the temperature. The depletion capacitance per unit area associated with the depletion layer in a metal-semiconductor system is given by [5]

$$C_{dep} = \sqrt{\frac{q\varepsilon_r \varepsilon_0 N_D}{2\left(V_{bi} - V - \frac{k_B T}{q}\right)}}. \tag{11.4}$$

Once again, inclusion of lumped elements in addition to C_{dep}, such as the stray capacitance, tip resistance, and semiconductor sheet resistance, is necessary to complete the full tip-sample model. The form of Equation (11.4) suggests that a measurement of the voltage-dependence of the depletion capacitance will allow determination of the dopant density, provided that material parameters (ε_r and V_{bi}) are known or measurable. Indeed, as we will see in the following text, local capacitance versus voltage (C-V) measurements (and differential dC/dV versus voltage measurements) are central to dopant profiling with NSMM and related techniques.

Experimentally, C-V tests require that a small AC voltage V_{ac} is added to the swept DC bias V. In the context of a scanning probe microscope where a DC bias is applied to the probe tip, V_{ac} can be added to the tip bias as a modulation signal. For dopant profiling applications, the amplitude of V_{ac} is typically on the order of millivolts while the tip bias is typically on the order of volts. In the presence of V_{ac}, an AC current will flow between the electrodes of the capacitor, leading to a net change in the charge Q at the capacitor's electrodes. The AC current can be integrated over time to find Q and, in turn, the capacitance can be determined from the definition: $C = dQ/dV$. As we will see in this chapter, in certain cases the sample capacitance may be found by measurement of the relative change in the reflection coefficient in an NSMM. In such cases, modulation of the tip bias is still useful, as it enables dC/dV (and higher order derivatives) to be determined by use of a lock-in technique.

11.2.3 Metal-Oxide-Semiconductor Model

Metal-oxide-semiconductor or, more generally, metal-insulator-semiconductor (MIS) systems, are more likely to be encountered in scanning probe measurements of semiconductor systems. This is due largely to the utility and prevalence of silicon-based devices and the convenience of characterization under ambient conditions. Well-established MOS diode theory provides a foundation for modeling the probe tip-oxide-semiconductor system [5]–[7]. As in the metal-semiconductor system, the formation of a depletion layer in the semiconductor is characterized by a voltage-dependent depletion capacitance. In an MOS system, there is an additional contribution from the oxide capacitance. Coupling between the tip and trapped charges, as well as quantum capacitive effects, also influence the total tip-sample capacitance observed in the junction of an RF scanning probe microscope.

The curve can be conceptually divided into three modes of operation: accumulation, depletion, and inversion. The accumulation regime for an n-type device corresponds to large, positive voltages applied to the metal contact (or probe). In this range of voltages, negatively charged electrons (the majority carriers) in the semiconductor will accumulate at the semiconductor-oxide interface. For devices with sufficiently thick oxide layers, the C-V curve will fully flatten at sufficiently large positive voltages. In accumulation, the measured capacitance is dominated by the oxide capacitance. An example of a C-V curve measured by use of an NSMM is shown in Fig. 11.2 for an n-type silicon sample [8]. The accumulation mode corresponds to the far-right side of the figure. As the bias voltage is decreased toward and past zero volts, the electrons are displaced from the semiconductor-oxide interface,

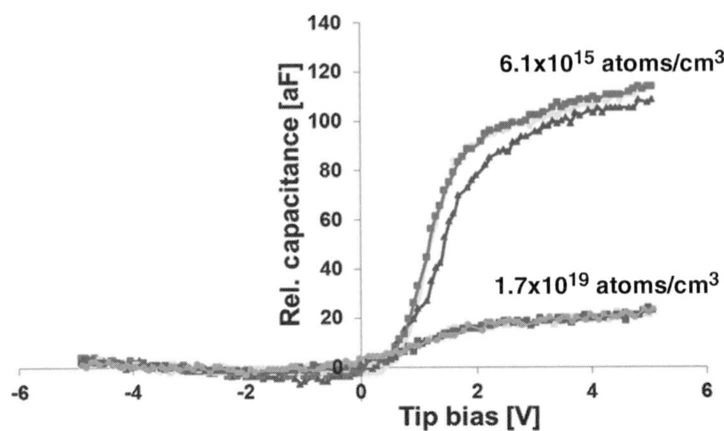

Figure 11.2.   Capacitance-voltage curve for n-type silicon.
The relative capacitance of an n-type MOS device is shown as a function of the bias voltage on the metal contact. The measurements were made by use of an NSMM and calibrated by use of the techniques described in the text. The tip-sample geometry is as shown in Fig. 11.1 with $z = 0$ (contact mode). Measurements were made with the probe tip positioned over two areas of known dopant concentration, as indicated in the figure. Multiple measurements are shown for each probe position. Reprinted from H. P. Huber et al., *Journal of Applied Physics* 111 (2012), art. no. 014301, with permission from AIP Publishing.

creating a depletion region near the surface of the semiconductor that effectively acts as an insulating layer. Thus, when an MOS device is operated in the depletion mode, the total measured capacitance consists of the oxide capacitance in series with the depletion capacitance. Note that in depletion, the depletion width and depletion capacitance are once again voltage-dependent, as was the case in metal semiconductor junctions. Finally, for sufficiently large negative voltages, electron-hole pairs will be generated in the semiconductor and move toward the metal contact, accumulating near the semiconductor-oxide interface. Thus an inversion layer is formed near the surface of the semiconductor within which the holes are the dominant carrier type. At extreme voltages, the C-V curve once again flattens out and the measured capacitance is the series combination of the oxide capacitance and the maximum depletion capacitance. For a p-type device, similar modes of operation exist, albeit with reversed voltage polarity.

A number of characteristic geometric and material parameters may be extracted from a C-V measurement. For example, if a device is operated in accumulation mode, the oxide thickness can be obtained in a straightforward way. In accumulation, the measured capacitance is the oxide capacitance and may be modeled by a simple expression such as Equation (11.1). Provided that the relative permittivity of the oxide and the metal contact area are known, the oxide thickness is easily determined by use of Equation (11.1). In a scanning probe microscope, topographic images can often provide values for the contact area and thickness. In that case, the relative permittivity of the oxide layer can be determined from the C-V measurement. Equation (11.2) can provide a more precise representation of the tip geometry, but there are additional unknowns (the effective tip radius and cone angle) that can't be determined from topographic images. In this case, C-V measurements performed on a series of oxide steps of different thicknesses allows us to determine the permittivity in addition to the geometric parameters of the tip [4]. While electromagnetic material parameters determined by use of macroscopic C-V measurements represent average quantities in MOS systems, the corresponding values extracted from C-V measurements made with a scanning probe microscope will be local with a lateral spatial resolution on the order of the probe dimensions. In general, appropriate analysis of the measured C-V curve for an MOS capacitor enables estimation of several material and device parameters, including the flatband voltage, the threshold voltage, and the difference between the metal and semiconductor work functions.

Here, the primary goal is to determine the semiconductor dopant concentration. As in the metal-semiconductor model, this objective requires calculation of the semiconductor depletion layer capacitance. In the standard approach [5]–[7], C_{dep} is given by the derivative of the surface charge density Q_S with respect to the surface potential Ψ_S. Initially, the potential in the semiconductor Ψ is introduced and defined with respect to the intrinsic Fermi level within the semiconductor. The value of Ψ at the surface is Ψ_S. If the space charge density varies only in one direction, as in the center of a parallel-plate-like geometry, then the potential Ψ may be determined from a one-dimensional Poisson equation of the form

$$\frac{d^2\Psi}{dz^2} = -\frac{\rho(z)}{\varepsilon_r\varepsilon_0}, \tag{11.5}$$

where $\rho(z)$ is the one-dimensional space charge density distribution and z is normal to the semiconductor-oxide interface. Equation (11.5) can be solved analytically, though the existence of solutions depends on the form of $\rho(z)$. For example, in Reference [5], this approach is used to find the depletion capacitance for an ideal MOS structure under flat-band conditions ($\Psi_S = 0$):

$$C_{dep} = \sqrt{\frac{\varepsilon_r\varepsilon_0 p_o q^2}{k_B T}}, \tag{11.6}$$

where p_o is the equilibrium density of holes in the bulk of the semiconductor. Note that the total capacitance is the series combination of the depletion capacitance and the oxide capacitance. In the context of a scanning probe microscope, the one-dimensional approximation is of limited value, as carriers are likely to be redistributed in the lateral dimensions, as well as z [9]. Given the complexity of the charge distribution in the probe-sample junction of a scanning probe system, numerical solutions may complement or even supersede approximate analytical solutions to the Poisson equation.

Before proceeding to discussion of scanning-probe-based dopant profiling techniques, it is vital to note limitations of the simplified MOS picture that we have introduced. To date, scanning-probe-based dopant profiling has primarily been applied to silicon. However, the silicon-silicon oxide interface is unusually free of trapped charge and interface states. In contrast, most compound semiconductor surfaces and oxide interfaces contain high densities of mid-gap states whose charge contributes significantly to the magnitude of surface band bending and the formation of the depletion region. Though these states tend to be deep within the gap and thus unresponsive to high-frequency microwave signals, they will alter the dependence of the depletion capacitance upon DC and low-frequency tip bias voltages. As a result, extension of scanning-probe-based dopant profiling techniques to material systems beyond silicon will require refinement of the tip-sample models to account for trapped charge and interface states. Additional details on interface states can be found in References [2], [10], and [11].

11.3   Dopant Profiling with Scanning Capacitance Microscopy

Scanning capacitance microscopy is similar in many respects to other scanned probe techniques, but the SCM probe detects changes in local capacitance as opposed to current, force, or other aspects of the probe–sample interaction [12]. The SCM is an RF measurement tool in that the capacitive sensor that is integrated into the probe structure typically operates at a frequency near 1 GHz. Often, the capacitance may be detected by integrating the sensor with a resonant LCR circuit and measuring changes in the resonant behavior, reminiscent of a resonant NSMM. The sensor measures the differential capacitance ΔC induced by an alternating voltage signal of amplitude V_{ac} that is applied to either the probe tip or the sample. The alternating voltage signal, oscillating at a frequency ω_{ac}, also serves as the reference for a lock-in amplifier, which is used to perform the measurement of ΔC. A slowly swept DC bias V is applied, in addition to V_{ac}, in order to measure the ΔC–V relationship. The SCM has long been applied to two-dimensional dopant profiling problems [13]–[23].

As we saw in Chapter 9, the extraction of electromagnetic material parameters by use of high-frequency scanning probe microscopy relies on accurate modeling of the probe–sample interaction. Likewise, interpretation of SCM data requires reliable modeling in order to extract parameters from the measured ΔC-V data [15]. Often, these models require complex numerical simulation involving a large number of parameters, including probe tip shape, oxide layer thickness, oxide permittivity, and the SCM operating voltage. In many cases, the numerical simulations are implemented in software packages. One example of a model for two-dimensional dopant profiling with an SCM is a quasi-three-dimensional model in which the probe-sample capacitance is represented by a set of capacitive MOS ring-shaped capacitors arranged in a concentric geometry that corresponds to the probe tip geometry [13], [15]. Each MOS capacitance is calculated by use of Equation (11.5),

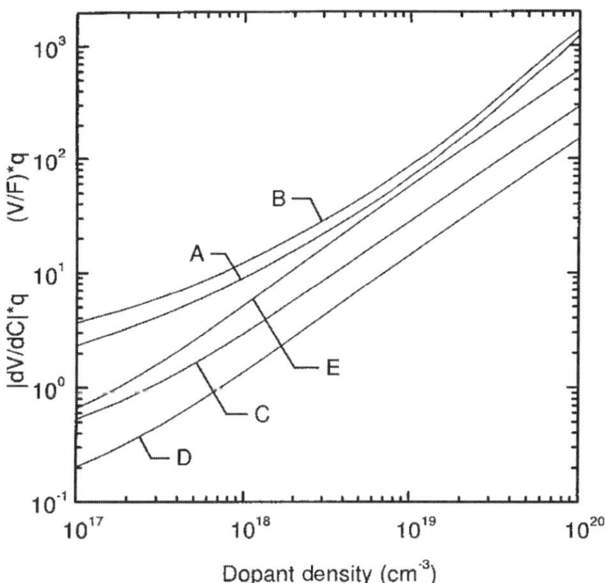

Figure 11.3. Comparison of dopant density conversion curves for different models. Calculated curves for converting dC/dV measurements to estimates of dopant density for five different models of an MOS capacitor. The models are denoted by the labels (A) through (E). Details of the models can be found in Reference [16] and the citations within. Reprinted with permission from J. J. Kopanski, J. F. Marchiando, and B. G. Rennex, *Journal of Vacuum Science and Technology B* 20 (2002) pp. 2101–2107. © 2002, American Vacuum Society.

with the space charge density distribution approximated by either Boltzmann or Fermi-Dirac statistics. The use of an approximation for the space charge density and other simplifying assumptions in the solution of Poisson's equation can speed computation time, but unfortunately this can also compromise the accuracy of the extracted dopant concentration [16]. Provided that one can tolerate the increased calculation time, a three-dimensional, finite-element solver can provide more accurate solutions of the three-dimensional Poisson's equation [17]. Note that in some cases, the effects of air gaps between the tip and sample, and the effects of condensed water on the sample, were needed in the model to improve agreement between SCM and measurements by other macroscopic techniques such as secondary ion-mass spectrometry [13]. In Fig. 11.3, several different models were used to calculate the conversion of measured dC/dV curves into estimates of dopant density. Clearly, the estimates are sensitive to the model used. In order to build confidence in both the interpretation of measurements and parameter extraction by use of a given model, calibration is required.

Calibration techniques and reference samples are critical for establishing accurate, absolute dopant profiling. Without calibration, SCM and other RF scanning probe techniques are limited to measurement of relative dopant concentration. The ideal reference sample for dopant profiling will be topographically smooth,

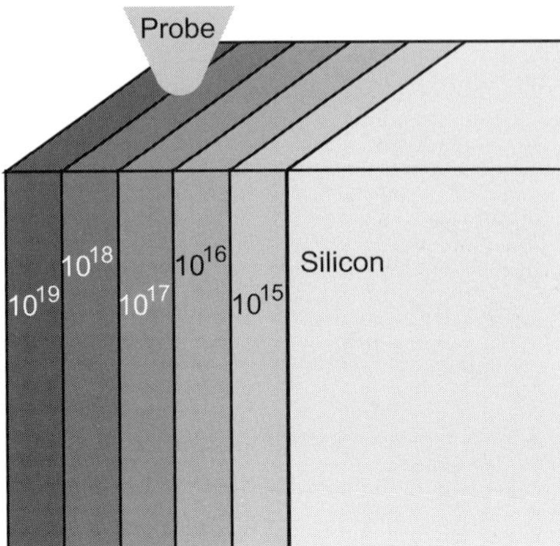

Figure 11.4.   Reference sample for dopant density calibration.
The reference sample is fabricated by depositing a series of semiconducting layers with known, uniform dopant concentration in a multilayer geometry. When this sample is imaged in cross section with a scanned probe, the sample appears as a series of stripes of varying dopant concentrations (e.g., 10^{14} cm^{-3}, 10^{15} cm^{-3}... 10^{19} cm^{-3}). In practice, separate p-doped and n-doped samples of this type have been utilized for calibration of dopant profiles.

minimizing the contribution of geometric effects to the probe-sample capacitance. In addition, the ideal reference sample should provide distinct, contrasting regions of varying dopant concentrations across the dynamic range of interest: roughly speaking, between 10^{14} cm^{-3} and 10^{20} cm^{-3}. Finally, the ideal reference sample will include both p-doped and n-doped regions, though in practice separately realized p-doped and n-doped reference samples have often been utilized. One approach is to create a multilayer structure in which each layer is uniformly doped at a known concentration. As needed, buffer layers may be introduced between the uniformly doped layers. In cross section, this sample will appear as a series of uniformly doped stripes, e.g., 10^{14} cm^{-3}, 10^{15} cm^{-3}, 10^{16} cm^{-3}, and so forth [18], [19]. A conceptual illustration of such a reference sample is shown in Fig. 11.4. This type of reference sample was also described in Chapter 9 in relation to material characterization techniques.

Figure 11.5 shows an SCM image of a variably doped n-type gallium nitride sample [20]. The stripes corresponding to different doping levels are clearly distinguished in the raw SCM image, shown in Fig. 11.5(a). Comparisons of SIMS measurements of a reference sample are compared to the SCM measurements in Fig. 11.5(b), providing a calibration approach that converts SCM measurements to estimates of the absolute dopant concentration. Systematic, cross-sectional SCM imaging of multilayered reference structures with varying layer thicknesses reveal best-case lateral resolution on the order of 10 nm for SCM imaging [20], [21]. Note, however, that both the sensitivity and resolution of SCM images are strongly dependent on tip geometry.

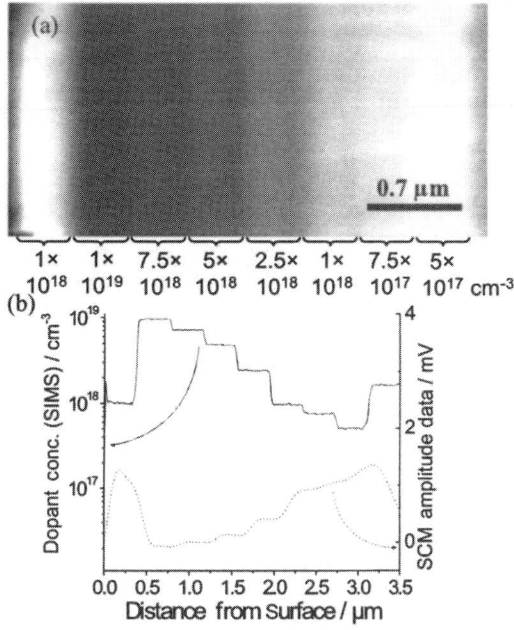

Figure 11.5. Scanning capacitance microscopy of variably doped, n-type gallium nitride. (a) Cross-sectional SCM amplitude image of the multilayer gallium nitride sample. Dopant concentrations for each layer are labeled. (b) Processed SCM data (dashed line) are compared to SIMS data (solid line). Reprinted with permission from J. Sumner, R. A. Oliver, M. J. Kappers, and C. J. Humphreys, *Journal of Vacuum Science and Technology B* 26 (2008), pp. 611–617. © 2008, American Vacuum Society.

11.4 Dopant Profiling with Near-Field Scanning Microwave Microscopy

As we have described earlier, measurement of the depletion layer capacitance–voltage relationship provides an avenue for estimating dopant concentration in metal-semiconductor and MOS structures. Since NSMM measurements of such structures include a significant contribution from the local depletion layer capacitance, NSMM provides a useful capability for local capacitance-voltage spectroscopy and in turn, dopant profiling. These capabilities are only realized when other contributions to the tip–sample interaction, including any parasitic reactance, are accounted for by use of calibration and modeling. In this section, we introduce approaches to dopant profiling by use of NSMM. In the first approach, the dopant profiling procedure has three main components: (1) measurement of the voltage dependence of the reflection coefficient, (2) calibration of the instrument so that the reflection coefficient can be converted to quantitative values of the capacitance, and (3) extraction of the dopant concentration from the capacitance–voltage relationship. We focus on the determination of the capacitance–voltage relationship in order to connect the procedure to the underlying physics, standard semiconductor characterization approaches, and dopant profiling with other scanned probe techniques such as scanning capacitance microscopy.

The following discussion assumes an experimental setup similar that of References [4] and [8] and generally follows the methods laid out in those references. Naturally, alternative test platforms will require corresponding modifications to the dopant profiling procedure, though the general concepts will remain intact. Measurements are made by use of a resonant NSMM operating in a reflection mode. The operating frequency is near, but not equal to, the resonant frequency. Further, the NSMM is assumed to be based on a contact-mode AFM in which the probe tip is in direct mechanical contact with the sample. During measurements, a DC bias is applied to the tip. The tip is laterally positioned above a point of interest and the complex reflection coefficient S_{11} is measured as a function of the DC tip bias by use of a VNA. The effects of the measurement platform may be partially removed by calibrating the network analyzer at coaxial reference planes located outside of the microscope head. Alternatively, the system may be calibrated such that the complex impedance is obtained in place of S_{11}, as described in Chapter 7 [24], [25]. Here, our approach is based on measurements of the raw, uncalibrated, complex reflection coefficient S_{11}.

The conversion of the raw S_{11} data to quantitative capacitance measurements requires a known reference sample, an accurate model of the tip–sample interaction, and a calibration procedure. The reference sample must incorporate features with a known impedance. Note that if a three-term error model is used for impedance calibration of NSMMs that operate in a reflection mode, then a minimum of three reference features are required for a one-port calibration, though the inclusion of additional reference features will decrease the statistical uncertainty in the calibrated measurements. Ideally, all of the features will be located close together such that the calibration requires only a single NSMM calibration image to be acquired and subsequently analyzed. It is also useful to incorporate the reference features into the same wafer or substrate as the SUT, if possible. The following calibration procedure requires that the reference features are on either the same device substrate or a similar substrate that can be assumed to have similar electromagnetic material properties as the device substrate. Specifically, the model of the tip–sample interaction must be valid across both the reference sample and the SUT. Here, we will use a set of microcapacitors, illustrated in Fig. 11.6(a), as the reference sample [3]. This reference sample was fabricated using standard microfabrication techniques. The sample comprises a series of circular gold pads patterned upon a series of steps in a SiO_2 staircase on a doped silicon substrate. The NSMM tip is in direct electrical and mechanical contact with one of the gold contacts, forming the top electrode of a microcapacitor while the doped substrate forms the bottom electrode. The capacitance C_{diel} is calculated by use of Equation (11.1) with the geometric parameters (pad area A and oxide thickness d) determined from topographic images of the sample. In addition, the oxide is assumed to be uniform and have a known relative permittivity. Assuming that the leakage current through the oxide is negligible, the impedance of the microcapacitor is given by the standard definition $Z = 1/j\omega C_{tot}$, where $\omega = 2\pi f$ and f is the operating frequency.

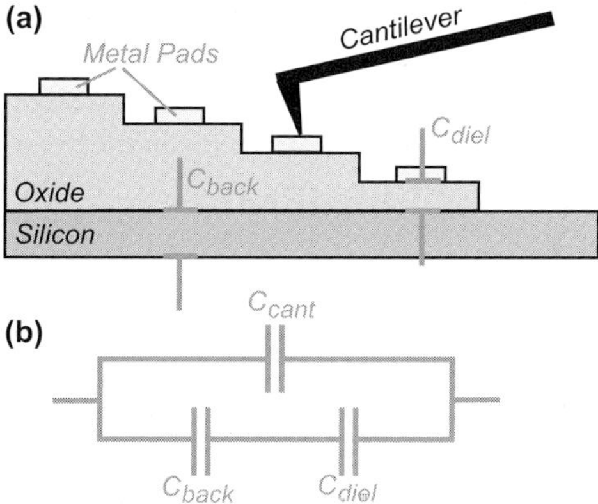

Figure 11.6. Microcapacitor reference sample.
(a) A side-view schematic of the microcapacitor reference sample. The oxide layer is patterned to create a staircase with step thicknesses of 50 nm, 100 nm, 150 nm, and 200 nm, for example. Circular gold pads are patterned on top of the steps. On each step, there is a sequence of pads with diameters equal to 1 μm, 2 μm, 3 μm, and 4 μm. (b) A lumped-element model of the tip–sample interaction, including the capacitance C_{diel}, a parasitic background capacitance C_{back}, and the cantilever-sample capacitance C_{cant}. Reprinted from H. P. Huber et al., "Calibrated Nanoscale Capacitance Measurements Using a Scanning Microwave Microscope," *Review of Scientific Instruments* 81 (2010) art. no. 113701, with permission from AIP Publishing.

A simple, lumped-element circuit model is used to represent the reference sample, as shown in Fig.11.6(b). Following a similar analysis as was presented in Chapter 8, the total microcapacitor capacitance C_{tot} is defined by the parallel combination of the oxide capacitance C_{diel} and the substrate or back capacitance C_{back}:

$$\frac{1}{C_{tot}} = \frac{1}{C_{diel}} + \frac{1}{C_{back}}. \tag{11.7}$$

To find the calibrated capacitance measured by the probe, two measurements are made for each microcapacitor: the reflection coefficient with the probe positioned on the gold pad $S_{11}{}^{Au}$ and the reflection coefficient with the probe positioned on the oxide surface $S_{11}{}^{Ox}$. The relative amplitude signal ΔS_{11} is defined simply as the difference between the scattering parameters

$$\Delta S_{11} = S_{11}{}^{Au} - S_{11}{}^{Ox}. \tag{11.8}$$

The use of a relative value is necessitated by the fact that the absolute reflection coefficient amplitude will depend on the test platform, the nature of the NSMM resonance, and the selected operating frequency. Further, by performing this subtraction, we are excluding the effects of the parallel stray capacitance C_{cant} from the cantilever body, albeit under the assumption that C_{cant} is unchanged by movement from the gold pad to the oxide. Note that sensitivity of C_{cant} to changes in probe position can be reduced by increasing the aspect ratio of the asperity on which the probe tip is formed. This increased aspect ratio reduces the capacitive coupling between the cantilever body and other parts of the probe platform with the sample. In Reference [4], it has been shown that there is a linear relationship between the relative amplitude signal and the total microcapacitor capacitance:

$$C_{tot} = \alpha |\Delta S_{11}|, \tag{11.9}$$

where α is a calibration constant. The two unknown calibration parameters, C_{back} and α, in Equations (11.7) and (11.9) can now be determined from measurements of reference microcapacitors.

Once C_{back} and α are known, the absolute capacitance as a function of voltage and tip position can be calculated from voltage- and position-dependent measurements of ΔS_{11}. Once this dependence is known, the local dopant concentration can be determined from numerical models or from analytical solutions for the semiconductor potential Ψ in Poisson's equation. Essentially, once the NSMM measurement has been calibrated and converted to a voltage-dependent capacitance (or voltage-dependent dC/dV) measurement, estimation of the dopant concentration from the C-V curve follows the same procedure as SCM, as described in the previous section. As the NSMM capacitance calibration relies on a relative measurement referenced to the reflection coefficient on the oxide surface S_{11}^{Ox}, it is limited to applications in which there is an oxide or other suitable reference surface. Given the widespread use of silicon substrates and the presence of native silicon oxides, there are many applications that meet this requirement. Additional requirements for successful and sustainable capacitance calibration include a stable test platform. Given the current state of commercial network analyzers, test cabling, and other hardware, it is generally straightforward to assemble a nominally stable test platform. However, microwave measurements in general and resonant NSMMs in particular are highly sensitive to environmental factors, such as temperature and humidity. A stable tip shape is also a necessity for sustainable, calibrated NSMM capacitance measurements, as capacitance is inherently geometry dependent. In the case of a contact-mode, AFM-based NSMM, significant mechanical wear can occur in a short period of time, thus invalidating the calibration procedure. Custom, metal-coated nanowire probes have demonstrated reduced mechanical wear over large numbers of scans [26].

Calibrated, spectroscopic dC/dV measurements are compared to modeled data in Fig. 11.7. The different curves in Fig. 11.7(a) correspond to different regions of known dopant concentration on an n-type reference sample similar to the one shown in Fig. 11.4. The numerical simulation package, FASTC2D [27], initially calculates the spectroscopic curves based on best-known values of measurement parameters and then adjusts a set of variable parameters, including tip geometry, surface insulator thickness t_{ox}, and the relative permittivity of the oxide ε_r, to fit the experimental data. To achieve reasonable agreement, the voltage amplitude had to be significantly increased from a nominal value of 800 mV to a value of 3000 mV, broadening the modeled response. Additionally, the flatband voltage in the

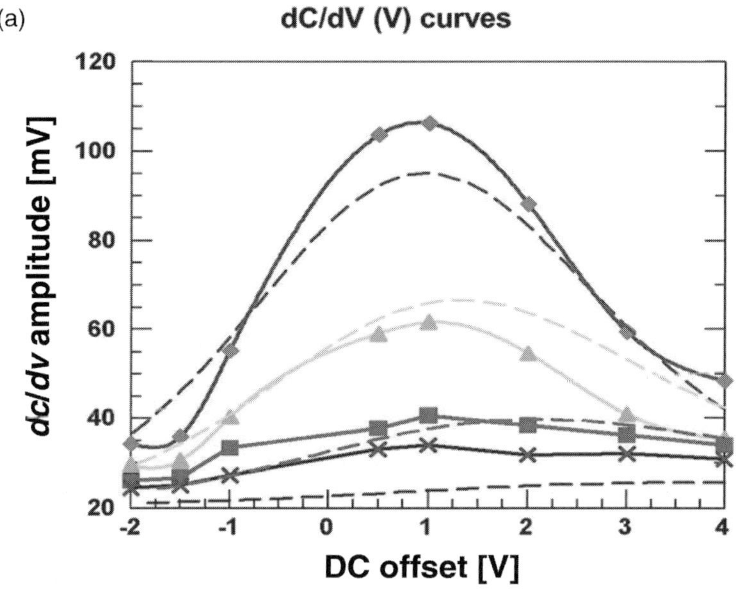

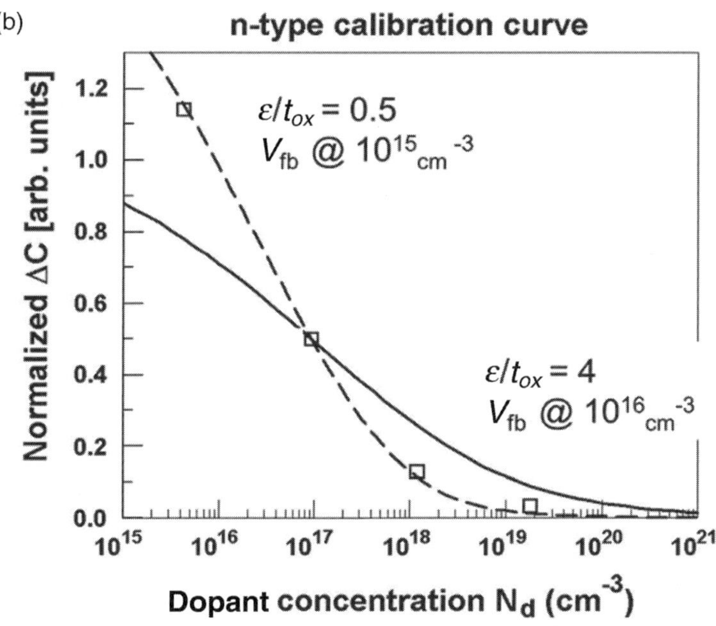

Figure 11.7. Modeling of capacitance voltage measurements. (a) Experimentally measured, voltage-dependent dC/dV curves (markers and solid lines) are compared to numerically modeled values (dashed lines). (b) Experimentally determined calibration curves for an n-type calibration sample. Reprinted from H. P. Huber et al., *Journal of Applied Physics* 111 (2012), art. no. 014301, with permission from AIP Publishing.

model had to be adjusted by + 0.8 V to align with the measured NSMM data. Overall, approximate agreement between the model and measured data is seen in Fig. 11.7(a). Finally, the FASTC2D package is also used to model the n-type calibration curve shown in Fig. 11.7(b). The solid black line is an initial calculation, based on nominal values of t_{ox} (1 nm) and ε_r (3.9), with the DC bias set in the model to produce the maximum differential capacitance signal at a concentration of 10^{16} cm^{-3}. The dashed line represents a fit found by reducing the ratio ε_r/t_{ox} to 0.5 and setting the DC bias to produce the maximum differential capacitance signal at a concentration of 10^{15} cm^{-3}.

11.5   Dopant Characterization with Other Microscopy Techniques

In addition to the SCM and NSMM, several additional scanning probe tools are commonly applied to local characterization of defects and dopants in semiconductors [28], [29]. As these additional tools are based on DC or low-frequency techniques, they are not, strictly speaking, applicable to microwave and RF measurements. However, they provide complementary insight into semiconductor micro- and nanoelectronics in general and RF nanoelectronics in particular. Here we will briefly review three such techniques: scanning spreading resistance microscopy, scanning kelvin probe microscopy, and scanning tunneling microscopy. In general, these techniques are two-dimensional and surface-sensitive. Thus, sample cross-sectioning is required for subsurface, depth-dependent profiling. As in the case of SCM and NSMM, extraction of material parameters such as dopant concentration requires accurate physical models, reliable calibration techniques, and well-known reference samples.

From an instrumentation point of view, the SSRM is similar to a contact-mode, conductive AFM [30]. During SSRM measurements, a sharp, conductive cantilever probe is in direct mechanical contact with the sample. With a small (~100 mV) bias applied to the probe, the SSRM measures the DC resistance between the probe and a backside contact on the sample. If the force applied by the cantilever probe on the sample exceeds a threshold value, then the measured resistance will be approximately equal to the local spreading resistance, which is, in turn, inversely proportional to the local carrier concentration. Because the SSRM applies high forces, a mechanically hard probe such as a diamond-coated tip is usually used [31].

Often, SKPMs are based on noncontact, cantilever-based AFMs [32]. A DC voltage is applied to the probe tip, inducing an electrostatic force between the probe and the sample. If the DC bias voltage is equal to the surface potential, then the electrostatic force will be zero. Thus, the surface potential can be measured by adjusting the tip voltage such that the electrostatic force is nulled. In practice, the force can be detected by modulating the bias voltage at a mechanical resonance frequency of the cantilever and using a lock-in technique to detect the resulting oscillations of the cantilever. Note that if a noncontact, dynamic AFM mode is used for distance following, then the fundamental resonance frequency may be unavailable for modulating the tip bias and a higher order resonance frequency will be required. The work function of the sample may be determined from the surface potential, provided that the work function of the tip is known. The sample carrier concentration may be extracted from the work function. This method has been used to perform dopant profiling on Si samples, for example [33].

At the limit of device scaling, as device dimensions become comparable to a few atomic diameters, device performance depends upon the identity and placement of individual atoms. At this scale, the STM's capability to perform local electronic spectroscopy on individual atoms is a powerful tool [34], [35]. In an STM, a conducting tip is placed within a nanometer of a conducting sample. When there is a potential difference between the tip and the sample, a small, but measurable current will flow due to quantum mechanical tunneling. The magnitude of the tunneling current will depend upon the tip-sample separation, the potential difference, and the local electronic DOS of the sample. The latter dependence is the basis in scanning tunneling spectroscopy (STS), a technique in which the tunneling conductivity is measured as a function of the potential difference between the tip and the sample. If the tip-sample height is fixed, then STS provides a method for measuring the DOS with atomic-scale resolution. This enables the identification and electronic characterization of individual atoms, enabling the measurement of the role of individual dopants in nanoelectronic devices [36].

References

[1] *The International Technology Roadmap for Semiconductors 2.0: 2015.*

[2] B. S. Eller, J. Yang, and R. J. Nemanich, "Electronic Surface and Dielectric Interface States on GaN and AlGaN," *Journal of Vacuum Science and Technology A* 31 (2013) art. no. 050807.

[3] G. Gomila, J. Toset, and L. Fumagalli, "Nanoscale Capacitance Microscopy of Thin Dielectric Films," *Journal of Applied Physics* 104 (2008) art. no. 024315.

[4] H. P. Huber, M Moertelmaier, T. M. Wallis, C. J. Chiang, M. Hochleitner, A. Imtiaz, Y. J. Oh, K. Schilcher, M. Dieudonne, J. Smoliner, P. Hinterdorfer, S. J. Rosner, H. Tanbakuchi, P. Kabos, and F. Kienberger, "Calibrated Nanoscale Capacitance Measurements Using a Scanning Microwave Microscope," *Review of Scientific Instruments* 81 (2010) art. no. 113701.

[5] S. M. Sze, *The Physics of Semiconductor Devices* (Wiley, 1981).

[6] Y. Taur and T. H. Ning, *Fundamentals of Modern VLSI Devices* (Cambridge University Press, 1998).

[7] E. H. Nicollian and J. R. Brews, *MOS (Metal Oxide Semiconductor) Physics and Technology* (Wiley, 1982).

[8] H. P. Huber, I. Humer, M. Hochleitner, M. Fenner, M. Moertelmaier, C. Rankl, A. Imtiaz, T. M. Wallis, H. Tanbakuchi, P. Hinterdorfer, P. Kabos, J. Smoliner, J. J. Kopanski, and F. Kienberger, "Calibrated Nanoscale Dopant Profiling Using a Scanning Microwave Microscope," *Journal of Applied Physics* 111 (2012) art. no. 014301.

[9] S. Friedman, Y. Yang, O. Amster, and F. Stanke, "Characterizing Non-Linear Microwave Behavior of Semiconductor Materials with Scanning Microwave Impedance

Microscopy," *2016 IEEE MTT-S International Microwave Symposium Digest* (2016) pp. 1–3.

[10] W. Mönch, *Semiconductor Surfaces and Interfaces* (Springer, 2001).

[11] D. Colleoni, G. Miceli, and A. Pasquarello, "Origin of Fermi-Level Pinning at GaAs Surfaces and Interfaces," *Journal of Physics: Condensed Matter* 26 (2014) art. no. 492202.

[12] J. R. Matey and J. Blanc, "Scanning Capacitance Microscopy," *Journal of Applied Physics* 57 (1985) pp. 1437–1444.

[13] Y. Huang, C. C. Williams, and H. Smith, "Direct Comparison of Cross-sectional Scanning Capacitance Microscope Dopant Profile and Vertical Secondary Ion-Mass Spectroscopy Profile," *Journal of Vacuum Science and Technology B* 14 (1996) pp. 433–436.

[14] Y. Huang, C. C. Williams, and J. Slinkman, *Applied Physics Letters* 66 (1995) p. 344.

[15] J. J. Kopanski, J. F. Marchiando, and B. G. Rennex, "Comparison of Experimental and Theoretical Scanning Capacitance Microscope Signals and Their Impact on the Accuracy of Determined Two-Dimensional Carrier Profiles," *Journal of Vacuum Science and Technology B* 20 (2002) pp. 2101–2107.

[16] J. F. Marchiando and J. J. Kopanski, "Regression Procedure for Determining the Dopant Profile in Semiconductors from Scanning Capacitance Microscopy Data," *Journal of Applied Physics* 92 (2002) pp. 5798–5809.

[17] J. F. Marchiando, J. J. Kopanski, and J. R. Lowney, "Model Database for Determining Dopant Profiles from Scanning Capacitance Microscope Measurements," *Journal of Vacuum Science and Technology B* 16 (1998) pp. 463–470.

[18] R. Stephenson, A. Verhulst, P. De Wolf, M. Caymax, and W. Vandervorst, "Nonmonotonic Behavior of the Scanning Capacitance Microscope for Large Dynamic Range Samples," *Journal of Vacuum Science and Technology B* 18 (2000) pp. 405–408.

[19] L. Wang, J. Laurent, J. M. Chauveau, V. Sallet, F. Jomard, and G. Bremond, "Nanoscale Calibration of n-type ZnO Staircase Structures by Scanning Capacitance Microscopy," *Applied Physics Letters* 107 (2015) art. no. 192101.

[20] J. Sumner, R. A. Oliver, M. J. Kappers, and C. J. Humphreys, "Assessment of Performance of Scanning Capacitance Microscopy for n-type Gallium Nitride," *Journal of Vacuum Science and Technology B* 26 (2008) pp. 611–617.

[21] F. Giannazzo, V. Raineri, S. Mirabella, G. Impellizzeri, F. Priolo, M. Fedele, and R. Mucciato, "Scanning Capacitance Microscopy: Quantitative Carrier Profiling Down to Nanostructures," *Journal of Vacuum Science and Technology B* 24 (2006) pp. 370–374.

[22] N. Duhayon, et al. "Assessing the Performance of Two-Dimensional Dopant Profiling Techniques," *Journal of Vacuum Science and Technology B* 22 (2004) pp. 385–393.

[23] R. A. Oliver, "Advances in AFM for the Electrical Characterization of Semiconductors," *Reports on Progress in Physics* 71 (2008) art. no. 076501.

[24] G. Gramse, M. Kasper, L. Fumagalli, G. Gomila, P. Hinterdorfer, and F. Kienberger, "Calibrated Complex Impedance and Permittivity Measurements with Scanning Microwave Microscopy," *Nanotechnology* 26 (2015) art. no. 149501.

[25] M. Farina, D. Mencarelli, A. Di Donato, G. Venanzoni, and A. Morini, "Calibration Protocol for Broadband Near-Field Microwave Microscopy," *IEEE Transactions on Microwave Theory and Techniques* 59 (2011) pp. 2769–2776.

[26] J. C. Weber, P. T. Blanchard, A. W. Sanders, A. Imtiaz, T. M. Wallis, K. J. Coakley, K. A. Bertness, P. Kabos, N. A. Sanford, and V. M. Bright, "Gallium Nitride Nanowire Probe for Near Field Scanning Microwave Microscopy," *Applied Physics Letters* 104 (2014) art. no. 023113.

[27] J. J. Kopanski, J. F.Marchiando, and J. R. Lowney, "Scanning Capacitance Microscopy Applied to Two-Dimensional Dopant Profiling of Semiconductors," *Materials Science and Engineering B – Solid State Materials for Advanced Technology* 44 (1997) pp. 46–51.

[28] P. De Wolf, R. Stephenson, T. Trenkler, T. Clarysse, T. Hantschel, and W. Vandervorst, "Status and Review of Two-Dimensional Carrier and Dopant Profiling Using Scanning Probe Microscopy," *Journal of Vacuum Science and Technology B* 18 (2000) pp. 361–368.

[29] D. Y. Ban, B. Y. Wen, R. S. Dhar, S. G. Razavipour, X. Cu, X. R. Wang, Z. Wasilewski, and S. Dixon-Warren, "Electrical Scanning Probe Microscopy of Electronic and Photonic Devices: Connecting Internal Mechanisms with External Measures," *Nanotechnology Reviews* 5 (2016) pp. 279–300.

[30] D. T. Lee, J. P. Pelz, and B. Bhushan, "Instrumentation for Direct Low Frequency Scanning Capacitance Microscopy, and Analysis of Position Dependent Stray Capacitance," *Review of Scientific Instruments* 73 (2002) pp. 3525–3533.

[31] P. De Wolf, T. Clarysse, W. Vandervorst, and L. Hellemans, "Low Weight Spreading Resistance Profiling of Ultrashallow Dopant Profiles," *Journal of Vacuum Science and Technology B* 16 (1998) pp. 401–405.

[32] M. Nonnenmacher, M. P. Oboyle, and H. K. Wickramasinghe, "Kelvin Probe Force Microscopy," *Applied Physics Letters* 58 (1991) pp. 2921–2923.

[33] M. Tanimoto and O. Vatel, "Kelvin Probe Force Microscopy for Characterization of Semiconductor Devices and Processes," *Journal of Vacuum Science and Technology* 14 (1996) pp. 1547–1551.

[34] G. Binnig, H. Rohrer, C. Gerber, and E. Weibel, "Tunneling through a Controllable Vacuum Gap," *Applied Physics Letters* 40 (1982) pp. 178–180.

[35] J. A. Stroscio, R. M. Feenstra, and A. P. Fein, "Electronic Structure of the Si(111) 2×1 Surface by Scanning-Tunneling Microscopy," *Physics Review Letters* 57 (1986) pp. 2579–2582.

[36] P. M. Koenraad and M. E. Flatte, "Single Dopants in Semiconductors," *Nature Materials* 10 (2011) pp. 91–100.

12 Depth Profiling

12.1 Introduction to Nanoscale Depth Profiling

Microwave tomography is an active, developing research area. The objective is to visualize hidden, subsurface features through application of microwave radiation. Applications include ground penetrating radars (GPR) [1], defect spectroscopy in materials, fault detection in construction and manufacturing, and the detection of hidden objects at secure access points. Recently, significant effort has been devoted to microwave tomography for medical applications [2]–[4] and frequency-dependent, mechanical metrology of defects [5]. In spite of this broad effort, microwave tomography at the nanoscale is in its infancy. One significant challenge is that measurements at the nanoscale are done mostly in the near-field, as is also the case in GPR and some biological applications, but with the additional requirement that the nanoscale, electromagnetic interaction of microwave radiation with materials must be understood. These unique interactions with nanoscale systems usually differ significantly from corresponding interactions with macroscopic objects. In this chapter, we will discuss basic principles of nanoscale microwave tomography and some initial tomographic measurements with near-field scanning probe microscopes.

Note that we use "three-dimensional" to describe measurements of a physical variable such as permittivity or permeability as a function of three spatial coordinates. The resulting dataset is often referred to as a "tomograph." Similarly, we use "two-dimensional" to describe measurements of a physical variable as a function of two spatial coordinates. Here, we will refer to the resulting dataset as an "image."

In order for progress to continue in this area, existing macroscopic methods for tomography of buried materials and interfaces have to be modified. Specifically, the material-dependent, effective penetration depth of the near-field signal must be understood and controlled. Furthermore, both reflection and transmission measurements have to be considered. In principle, the combination of reflection mode and transmission mode measurements will enable detection of embedded objects and lead to quantitative measurement of their position, geometry, and electromagnetic material properties. We will begin with a detailed introduction of basic theoretical approaches to three-dimensional tomography before reviewing experimental demonstrations of nanoscale depth profiling with scanning probe microscopy. To date, the development of the theoretical descriptions has outpaced experiment demonstrations. At the core of the theoretical treatment lies the task of solving a complicated inverse problem. As a result, some complex mathematical details will necessarily be introduced in this chapter.

12.2 Theoretical Foundation of Depth Profiling

12.2.1 Near-Fields and Tomography

One central, ongoing challenge for quantitative near-field imaging, as well as three-dimensional tomography, is the convolution of the topography with the material properties of the sample. There are a few exceptions, such as two-dimensional imaging of truly flat surfaces for which observed contrast in images is entirely due to variations in the material properties of the system surface, but in most cases topographic contrast is difficult to distinguish from material contrast. This applies especially to near-field microwave scanning probe measurements as the measured reflection coefficient is determined largely by capacitive interactions. How does one de-embed the topographic contribution from the measured reflection coefficient image? Some practical, experimental approaches to this problem for the case of near-field microwave imaging will be discussed in Chapter 14. Here, we focus on the development of the theory of near-field scattering. The solution to the inverse scattering problem offers the enticing possibility to nondestructively retrieve the three-dimensional spatial distribution of charge, permittivity, permeability, and other physical parameters. Near-field, three-dimensional tomography as carried out by a near-field microscope is an electromagnetic problem with similar analytical approaches for both near-field optical and microwave microscopes. From the experimental point of view, microwave microscopy has an advantage in that it is easier to simultaneously measure the amplitude and phase of microwave signals. The measured phase response is an important input for the solution of the inverse scattering problem.

Analytically, three-dimensional tomography is a difficult inverse scattering problem, that is further complicated by the fact that electromagnetic fields are difficult to calculate in complex environments. Therefore, a general, electromagnetic, near-field theory of three-dimensional tomography is elusive. We will primarily address aspects of the problem that may prove to be important for the specific case of NSMM measurements, which are done by planar scanning above the sample surface. Thus, to obtain three-dimensional tomographic data, spatially resolved NSMM images must be complemented by changing at least one parameter in addition to the lateral probe position. From the outset, it is important to note that it is only possible to measure buried structures within sample volumes in which there is efficient formation of the near-field signals. As the depth of the buried feature increases, the resolution of a near-field microscope deteriorates rapidly.

12.2.2 Near-Field of an Elementary Dipole

As a starting point to describe the electric field for subsurface NSMM measurement applications, consider the field from a unit point source. The complex potential from a unit point source in an infinite, uniform medium is

$$\phi(r) = \frac{e^{-jkr}}{4\pi r},$$ (12.1)

where r is the distance from the source and k is the wave vector. Harmonic time dependence of the form $e^{j\omega t}$ is assumed, where ω is the radial frequency. Assume that the point source is an elementary dipole current source of the form $u_z I_0 \Delta l$, where u_z is a unit vector in the z direction, Δl is the length of the dipole, and I_0 is the current amplitude. The field of an elementary electric dipole of the length Δl at arbitrary distance $r u_r = r(sin\theta cos\phi u_x + sin\theta sin\phi u_y + cos\theta u_z)$, where angle θ is with respect to the z axis, can be expressed as

$$H_\phi \approx \frac{I_0 \Delta l e^{-jkr}}{4\pi r^2} sin\theta = j\frac{q\omega \Delta l e^{-jkr}}{4\pi r^2} sin\theta,$$ (12.2)

$$E_r \approx -jZ_0 \frac{I_0 \Delta l e^{-jkr}}{2\pi kr^3} cos\theta = \frac{q\Delta l e^{-jkr}}{2\pi \varepsilon r^3} cos\theta,$$ (12.3)

$$E_\theta \approx -jZ_0 \frac{I_0 \Delta l e^{-jkr}}{4\pi kr^3} sin\theta = \frac{q\Delta l e^{-jkr}}{4\pi \varepsilon r^3} sin\theta.$$ (12.4)

Here,

$$k = \frac{2\pi}{\lambda} = \omega\sqrt{\mu\varepsilon},$$ (12.5a)

$$Z_0 = \sqrt{\frac{\mu}{\varepsilon}},$$ (12.5b)

and

$$I_0 = \frac{\partial q}{\partial t} = j\omega q.$$ (12.5c)

H_ϕ is the polar component of the magnetic field. E_r and E_θ are the radial and azimuthal components of the electric field. The permittivity and permeability are represented by μ and ε, respectively. For near-field ($kr \ll 1$), the exponential term e^{-jkr} in the complex potential can be neglected. Note that we are taking advantage of the near-field nature of the problem, implementing a quasi-electrostatic solution. This quasi-static approximation is used for many near-field problems.

In order to solve inverse problems, it is useful to introduce the spectral decomposition (or plane wave expansion) of these fields. One can express

$$\phi(r) = \frac{e^{-jkr}}{4\pi r} = -j\int_0^\infty dk_\rho \frac{k_\rho}{k_z} J_0\left(k_\rho\rho\right)e^{-jk_z|z|} = \frac{-j}{2}\int_{-\infty}^\infty dk_\rho \frac{k_\rho}{k_z} H_0^1(k_\rho\rho)e^{-jk_z|z|} \quad (12.6)$$

with $k_\rho \cdot \rho = k_x x + k_y y$ and $k_z = \left(k_0^2 - k_x^2 - k_y^2\right)^{\frac{1}{2}} = (k_0^2 - k_\rho^2)^{1/2}$. $J_0(.)$ and $H_0^1(.)$ are Bessel (zero order) and Hankel (zero order first kind) functions, respectively. Equation (12.6) represents the Sommerfeld identity for spherical waves [6]. The wave vectors can be considered to be complex and include some loss. The spectral decomposition of the electromagnetic field components then can be expressed using the Sommerfeld identity (12.6) and realizing the fact that these components can be obtained from differentiation of $\phi(r)$ with respect to r.

12.2.3 Near-Field Scattering at a Sub-wavelength Aperture

The initial experimental approaches to near-field microwave microscopy were based on making small holes (apertures) into microwave waveguides. The end of the tip in near-field microwave microscopy can be considered, in first approximation, from the theoretical point of view as a small aperture. Therefore, it is important to start with the work related to diffraction of electromagnetic waves on small aperture as the initial approach to solution of the near-field microwave microscopy problem. Since the original solution of the problem by Bethe [7], the near-field physics of an electromagnetic wave scattered from an aperture has been intensively investigated by Levine and Schwinger [8], Bouwkamp [9], and many others. The total electromagnetic field components close to a small aperture are expressed as a superposition of the fields that would be present without the aperture and the scattered fields: $E = E_o + E_s$, $H = H_0 + H_s$. E_o and H_o are the electric and magnetic fields, respectively that would be present without the aperture. E_s and H_s are the scattered fields [6].

From here, we follow the approach of Reference [9]. The boundary value problem at the aperture is solved in terms of magnetic currents and charge densities in the aperture by use of a Green's function. We assume that the aperture is small with respect to incident wavelength. If we consider the case of normal incidence of a plane, monochromatic wave on a circular aperture of diameter D, then the magnetic charge density can be expressed as [10]

$$\sigma_m(r) = -\frac{H_0 \cdot r}{\pi^2\left(\left(\frac{D}{2}\right)^2 - r^2\right)^{1/2}} \quad (12.7)$$

and the magnetic current density is

$$J_m(r) = \frac{-jk_0}{3\pi^2}\left[2H_0\left(\left(\frac{D}{2}\right)^2 - r^2\right)^{1/2} + \frac{H_0(r\cdot r) - r(H_0\cdot r)}{\left(\left(\frac{D}{2}\right)^2 - r^2\right)^{1/2}}\right]. \tag{12.8}$$

The magnetic charge and current densities lead to expressions for the scattered field outside the aperture [9] in the form

$$E_{s,z}(r) = -\frac{2jk_0\left(\frac{D}{2}\right)^3}{3\pi}\frac{(r\times H_0)_z}{r^2\left(r^2 - \left(\frac{D}{2}\right)^2\right)^{1/2}} \tag{12.9}$$

and

$$H_{s,\tan}(r) = \frac{H_0}{\pi}\left[arcsin(D/2r) - D/2r^2\left(r^2 - \left(\frac{D}{2}\right)^2\right)^{1/2}\right] - \frac{D^3 r(H_0\cdot r)}{4\pi r^4\left(r^2 - \left(\frac{D}{2}\right)^2\right)^{1/2}}. \tag{12.10}$$

Note that the coordinate z is perpendicular to the plane of the aperture. These scattering fields represent the excitation fields that enable sub-wavelength resolution in the near-field and are thus the foundation for NSMM calculations.

Presently, inverse, near-field problems are well addressed for acoustic waves, but much less so for electromagnetic waves. Thus, inverse, near-field solutions for scanning electromagnetic probe microscopy applications are in the early stages of development. The development of calibration procedures for the corresponding measurements lags even farther behind. Therefore, we will limit our discussion to a few existing, theoretical treatments that are applicable to NSMM tomography. Most NSMM measurements are performed in a configuration in which the incident electromagnetic wave interrogates the material and the receiver detects the reflected or transmitted signal. The measured quantities are either the change of the reflection/transmission coefficients or the frequency shift of a resonant cavity. As discussed in previous chapters, these measurands reflect variations in the material parameters of the sample. This applies to both surface and subsurface characterization, but the subsurface problem is more complicated.

For three-dimensional electromagnetic tomography, two steps are required to find a solution. First the forward problem is solved: the electromagnetic fields are determined for a known configuration and known material parameters. In the second step, the inverse problem is treated based on the forward problem solutions. (The aforesaid solutions for the cases of an elementary dipole and scattered waves at an aperture are examples of forward problem solutions.) More specifically, the philosophy of this approach is as follows. The solution of the forward problem determines the signals measured by the NSMM or other detectors for known distributions of the conductivity, permittivity, and permeability in the object space. The solution to the inverse problem determines the distributions of the electromagnetic materials parameters within the object space from the detected signal, resulting in image reconstruction. For NSMM, work in this area is ongoing. It is necessary to keep in mind that inverse problems are inherently nonlinear and from the point of view of numerical computation ill-posed. As a reminder to the reader, a problem is ill-posed when one of the following conditions is satisfied: (a) a solution does not exist; (b) solutions exist, but are not unique; or (c) the solution is unstable or overly sensitive to small changes in measured data. The latter condition may result in solutions that do not remain valid after small perturbations. Therefore, regularization methods have to be implemented. There are several critical issues such as confirmation of the uniqueness of the solution, stabilization of the problem, and determination of the approximate solution to such a stabilized problem. These are not simple tasks [11].

12.2.4 Solution to the Forward Problem

In general, electromagnetic tomography requires the solution of a nonlinear, multidimensional problem for a particular configuration. Therefore, few if any fundamental theoretical approaches can be introduced that have universal validity. In the broadest sense, most cases require the solution of a basic equation, taking into account specific electromagnetic field distributions and boundary conditions. In finding this solution, the objective is to predict the values that will be measured by a given detection system for a known distribution of conductivity, permittivity and permeability in the object space. Information about the field distribution can be obtained from an appropriate analytical model of the system or through numerical calculations such as finite-element simulations. From the experimental point of view, there are important questions related to the feasibility of direct measurement of the fields. Can one accurately and directly measure two-dimensional distributions of electric and magnetic fields as a function of the frequency and current distributions within the source and the receiver? Alternately, can one indirectly measure the re-radiated fields from material samples or changes of the impedance, e.g., by use of a scanning antenna? The forward problem solution must be thoroughly studied over a wide parameter space in order to develop as complete an understanding of the near-field detection system as possible. Insight is required into physical principles, detectability limits and sensitivity maps in order to provide a sound basis for sensitivity analysis and quantitative image reconstruction algorithms of embedded, subsurface objects.

The fields for the specific configurations and the corresponding boundary conditions are naturally derived from solution of Maxwell's equations. Previously, we already introduced the near-field components that follow from the field of an elementary antenna and a small aperture that are relevant for the specific case of NSMM applications. In general, the electric and magnetic fields can be expressed following Reference [6] through integral equations. These expressions apply to a finite, inhomogeneous scatterer and can be expressed for electric field component as

$$E(r) = E_{inc}(r) + E_{scat}(r) = E_{inc}(r) + \int_V dr' G_E(r,r') \left[k_s^2(r') - k_0^2 \right] E(r')$$
$$- \int_V dr' G_E(r,r') \left[k_s^2(r') - k_0^2 \right] (\mu \nabla \mu^{-1}) \times [\nabla \times E(r')],$$

(12.11)

where wavenumber $k_s = \sqrt{\omega^2 \mu \varepsilon}$ is the function of the position within the inhomogeneous region and k_0 is the wavenumber of the surrounding medium. The permeability μ and the permittivity ε can be in general tensors and are also functions of positions inside the inhomogeneous region. Similarly, the corresponding expression for the magnetic field is obtained by replacing the electric field Green's function with a corresponding magnetic field Green's function and by modifying the coefficient in front of the integral. The dyadic Green's function $G(r,r')$ is a solution of the equation

$$\nabla \times \nabla \times G(r,r') - k_0^2 \, G(r,r') = I\delta(r - r')$$

(12.12)

with I the identity dyadic. The solution is

$$G(r,r') = \left[I + \frac{\nabla \nabla}{k_0^2} \right] g(r,r'); \; g(r,r') = \frac{e^{-jk_0|r-r'|}}{4|r-r'|} = \phi(|r-r'|).$$

(12.13)

For a homogeneous medium, the Green's function in Equation (12.13) is reduced to $g(r,r')$. This Green's function can also be expressed in terms of Bessel and Hankel functions as it was introduced earlier in Equation (12.6). For constant permeability μ, Equation (12.11) is reduced to

$$E(r) = E_{inc}(r) + \int_V dr' G_E(r,r') \left[k_s^2(r') - k_0^2 \right] E(r').$$

(12.14)

In the preceding equations, $E_{inc}(r)$ is usually known (see the introduced near-fields applicable to NSMM problems), but the total field is unknown and as follows from Equation (12.14) is a part of the integral as well. In the case when $| k_s^2(r') - k_0^2 |$ is small, one can approximate $E(r) \cong E_{inc}(r)$ and the field can be expressed as

$$E(r) = E_{inc}(r) + \int_V dr' G_E(r,r') \left[k_s^2(r') - k_0^2 \right] E_{inc}(r') =$$
$$= E_{inc}(r) + \int_V K_E(r,r') E_{inc}(r') dr' = E_{inc}(r) + E_{scat}(r).$$

(12.15)

This is known as the first Born approximation and represents the first correction term that is obtained by use of perturbation theory and usually it is justified in long wavelength solutions. For the solution of near-field problems where they are treated as electrostatic problems, its use is usually justified. Note that we introduced the electric field kernel of the integral $K_E(r,r')$. If necessary, higher order

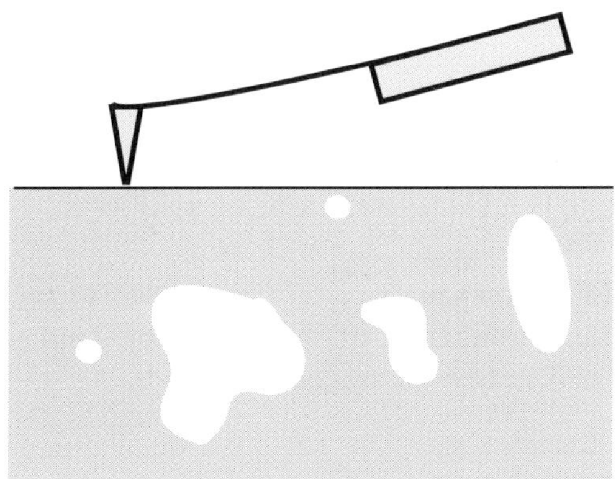

Figure 12.1. Near-field tomography performed by use of a scanning probe system. A schematic of a microcantilever probe positioned above an SUT. Assorted subsurface, embedded structures are shown in white.

correction terms can be obtained through an iterative procedure. The solution for the magnetic field may obtained by an analogous approach.

As follows from Equation (12.15), the problem is reduced to the solution of a three-dimensional integral equation. Through a Fourier transform, the problem may be further simplified to a one-dimensional Fredholm integral equation of the first kind, as will be shown later in the text. An illustration of near-field tomography performed by use of a cantilever-based scanning probe system is shown in Fig. 12.1. The scanning probe is positioned over the half space with embedded nonuniformities [12]. Assume that the measured two-dimensional distribution of the measured signal is a function of a parameter h representing the distribution of the material properties in the scanned volume. In this case the relation of the measured signal R with respect to any field component and material parameters, represented, say, through permittivity, can be expressed as [12]

$$R(x,y,h) = \int_{V'} K(x-x',y-y',z',h)\,\varepsilon(x',y',z')\,dx',dy',dz', \qquad (12.16)$$

where kernel K contains the instrumental response and $\varepsilon(x',y',z')$ represents the unknown parameters of the material to be obtained. Equation (12.16) is the three-dimensional Fredholm's integral equation. This equation is known to be an ill-posed, complicated problem that usually requires additional a priori information about possible, correct solutions. The two-dimensional Fourier transform of Equation (12.16) leads to [12]

$$\tilde{R}(\kappa_x,\kappa_y,h) = 4\pi^2 \int_{-\infty}^{0} \tilde{K}(\kappa_x,\kappa_y,z',h)\cdot\tilde{\varepsilon}(\kappa_x,\kappa_y,z')\,dz', \qquad (12.17)$$

where

$$\tilde{R}(\kappa_x,\kappa_y,h) = \frac{1}{4\pi^2}\iint R(x,y,h)\exp(-j\kappa_x x - j\kappa_y y)\,dx\,dy, \qquad (12.18a)$$

$$\tilde{\varepsilon}\left(\kappa_x,\kappa_y,h\right) = \frac{1}{4\pi^2}\iint \varepsilon(x,y,h)\exp(-j\kappa_x x - j\kappa_y y)dxdy, \qquad (12.18b)$$

$$\tilde{K}\left(\kappa_x,\kappa_y,h\right) = \frac{1}{4\pi^2}\iint K(x,y,h)\exp(-j\kappa_x x - j\kappa_y y)dxdy. \qquad (12.18c)$$

Provided that the kernel and solution satisfy certain conditions of regularity, it is possible to define the periodic continuation of these functions and use the two-dimensional, fast Fourier transform to evaluate these integrals.

12.2.5   Solution to the Inverse Problem

In the next steps, Equation (12.17) is solved for each pair of κ_x,κ_y as a function of the depth profile of $\tilde{\varepsilon}(\kappa_x,\kappa_y,z')$ and finally inverse Fourier transformed into $\varepsilon_{scat}(x,y,z) = \iint \tilde{\varepsilon}_{scat}\left(\kappa_x,\kappa_y,z\right)\exp(j\kappa_x x + j\kappa_y y)d\kappa_x d\kappa_y$. Following the Tikhonov's regularization [13], [14], the approximate solution of Equation (12.17) is obtained from the minimum of the functional

$$M_\alpha[\tilde{\varepsilon}] = \left\|4\pi^2\tilde{K}_\upsilon\tilde{\varepsilon} - \tilde{R}^\delta\right\|^2_{L_2} + \alpha\|\tilde{\varepsilon}\|^2_{W_2^1}, \qquad (12.19)$$

where $\tilde{R}^\delta$ are the measured values with an error δ. The parameter υ is the error in the estimate of the kernel operator $\tilde{K}$. L_2 and W_2^1 are the corresponding metrics. The parameter α should be found under the additional condition of having knowledge of measurement errors and the operator errors

$$\left\|4\pi^2\tilde{K}_\upsilon\tilde{\varepsilon} - \tilde{R}^\delta\right\|^2_{L_2} = (\delta + h\|\tilde{\varepsilon}\|^2_{W_2^1})^2 \qquad (12.19a)$$

The quality of the inverse problem solution depends on the values of error estimates δ and υ. In the limit where $\delta,\upsilon \to 0$, the solution approaches the correct one. The error parameter υ is usually estimated from the numerical solution of the forward problem. The estimate of the experimental error δ can be obtained from [12]

$$\delta^2 = \frac{1}{\Delta h}\iint_{\kappa_x,\kappa_y} d\kappa_x d\kappa_y \int_h \left[\delta\tilde{R}(h,\kappa_x,\kappa_y)\right]^2 dh \qquad (12.20)$$

with $\delta\tilde{R}\left(h,\kappa_x,\kappa_y\right) = \tilde{R}\left(h,\kappa_x,\kappa_y\right) - \tilde{R}^\delta(h,\kappa_x,\kappa_y)$. The resulting parameter distribution is obtained from inverse Fourier transform of $\tilde{\varepsilon}$. At this point, it is necessary to stress once again, that though solutions can be found for specific problems, a general solution is not possible due to the complexity and nonlinearity of the problem. In addition, when calculating numerical solutions to these equations, it is necessary to implement the stingiest possible conditions on the discretization of the material domain.

When the wavelength of the incident wave is much smaller than the size of the subsurface inhomogeneity $k_0 a = 2\pi a / \lambda \gg 1$ (where a is the dimension of the scattering object), the electromagnetic problem can be reduced to the study of the scalar wave equation. Under these conditions, the problem can be reduced to Rytov approximations, i.e., the phase perturbation is a linear functional of the object. Because the Rytov approximation is not valid for NSMM, we are not going to discuss the details of this approach.

The aforesaid approximations were introduced because of their mathematical simplicity. However, it is necessary to remember that these assumptions essentially ignore the fact that there are multiple reflections of the electromagnetic radiation. To account for multiple reflections, it is necessary to reformulate the problem as a nonlinear optimization problem. This requires the solution of the direct scattering problem for different domains of the embedded structures by use of iterative protocols. At each step of the iterative procedure, a solution to the scattering problem must be obtained. Clearly, this is a tedious, time-consuming process. Some progress has been obtained by separating the problem into an ill-posed linear part and a well-posed nonlinear part. This strategy avoids the solution of the direct scattering problem at each iteration step [15], [16], [17]. For NSMM, one can safely assume the condition $k_0 a = 2\pi a / \lambda \leq 1$, leading to solution of diffraction problems within the so-called resonance regime. Mathematical methods in the resonance regime are significantly different from problems where the Rytov condition is valid. In particular, for NSMM it is not necessary to consider the possibility of existence of shadows behind the scattering object. We will return to this solution when discussing the multifrequency approach later in the text. In the following sections, we address two specific cases that apply to NSMM measurements.

12.2.6 Linear Inverse Problem Solutions from Frequency Shift Measurements

We start with a simple case, namely linear inverse problems. It is assumed that the only fields that are measureable lie outside of the scatterer, i.e., they lie outside of the material or the DUT. This is the situation in most experiments. The general nonlinear problem can be significantly simplified if the scattered field can be approximated as a linear functional of the object. The goals of inverse tomographic problems include global quantitative and qualitative image reconstruction of the internal constituents of an object. Additional objectives include reconstructions of the object boundary and localized, internal inclusions. Specific examples of these goals include mapping of embedded material properties such as complex permittivity and detection of subsurface voids in electrical circuits. In order to pinpoint a localized, subsurface object, the inverse problem requires as inputs a given set of measurements that are made as a function of parameters in addition to position. In NSMM, the frequency is the most common such parameter. The localization of a subsurface object may be described by a position-dependent, characteristic function. In the simplest form, this function is equal to one at positions where the object is present and equal to zero elsewhere. This function's form may be customized for certain types of problems, e.g., the function may take on any real value [18], [19].

Experimentally, the first step in subsurface mapping of electromagnetic material distributions is to measure the two-dimensional, transverse distribution of the scattered field as a function both of position along the media interface (x and y) and frequency (f). In this case, f serves as the depth-sensitive parameter, due to the skin depth effect. Recall that the skin depth is given by $\delta = \sqrt{2\rho/\omega\mu}$ with ρ being the resistivity of the conductor, ω the angular frequency, and μ the permeability of the conductor. As applied to NSMM, the details of the inverse method depend on how the measurement has been done. Common approaches include the measurement of the frequency shifts of the resonator or the measurement of the complex impedance via the reflection coefficient. For measurement of the frequency shifts, most of the corresponding inverse scattering problems can be formulated by the equation [20], [21]

$$[O_s][\epsilon] = [\Delta F]. \tag{12.21}$$

$[\Delta F] = (\Delta f_1, \Delta f_2, \ldots, \Delta f_N)$ is the vector of measured frequency shifts at a given position on the sample, and $[O_s]$ is the direct, forward problem vector operator that determines the frequency shifts from the vector of given material parameters, $[\epsilon]$. In the case of Reference [19], $[\epsilon]$ is a permittivity vector. The operator $[O_s]$ must be calculated using numerical or analytical physical models of the forward problem for the given subsurface geometry. If the problem is linearized, $[O_s]$ is a constant, ill-posed matrix, and therefore the solution can't be obtained by the standard algebraic procedure of simply multiplying the equation from the left with inverse matrix. To regularize the problem, an alternate vector $[\hat{\epsilon}]$ may be defined in place of $[\epsilon]$ by use of a minimization algorithm [18]:

$$[\hat{\epsilon}] = \arg min\left(\sum_{k=1}^{N}\left(\Delta f_k(\epsilon) - \Delta f_k^{meas}\right)^2\right). \tag{12.22}$$

Here $\Delta f_k(\epsilon)$ are calculated values of frequency shift and Δf_k^{meas} are the corresponding measured values. The authors of Reference [18] used a Levenberg-Marquardt algorithm in the iteration procedure

$$[\hat{\epsilon}]_{i+1} = [\hat{\epsilon}]_i + ([A]_i^T [A]_i + \alpha[I])^{-1}[A]_i^T ([S][\hat{\epsilon}]_i - [\Delta F]^{meas}). \tag{12.23}$$

$[\Delta F]^{meas} = \left(\Delta f_1^{meas}, \Delta f_2^{meas}, \ldots, \Delta f_N^{meas}\right)$ is a vector of experimentally measured frequency shifts, $[A]$ is the matrix with elements $\dfrac{\partial \Delta f_k}{\partial \hat{\epsilon}_j}\big|_{\hat{\epsilon}_i}, k = 1, 2, \ldots, N; j = 1, 2, \ldots, M$, $[I]$ is the identity matrix and α is the Marquardt regularization parameter. The calculation of the $[A]$ matrix is not easy and therefore one can use the approximation $[A]_i = [A]_0$ to simplify the problem. In addition, within the iteration process the values of $\hat{\epsilon}_j$ could lead to unphysical values, such as a relative permittivity less than one. Therefore, it is necessary in the iteration process to bound the values $\hat{\epsilon}_j$ such that $\hat{\epsilon}_{min} < \hat{\epsilon}_j < \hat{\epsilon}_{max}$, where the limiting cases correspond to physically reasonable minimum and maximum values of the subsurface structure. In the specific case where $\hat{\epsilon}_j$ corresponds to permittivity, the regularization projection operator $\varepsilon^R = R_P(\epsilon)$ may be introduced:

$$\varepsilon_i^R = \begin{cases} \varepsilon_{min} & \varepsilon_i < 1 \\ \varepsilon_{max} & \varepsilon_i > \varepsilon_{max} \quad \text{for} \quad i = 1, 2, \ldots M. \\ \varepsilon_i & \text{otherwise} \end{cases} \tag{12.24}$$

Given the preceding analysis, including the introduced assumptions, the final iteration formula is:

$$[\hat{\epsilon}]_{i+1} = R_P \left([\hat{\epsilon}]_i + \left([A]_0^T [A]_0 + \alpha [I] \right)^{-1} [A]_0^T \left([S][\hat{\epsilon}]_i - [\Delta F]^{meas} \right) \right). \tag{12.25}$$

In solving the inverse problem, special treatment is required for the edges of the embedded structure. It may seem to be enough to fill the area within known surface with parameter (value of permittivity in the case that the object is dielectric) and reduce the problem to selection of finding the optimal ε values. In the case that not all edges are known a priori, this approach would fail, but it can be made to behave correctly if the regularization coefficients α in Equation (12.23) are properly selected. The greater values of α provide increased smoothing within the given area. Note that an additional, implicit parameter representing the effective probe tip (or antenna) diameter also enters the calculations.

12.2.7 Inverse Problem Solutions from Multifrequency or Multipoint Scattering Field Data

The preceding discussion relates to the reconstruction of subsurface information from the measured frequency shift at each probe position on a grid in two-dimensional space. Another experimental approach to the inverse problem is to measure the reflection coefficient within the experimental grid using a multifrequency approach. To illustrate this approach, consider NSMM imaging of an object confined in a subsurface volume that is surrounded by a uniform domain. Frequency is an easily controlled parameter in NSMM. Therefore, one natural approach to this imaging problem is measurement of two-dimensional, scattered signal for a number of incident fields of different frequencies, indexed by $i = 1, 2, \ldots, N$. For an NSMM probe, this can be modeled through elementary dipoles that are expressed through one of the field components of a TE or TM field. In such a case, we can reduce the calculation to a formulation component represented in terms of a scalar variable u alone [18]. As in previous cases, one has to start with the solution of the forward problem. For each incident excitation, indexed by j, the forward problem can be expressed as [18]

$$u_i(r) = u_i^{inc} + k_i^2 \int_{V_D} dr' G_i(r,r') \chi_i(r') u_i(r') \tag{12.26}$$

The integral is over the bounded, simply connected domain V_D within which the object is embedded. $G_i(r,r')$ is the Green's function of the uniform background medium. $G_i(r,r') = \frac{j}{4} H_0^{(1)}(k_i|r-r'|)$ at frequency i and

$$\chi_i(r') = \left[\frac{k^2(r',k_i)}{k_i^2} - 1 \right], \tag{12.27}$$

with $k_i = \omega_i \sqrt{\varepsilon \mu}$. In a simple, practical case where the scatterer is a nonmagnetic medium, the material properties may be represented by a function of the form $\chi_i(r')$. This function incorporates both permittivity and conductivity and can be expressed as

$$\chi_i(r') = \frac{\varepsilon(r') - \varepsilon}{\varepsilon} - j \frac{\sigma(r')}{\omega_i \varepsilon}, \tag{12.28}$$

where ε, μ are the permittivity and permeability of the lossless host material and $\varepsilon(r')$ and $\sigma(r')$ are the permittivity and conductivity of the scatterer. From the definition of $\chi_i(r')$, it follows that if r is not within the volume of the scatterer, then χ_i vanishes. It is important to remember that the position and composition of the scatterer is not known a priori. All that is known is that χ_i vanishes outside of the scatterer. The scattered field outside of the volume V_D [17], measured at a discrete number of points at each frequency, is given by

$$M_i(r) = k_i^2 \int_{V_D} dr' G_i(r,r') \chi_i(r') u_i(r') \qquad r \in C \tag{12.29}$$

where r is from the measurement domain C representing surface, curve or a discrete collection of points outside of volume V_D where the scattered field $M_i(r)$ is measured.

Strictly, this relation applies only in an ideal case free from noise and errors. Next, following the shortened operator annotation of Reference [17], we will rewrite the equations in a form

$$u_i = u_i^{inc} + G_i^{V_D} \chi_i u_i \qquad r \in V_D. \tag{12.30}$$

The corresponding equation for the measured fields takes the form

$$M_i = G_i^C \chi_i u_i \qquad r \in C. \tag{12.31}$$

The only known variables in these equations are the incident fields u_i^{inc}, the measured data M_i, and wavevector k_i. Equation (12.29) can be solved for the unknown variable u_i:

$$u_i = (I - G_i^{V_D} \chi_i)^{-1} u_i^{inc}. \tag{12.32}$$

By inserting (12.32) into (12.31), it becomes

$$M_i = G_i^C [\chi_i (I - G_i^{V_D} \chi_i)^{-1} u_i^{inc}]. \tag{12.33}$$

Equation (12.33) is a nonlinear equation. Approximating

$$(I - G_i^{V_D} \chi_i)^{-1} \approx I \tag{12.34}$$

leads to a linearized form in the Born approximation [17]. Alternately, one may use an iterative approach

$$(I - G_i^{V_D} \chi_{i,n})^{-1} \approx (I - G_i^{V_D} \chi_{i,n-1})^{-1}, \tag{12.35}$$

which represents the iterative Born method. Meanwhile, the first order Taylor expansion (linearization) in terms of $\Delta\chi_{i,n} = \chi_{i,n} - \chi_{i,n-1}$:

$$(I - G_i^{V_D} \chi_{i,n})^{-1} \approx [I + \left(I - G_i^{V_D} \chi_{i,n-1}\right)^{-1} G_i^{V_D} \Delta\chi_{i,n}](I - G_i^{V_D} \chi_{i,n-1})^{-1} \tag{12.36}$$

leads to Newton-Kantorovich method [18].

Note that the forward problem has to be solved at each iterative step in the methods described so far. Significant effort has been made in order to introduce approaches that do not require the solution of the forward problem. One approach is to introduce a variable w_i: $\chi_i u_i = w_i$ and rewrite Equations (12.31) and (12.32) in terms of w_i. Then

$$\chi_i u^{inc} = w_i - \chi_i G_i^{V_D} w_i. \tag{12.37}$$

The next step is to define the cost functional

$$F = \frac{\sum_i \left\| M_i - G_i^C w_i \right\|_C^2}{\sum_i \left\| M_i \right\|_C^2} + \frac{\sum_i \left\| \chi_i u^{inc} - w_i + \chi_i G_i^{V_D} w_i \right\|_{V_D}^2}{\sum_i \left\| \chi_i u^{inc} \right\|_{V_D}^2} \tag{12.38}$$

and minimize it. The operation $\| \ \|$ denotes the norm on the respective domains with the normalization chosen in a way that $w_i = 0$. The first term in Equation (12.38) represents the error in the measurement equations and the second term represents the error in the equations due to change of the variable to w_i. This equation is highly nonlinear and the minimization of the cost function is found iteratively. The process is quite involved and the interested reader can find further details in Reference [18].

12.2.8 Inverse Problem Solutions from Multifrequency Scattering Field Data

In Reference [22], the three-dimensional permittivity distribution was determined from multifrequency data. The inverse problem was solved by use of two-dimensional, lateral plane decomposition of corresponding Green's functions.

This strategy effectively reduces the three-dimensional integral equation to one-dimensional Fredholm integral equation of the first kind relative to the depth profile of the lateral permittivity spectrum. The foundation for this approach was introduced in Reference [23] and complements the general case, introduced earlier (see Equations (12.16) through (12.20)) [12]. One way to treat complex, electromagnetic, three-dimensional reconstruction problems is plane wave decomposition in k-space and the use of evanescent waves. Here, it turns out that it is advantageous to use the probing field of a near-field source in place of the decomposition into plane evanescent waves. The procedure is inherently dependent upon data acquisition. Two methods of data acquisition are proposed: a multifrequency approach and a multilevel approach. The former is two-dimensional, lateral scanning at several frequencies. The latter is two-dimensional scanning at several heights above the investigated area. Here, it is worth mentioning that the depth sensitivity of near-field measurements further depends on transfer functions of the source (determined mainly by their dimensions) and on the source–receiver distance.

As before, the solution of the inverse problem starts with Born approximation followed by an iterative solution based on one-dimensional equations. The solution is similar to one expressed in general in Equations (12.12) through (12.16) and follows from the convolution of the current distribution of the source and the Green's function in Equation (12.13). A textbook vector potential can be expressed inside a volume V in the k-space representation in plane wave expansion as

$$A(x,y,z) = \mu \int_{V'} J(r') \frac{j}{2\pi} \int_{-\infty}^{\infty} \int_{-\infty}^{\infty} e^{jk_x(x-x')+jk_y(y-y')} \frac{exp\left\{\pm j\sqrt{k^2 - k_x^2 - k_y^2}\,(z-z')\right\}}{\sqrt{k^2 - k_x^2 - k_y^2}} dk_x dk_y dr'. \quad (12.39)$$

The sign in the exponential function is positive for $(z-z') > 0$ and negative at $(z-z') < 0$. In addition, $k_z = \sqrt{k^2 - k_x^2 - k_y^2}$. From the vector potential, as before, it is possible to calculate the electric field in the form of a plane wave decomposition as a sum of the TM and TE wave field components following Equation (12.15) and introducing an effective current source $J_{eff}(r) = j\omega\varepsilon_{scat}(r)E(r)$. In general, $E(r)$ is the total electric field that is in this case composed from the incident and scattered fields. The situation simplifies if the total field can be replaced by the incident field $E_0(r)$. Additional simplification follows from the near-field assumption by neglecting the e^{jkr} component for $kr << 1$ and using the elementary dipole field distributions as the incident field. The Fourier transformation of the scattered fields in the first order Born approximation can be expressed as [12]

$$E_{scat}(\kappa_x, \kappa_y, z) = \frac{1}{j\omega\varepsilon_0} \int_{-\infty}^{0} \tilde{G}_E(\kappa_x, \kappa_y, z, z') J_{eff}(\kappa_x, \kappa_y, z') dz' \quad (12.40)$$

and

$$H_{scat}(\kappa_x, \kappa_y, z) = \int_{-\infty}^{0} \tilde{G}_H(\kappa_x, \kappa_y, z, z') J_{eff}(\kappa_x, \kappa_y, z') dz'. \quad (12.41)$$

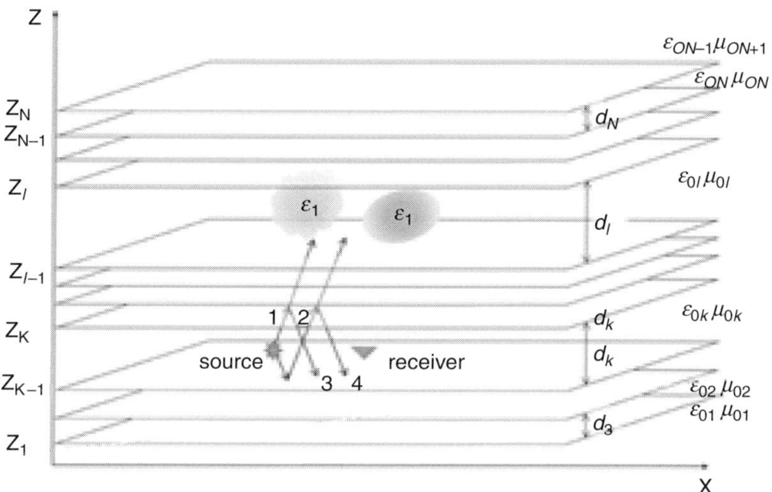

Figure 12.2. Schematic of the inverse scattering problem in a multilayered medium. The inhomogeneities in the *l*-th layer are probed with the field generated by the source and detected by the receiver in the *k*-th layer [23].
© IOP Publishing. Reproduced with permission. All rights reserved.

Here, ε_0 is the constant permittivity of the host material. For a multilayer medium, as shown in Fig. 12.2, the detailed expressions and the corresponding derivation of the fields in the different layers of the multilayer medium may be found in the original papers in References [22], [23]. Here, we will present just the final results for the components of the electric field expressed through the Green's functions for the total field in layer *l* as generated by a source in layer *k* of the planar multilayer medium as

$$E_n^l(\mathbf{r}) = \sum_m \int_{V'} J_m^k(\mathbf{r}') \int_{-\infty}^{\infty} \int_{-\infty}^{\infty} e^{jk_x(x-x')+jk_y(y-y')} \tilde{G}_{mn}^{kl}\left(k_x, k_y, z, z'\right) dk_x dk_y d\mathbf{r}'.$$
$$= \sum_m \int_{V'} J_m^k(\mathbf{r}') G_{mn}^{kl}\left(x-x', y-y', z, z'\right) d\mathbf{r}'$$

$$(12.42)$$

where *n, m* = *x, y, z* are the field and current components of the corresponding vectors and notice the change of the integration variables from κ_i to k_i. This total field can be expressed similarly as before as a sum of the probing (incident) and scattered fields (see Equation (12.15)) in the *k*-th layer from the target in the *l*-th layer in the form

$$E_{scatn}^k(\mathbf{r}) = j\omega \sum_m \int_{V'} \varepsilon_{scat}(\mathbf{r}') E_{0m}^l(\mathbf{r}') G_{mn}^{kl}\left(x-x', y-y', z, z'\right) d\mathbf{r}', \quad (12.43)$$

where

$$E_{0n}^l(\mathbf{r}) = \sum_m \int_{V'} J_m^k(\mathbf{r}') G_{mn}^{kl}\left(x-x', y-y', z, z'\right) d\mathbf{r}' \quad (12.44)$$

is the source field at the l-th layer and

$$E_{0n}^{k}(r) = j\omega \sum_{m} \int_{V'} J_{m}^{k}(r') G_{mn}^{kl}(x - x', y - y', z, z') dr' \tag{12.45}$$

is the source field in the k-th layer of the multilayered medium. Equation (12.43) can be reduced to a one-dimensional Fredholm integral equation of the first kind in frequency space by fixing the reference field around the receiving point and by fixing the source-receiver vector δr as [23]

$$E_{scatn}^{k}\left(k_x, k_y, z, \delta r\right) = j\omega \sum_{m} \int_{z'} \varepsilon_{scat}(k_x, k_y, z')$$
$$\left[\int_{-\infty}^{\infty} \int_{-\infty}^{\infty} e^{-j\kappa_x \delta x - j\kappa_y \delta y} \int_{z''} J_{n}^{k}\left(\kappa_x, \kappa_y, z'' - z - \delta z\right) G_{nm}^{kl}(\kappa_x, \kappa_y, z'z'') \right]$$
$$G_{mn}^{lk}(\kappa_x + k_x, \kappa_y + k_y, z, z') d\kappa_x d\kappa_y dz'' dz'$$

$$\tag{12.46}$$

Equation (12.46) can be used in the solution of the multifrequency scheme of the inverse scattering problem. In particular, Equation (12.46) can be expressed as

$$E_{scat}\left(k_x, k_y, \omega\right) = \int_{z'} \varepsilon_{scat}(k_x, k_y, z') K(k_x, k_y, z', \omega) dz', \tag{12.47}$$

where the depth sensitivity is determined by the frequency dependence of kernel K. The data acquisition is once again performed with the multilevel approach, i.e., two-dimensional, transverse scanning measurements done at different vertical positions z with respect to the k-th layer (please note the difference between the index k and the wave vector components k_x etc.). In the case of NSMM, the depth sensitivity is related to the strong dependence of near-field components of emitted and scattered field upon source/receiver-target (probe-sample) distance. The effective normalized scattering permittivity can be expressed as [12]

$$\varepsilon_{scat}^{eff}\left(k_x, k_y, z\right) = \int_{-\infty}^{0} \varepsilon_{scat}(k_x, k_y, z') K(k_x, k_y, z', z) dz'. \tag{12.48}$$

Equations (12.47) and (12.48) should be solved for each pair of spectral components k_x, k_y of the scattering permittivity spectrum. The details of the numerical procedure for selected properties of the embedded dielectric inclusion are beyond the scope of this chapter, but can be found in the literature [22, 23].

Now, we will express the source field in terms of the currents across the cross section of the aperture. For simplicity, these currents will be assumed to move in only one direction and to be localized at the distance $z = z_0$ from the surface. The surface currents are then

$$J^{s}(x, y, z) = J_{x}^{s}(x, y, z) \delta(z - z_0). \tag{12.49}$$

Using the boundary conditions at the surface of the medium ($z = 0$) and expressing the fields through the Fresnel transmission coefficients

$$T_{\parallel} = \frac{2\sqrt{\varepsilon_0}\sqrt{k_0^2 - \kappa_{\perp}^2}}{\varepsilon_0\sqrt{k_0^2 - \kappa_{\perp}^2} + \sqrt{k^2 - \kappa_{\perp}^2}} \qquad (12.50)$$

and

$$T_{\perp} = \frac{2\sqrt{k_0^2 - \kappa_{\perp}^2}}{\sqrt{k_0^2 - \kappa_{\perp}^2} + \sqrt{k^2 - \kappa_{\perp}^2}}, \qquad (12.51)$$

the kernel in (12.54) can be expressed as [12]

$$K\left(k_x, k_y, z\right) - \frac{\tilde{K}(k_x, k_y, z)}{\int_{-\infty}^{0}\iint \tilde{K}(k_x, k_y, z')dk_x dk_y dz'}, \qquad (12.52a)$$

$$\tilde{K}\left(k_x, k_y, z\right) = \iint_{-\infty}^{\infty} J_x^s\left(\kappa_x, \kappa_y, z_0\right) J_x^{s*}(\kappa_x + k_x, \kappa_y + k_y, z_0) \frac{\exp(\pm j\sqrt{k^2 - \kappa_{\perp}^2}\, z \pm j\sqrt{k_0^2 - \kappa_{\perp}^2}\, z_0)}{j\sqrt{k_0^2 - \kappa_{\perp}^2}}$$

$$\times \frac{\exp\{\left(\pm j\sqrt{k^2 - (\kappa_x + k_x)^2 - (\kappa_y + k_y)^2}\, z \pm j\sqrt{k_0^2 - (\kappa_x + k_x)^2 - (\kappa_y + k_y)^2}\, z_0\right)\}^*}{\left(j\sqrt{k_0^2 - (\kappa_x + k_x)^2 - (\kappa_y + k_y)^2}\right)^*}$$

$$\times \left[\begin{array}{l} f_x^{E_{inc}}\left(\kappa_x, \kappa_y, \varepsilon_0\right) f_x^{E_{inc}*}\left(\kappa_x + k_x, \kappa_y + k_y, \varepsilon_0\right) \\ + f_y^{E_{inc}}\left(\kappa_x, \kappa_y, \varepsilon_0\right) f_y^{E_{inc}*}\left(\kappa_x + k_x, \kappa_y + k_y, \varepsilon_0\right) \\ + f_z^{E_{inc}}\left(\kappa_x, \kappa_y, \varepsilon_0\right) f_z^{E_{inc}*}\left(\kappa_x + k_x, \kappa_y + k_y, \varepsilon_0\right) \end{array}\right] d\kappa_x d\kappa_y$$

$$(12.52b)$$

where

$$f_x^{E_{inc}} = \frac{1}{\kappa_{\perp}^2}\left[\kappa_x^2\left(k_0^2 - \kappa_{\perp}^2\right)T_{\parallel} + \kappa_y^2 k_0^2 T_{\perp}\right];$$

$$f_y^{E_{inc}} = \frac{\kappa_x \kappa_y}{\kappa_{\perp}^2}\left[\left(k_0^2 - \kappa_{\perp}^2\right)T_{\parallel} - k_0^2 T_{\perp}\right]; \qquad (12.53)$$

$$f_z^{E_{inc}} = \pm\kappa_x T_{\parallel}\sqrt{k_0^2 - \kappa_{\perp}^2}.$$

Further details of this approach are given in Reference [24]. The reflected field is again expressed in the same form as Equation (12.11). The complex amplitudes of the received signal are expressed as a convolution of the scattered field $E_{scat}(r)$ and the instrument function of the receiver F as [22]

$$s(r_r) = \int E_{scat}\left(r'\right) F\left(x_r - x', y_r - y', z_r, z'\right) dx'dy'dz', \qquad (12.54)$$

where r_r is the vector of the receiver position. Taking into account the expressions for the scattered fields due to change of the permittivity of the embedded object and the instrument function the transfer spectrum of the measured signal then can be expressed as

$$s\left(k_x,k_y,\omega\right)=\int_{z'}\varepsilon_{scat}\left(k_x,k_y,z'\right)K\left(k_x,k_y,z',\omega\right)dz'.$$ (12.55)

Following Reference [22], the multifrequency data is transformed to the synthetized pulse

$$s_{Re}\left(x,y,t\right)=\Re\left\{\int_0^\infty s(x,y,\omega)\exp\left(j\omega t\right)d\omega\right\}$$ (12.56)

that can be expressed as a function of the effective depth parameter, if one takes into account the velocity of the electromagnetic wave propagation in the medium. Clearly, the bounds of the integration must span the measured frequency band $\Delta\omega$. It is possible to try the same transformation in Equation (12.39) while including the time dependence t:

$$s_{Re}\left(k_x,k_y,z_{eff}\right)=\Re\left\{\int_0^\infty s\left(k_x,k_y,\omega\right)\exp\left(j\omega t\right)d\omega\right\}=\int_{z'}\varepsilon_{scat}\left(k_x,k_y,z'\right)K\left(k_x,k_y,z',z_{eff}\right)dz';$$
$$K\left(k_x,k_y,z',z_{eff}\right)=\int_0^\infty K\left(k_x,k_y,z',\omega\right)\exp\left(j\omega t\right)d\omega$$ (12.57)

where the kernel in Equation (12.57) is formed from Equation (12.46) after the summation over n,m and integration over κ_x and κ_y with addition of the instrument function in Equation (12.54). Subsequently, it is shown how this kernel can be obtained from the experiment. The solution of the Fredholm integral Equation (12.55) simplifies when it is known that the permittivity of the target ε_{scat}^0 is constant. In this case, the problem is reduced to shape retrieval. The shape is determined by two space-dependent functions $x_1(y,z)$ and $x_2(y,z)$. The k-space permittivity spectrum can be written as

$$\varepsilon_{scat}\left(k_x,k_y,z\right)=\int_{y_1}^{y_2}\int_{x_1(y)}^{x_2(y)}\varepsilon_{scat}^0e^{-jk_xx-jk_yy}dxdy=\varepsilon_{scat}^0\int_{y_1}^{y_2}\frac{e^{-jk_yy}}{jk_x}\left(e^{-jk_xx_1(y)}-e^{-jk_xx_2(y)}\right)dy.$$ (12.58)

After taking the inverse Fourier transform over k_y

$$\varepsilon_{scat}\left(k_x,y,z\right)=\frac{\varepsilon_{scat}^0}{jk_x}\left(e^{-jk_xx_1(y,z)}-e^{-jk_xx_2(y,z)}\right),$$ (12.59)

which is equivalent to a system of two real equations.

The kernel in Equation (12.55) can be determined experimentally from the measurement of weakly scattering, thin test samples with known shape and permittivity positioned at different depths z_0 of the investigated region. In this case $\varepsilon_{scat}\left(k_x,k_y,z\right)=\varepsilon_t(k_x,k_y)\delta(z-z_0)$ and from Equation (12.55), it follows

$$K\left(k_x,k_y,z_0,z_{eff}\right)=\frac{s(k_x,k_y,z_0,z_{eff})}{\varepsilon_t(k_x,k_y)}. \tag{12.60}$$

In this way, measurements of the known, thin test samples form the basis for a calibration procedure for subsurface permittivity profiling of weakly scattering objects with a constant scattering parameter.

In the case of strong inhomogeneity the Born approximation is not valid, but it can still be applied by an iterative algorithm [22], [23]

$$s^{(n)}\left(k_x,k_y,z_{eff}\right)=s\left(k_x,k_y,z_{eff}\right)-\Delta s\left(\varepsilon_{scat}^{(n-1)},k_x,k_y,z_{eff}\right)=\int_{z'}\varepsilon_{scat}^{(n)}(k_x,k_y,z')K\left(k_x,k_y,z',z_{eff}\right)dz' \tag{12.61}$$

with $\varepsilon_{scat}^{(0)}=0$. In this case it is also possible to obtain the kernel experimentally, but the instrumental transfer function has to be included at each step of the solution for signal correction.

12.2.9 Inverse Problem Solutions from Multifrequency or Multipoint Scattering Reflection Coefficient Data

The approach described earlier requires minimization of the cost functional based on the measured field values over a certain surface or line that lies outside of the volume of the embedded structure. In NSMM, most of the time the reflection coefficient or the change of the impedance is the measured, rather than the field. In Reference [25], the authors considered the method for NSMM in light of this distinction. As a first step, detailed formulas were derived for small antenna fields over multilayer media for TE and TM modes. Noting that the measured impedance depends on the complex permittivity, a parametric model of the permittivity of the nonmagnetic medium is introduced in the form

$$\varepsilon=\varepsilon_r\left(z,\boldsymbol{p}\right)-j\frac{\sigma(z,\boldsymbol{p})}{\omega\varepsilon_0}, \tag{12.62}$$

where z is the direction perpendicular to the multilayer stacking and $\boldsymbol{p}$ is a vector of parameters. For example, in a discrete, layered medium $\boldsymbol{p}$ might include the thicknesses t_i, permittivities ε_i and conductivities σ_i of the different layers:

$$\boldsymbol{p}=\left\{t_i,\varepsilon_i,\sigma_i,d_{eff}\right\}_{i=1}^{N}. \tag{12.63}$$

The effective tip (antenna) diameter d_{eff} is also included as it has to be considered in calculation of the impedance. Specifically, when modeling the NSMM interaction with an object through circuit models, the capacitance will depend on d_{eff}, assuming that the near-field of the aperture is consistent with Equations (12.9) and (12.10). One estimate of an effective tip (antenna) diameter d_{eff} for a small horizontal antenna at distance z_0 from the planar surface, where the magnetic charge and currents are replaced by current distribution on the aperture, is given by [26]

$$j(r,z) = \xi(r)\delta(z-h)u_y = \xi_0\exp(-4r^2/d_{eff}^2)\delta(z-h)u_y \tag{12.64}$$

where u_y is the unit vector parallel to the surface, z is the coordinate perpendicular to the surface, and ξ_0 is the maximum amplitude of the surface current.

In the case of a continuous medium with spatially varying conductivity in the form [27]

$$\sigma = \sigma_0 + (\sigma_m - \sigma_0)e^{-\frac{(z-z_0)^2}{t_0^2}}, \tag{12.65}$$

p may have a form

$$p = \left\{\left\{t_0, z_0, d_{eff}, \varepsilon_r, \sigma_0, \sigma_m\right\}\right\}. \tag{12.66}$$

σ_m is the maximum conductivity. The corresponding cost functional based on measured reflection coefficient values has the form

$$F(p) = \frac{\left\{\sum_{k=1}^{K}\left[(X_k(p) - X_k^m)^2 + (R_k(p) - R_k^m)^2\right]\right\}^{1/2}}{\left\{\sum_{k=1}^{K}\left((X_k^m)^2 + (R_k^m)^2\right)\right\}^{1/2}} \tag{12.67}$$

where $\left\{R_k^m, X_k^m\right\}_{k=1}^{K}$ are experimental data obtained from measurement of the impedance (as obtained from reflection coefficient) at K points and (R_k, X_k) are calculated data from the theoretical analysis for the given trial vector parameters p [25]. The values of the elements of p that minimize the functional Equation (12.67) are then the estimated parameters of the medium under investigation. For all the discussed inverse problem solutions, the number of frequencies or measurement points depends on the problem at hand. The lowest numbers leading to reasonable results can be 2 or 3 [25]. However, increasing the number of frequencies or measurement points will improve the precision of the extracted geometric and material parameters. The upper limit on the number of measurements is in practice controlled by the experimental and data processing capabilities.

12.3   Experimental Subsurface Tomography with Near-Field Microwave Microscopes

Recently, several experimental results have been published that demonstrate the subsurface tomographic capabilities of NSMM and related RF microscopy techniques.

One example of early work on subsurface imaging with NSMMs was introduced in Reference [28]. Although that experiment focused on the demonstration of spatial resolution beyond the diffraction limit, the authors studied the imaging of a steel surface as a function of both the frequency and probe sample height, necessarily exploring the frequency- and height-dependent inclusion of subsurface regions in the sampling volume. The system was modeled as a resonant coaxial line and the interpretation of the measurements was based on a standard transmission line theory. A more detailed transmission line approach tailored to near-field imaging applications was later developed [25, 26, 27]. Another early example introduced the application of scanning capacitance microscopy for subsurface characterization of metallic structures [29]. In this work, metallic structures were covered by 1 μm-thick, planarized SiO_2. Numerical modeling of the capacitance change due to subsurface metallic lines was done with a commercial software package. Specifically, the package was used to iteratively solve the static field equation $\nabla\left[\varepsilon_r\varepsilon_0\nabla\Phi(x,y)\right] = -\rho$, where $\Phi(x,y)$ is the electric potential and ρ the charge density. Given potential difference V between the probe and buried metallization, the capacitance is calculated from the standard electrostatic relation for the stored energy U_e in the electric field as $C = 2U_e / V^2$.

More recently, the application of NSMM to subsurface, nondestructive imaging of truly buried interfaces in metallic samples was introduced by Plassard et al. [30]. Note that depth profiling of high-loss materials is in practice much simpler than profiling of low-loss materials. As discussed theoretically earlier, this is due to the fact that the frequency serves as the depth-sensitive parameter and that the skin depth is more sensitive to frequency for high-loss materials. The authors of Reference [30] fabricated buried inclusions by patterning 20 nm-thick, well-defined aluminum shapes on a Si substrate and subsequently covering them with a continuous, planarized, 95 nm-thick Ni film. These samples were then studied by use of NSMM operating at frequencies between 1 GHz and 6 GHz. As the different constituent materials have different skin depths for a given operating frequency, properties of the buried interfaces were obtained by using a series of different operating frequencies. This approach effectively produces a different "cut" of the material at a given depth level corresponding to the selected NSMM frequency. The measurement results are shown in Fig. 12.3. The measured and displayed in the figure is the differential phase of the NSMM signal $\varphi = z / \delta$, where z is the position (depth) within the sample and δ is the classical skin depth. This technique was also used to investigate conventional, stainless steel samples [31], [32]. Stainless steel samples are often used in highly reactive, nonequilibrium environments that can cause changes in the crystalline structure and create oxides at grain boundaries with potential degradation of the material properties. The frequency-dependent measurements revealed that the NSMM is sensitive to the presence of subsurface oxide domains and enabled estimation of the size of these defects. In summary, these measurements demonstrated that NSMM is sensitive to buried interfaces in metallic samples with spatial resolution on the order of nanometers.

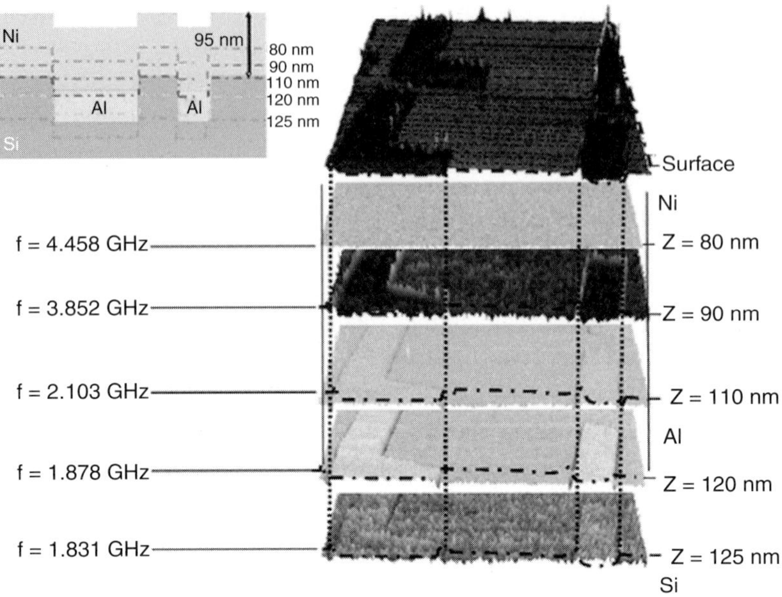

Figure 12.3. Images of buried Al patterns in a Ni matrix.
A schematic diagram of the fabricated structures is shown in the upper left corner.
The NSMM operating frequencies (f) are listed to the left of each image while the
corresponding depths (z) are shown on the right.
Reprinted figure with permission from C. Plassard, E. Bourillot, J. Rossignol, Y. Lacroute,
E. Lepleux, L. Pacheco, and E. Lesniewska, *Physical Review B* 83, (2011) art. no.
121409(R). © 2011, American Physical Society.

To date, experimental subsurface NSMM has been largely qualitative, as illustrated in the preceding examples. There are several ways forward to optimize subsurface imaging with NSMM and establish it as a quantitative technique. In Reference [33], the authors applied subsurface imaging to integrated circuits and nanoelectronics. Specifically, the focus of the work is on subsurface imaging of back-end-of-line processes in integrated circuits. Buried metal lines were imaged through the measurement of local capacitance. These measurements were carried out by use of NSMM and scanning Kelvin force microscopy. The procedure follows from the idea presented in Reference [29] and the skin depth of microwave radiation, introduced earlier in the chapter. To put buried line detection on a quantitative footing, the authors introduced two test chips with precisely calculable electromagnetic field distributions at the surface of the test chips. The first chip design included isolated metal lines and squares of Cr, Al, and Au. The first chip represents a calibration artifact with different work functions. The second chip design incorporated metallic interconnect lines buried at four different depths within a bulk SiO_2 insulator. The buried lines were 1.2 μm wide and separated by 1.2 μm. The second chip was designed to allow biasing of the buried interconnect lines in order to produce various patterns of electric and magnetic fields that could be imaged with the appropriate scanning probe microscope. The bond pads that provided electrical access to these buried interconnects were outside of the scan area, thus providing clearance for probing without mechanical interference. The electric field and surface potential distribution was calculated by use of a commercial multiphysics simulation package software. The simulations revealed that including the physical parameters of the tip significantly lowered the predicted resolution, suggesting that high aspect ratio probes are necessary for optimized resolution of lines buried a few micrometers below the surface of an insulator. The experimental results with NSMM proved that metal lines buried 2 μm under the surface of the dielectric can be imaged in reflection mode with reasonable resolution. Note that the measurements were done in contact mode.

This work was extended in Reference [34], in which a more detailed analysis is presented. In these latter experiments, the buried lines were 1.2 μm wide and covered with 0.8 μm or 2.3 μm SiO_2. The maximum contrast in the amplitude of the reflection coefficient between the areas with and without the buried metal lines was 0.006 dB for the deepest lines and 0.04 dB for the shallower lines. Note that the deeper line represented the limit of detection for this technique. To compare the observed experimental data with the theoretical prediction the authors used a simple lumped-element model of the tip-sample impedance that did not consider the geometry of the tip or changes in the reactive components of the tip-sample impedance. In spite of these simplifying assumptions, the results from this simple model were in qualitative agreement with the observed amplitude and phase changes. The model is based on a lumped-element description. Several similar models were discussed in more detail in Chapter 9. These measurements reaffirmed experimentally and qualitatively from a simple model calculation that NSMM has a reasonable sensitivity to metal lines buried under about 1 μm dielectric layer, but lacks the sensitivity to detect such lines buried more than a few micrometers below the surface.

The sensitivity of NSMM to structures buried under metallic layers was studied in Reference [35]. The use of NSMM at 1.8 GHz was made to measure CMOS test structures with eight metalized layers and two subsurface, 10 μm bus lines. The top two layers that covered the bus lines consisted of a 5 μm × 5 μm metallic square grid, serving as metal fill layers. The 15 μm diameter probe could detect the bus lines beneath the two metal grid layers.

The sensitivity of NSMM to subsurface features will necessarily depend on the frequency dependence of the signal-to-noise ratio of the instrument. Consider the resonance condition of a transmission line resonator can be expressed as [36]

$$\exp(-j2\gamma L)\, \Gamma\Gamma_0 = \exp(-j2\pi n) \tag{12.68}$$

where L is the resonator length, γ is the complex propagation constant of the transmission line, Γ is the reflection coefficient corresponding to the tip-sample impedance, Γ_0 is the reflection coefficient at the resonator input side, and n is the mode number of the resonator resonance. Expressing the propagation constant $\gamma = \omega\left(\varepsilon_0 \varepsilon_{eff} \mu_0\right)^{1/2} - j\gamma''$, with ε_{eff} equal to the effective permittivity of the transmission line, the frequency shift is

$$\frac{\delta f}{f} = \frac{Z_0}{L\sqrt{\varepsilon_0 \varepsilon_{eff} \mu_0}} \Delta C_t, \tag{12.69}$$

where ΔC_t is the change in the tip capacitance, or alternatively in terms of the phase

$$\delta\theta = 2\omega Z_0 \Delta C_t \tag{12.70}$$

where Z_0 is the characteristic reference impedance, often chosen to be 50 Ω. The signal-to-noise ratio of the reflection coefficient amplitude increases with frequency. For the phase of the reflection coefficient, the signal-to-noise ratio is more or less constant for frequencies up to about 10 GHz and decreases for frequencies above 10 GHz.

More general approaches to subsurface imaging exist, as described earlier in this chapter. Significant enhancement of the signal-to-noise ratio will require application of the more sophisticated theoretical approaches discussed earlier. One viable experimental approach is to apply a signal to the buried line with the same frequency as the NSMM operating frequency. This option further offers the opportunity to improve the signal-to-noise ratio through frequency mixing of the signals to the tip and sample. Additional options would be available if a second antenna is incorporated into the system.

Reference [37] introduced calibration techniques for quantitative, subsurface, noncontact NSMM imaging and tomography applications. The work was an extension of the capacitance calibration approach discussed in Chapter 7. The sensitivity of NSMM to subsurface elements was studied by used of calibrated experiments and theoretical finite-element modeling. The sensitivity was studied for both contact and noncontact microscope modes and as a function of tip radius and tip-sample separation. Subsurface imaging was demonstrated on a topographically flat semiconductor test sample with stripes of increasing dopant density from 1×10^{16} to 4×10^{19} atoms cm^{-3} separated by bulk interface layers. The stripes were buried by depositing additional 120 nm- and 200 nm-thick oxide layers. The sample flatness ensured that no cross-talk from the topography contributed to the S_{11} reflection coefficient image. The sample also included a bare surface that served as a reference for comparison to the buried doped silicon. The measurements made using NSMM revealed that the dopant-dependent contrast could be resolved through both oxide thicknesses, but quantitative measurement requires a reliable calibration approach. A calibration procedure established the tip-sample capacitance as a function of oxide thickness via measurements of a reference sample with 100 nm, 200 nm, and 400 nm oxide layers and a native oxide. From the calibrated tip-sample capacitance a reasonable estimate was found of the detection sensitivity to subsurface semiconducting artifacts.

In these experiments, the distance between the tip and subsurface feature of interest is established by the oxide thickness. As the oxide thickness increases, the lateral resolution decreases, as one would expect, assuming that the lateral resolution is related to tip diameter and the distance of the buried artifact from the tip. Furthermore, a thorough experimental investigation in Reference [37] showed that if the NSMM measurements are done in noncontact mode with the tip at a distance above the sample, the capability to resolve subsurface dopant contrast is lost when the tip height is on the order of the tip diameter. The lateral spatial resolution depends on the geometric structure of the feature of interest and the tip dimensions.

Another strategy for enhancement of the signal-to-noise ratio for materials buried in a dielectric environment is based on electric force microscopy [38]. This has been experimentally demonstrated by imaging of CNTs immersed in a polymer film. The measurement is done in two scanning passes. One pass is for measurement of the topography, including sample tilt, and the second pass is for measurement of the electrostatic force signal with the distance between the tip and the sample constant. The constant separation is maintained based on the topographic information acquired during the first pass. In addition, during the second pass a bias voltage can be applied to the conductive tip of the probe. The mechanical amplitude and phase of the cantilever probe are recorded as the electrostatic force signal. Recall that the electrostatic force is proportional to dC/dz and to the square of the tip bias voltage V_{tip}. The change of the phase shift of the AFM probe under the harmonic oscillator approximation can be expressed as [38]

$$\Delta\varnothing = \varnothing - \varnothing_0 \approx \tan(\varnothing - \varnothing_0) \approx \frac{Q}{2k}\left(C_1''(z) - C_2''(z)\right)V_{tip}^2, \tag{12.71}$$

where $\varnothing$ is the phase shift when the probe is over the area with the inclusion of interest (here, a CNT) and $\varnothing_0$ is the phase shift when the probe is over the plain polymer. Q is the quality factor of the AFM probe and k is the spring constant.

The underlying principle of the electrostatic force technique can be extended to NSMM. Like the electrostatic force, the microwave reflection coefficient is also a function of the capacitance between the tip and sample. Changes in the capacitance are not manifest through changes in the amplitude and phase of the vibrating cantilever, as it is in the case of electrostatic force microscopy, but rather through changes in the amplitude and phase of the reflection coefficient. Therefore, Equation (12.71) and the related analysis can be modified to apply to NSMM. For an electrostatic force microscope tip at a distance z_0 above a film surface, it follows from a simple plane capacitor model that the second derivative of the capacitance over portions of the sample with inclusions is [38]

$$C_2''(z) = 8\pi\varepsilon_0 \left[ln\left(\frac{1+\cos\theta}{1-\cos\theta}\right)\right]^{-2} \frac{1}{\left\{h + (D_1 + D_2)/\varepsilon_{film} + t/\varepsilon_{inclusion}\right\}^3} \tag{12.72}$$

and the second derivative of the capacitance over sample without the inclusions (plain polymer) is

$$C_1''(z) = 8\pi\varepsilon_0 \left[ln\left(\frac{1+cos\theta}{1-cos\theta}\right) \right]^{-2} \frac{1}{\left\{ h + (D_1 + D_2 + t)/\varepsilon_{film} \right\}^3}. \tag{12.73}$$

Here, θ is the conic half angle of the cantilever, D_1 is the depth of the inclusion, D_2 is the distance of the inclusion to the bottom of the film, t is the thickness of the inclusion, $\varepsilon_{inclusion}$ is the permittivity of the inclusion, and ε_{film} is the permittivity of the polymer film. Note that the phase contrast can be revealed only if the permittivity of the inclusion is significantly different from the permittivity of the polymer film. If the depth of the inclusion is not known a priori, then performing the measurements with the tip positioned at several heights over the sample surface enables a determination of D_1.

In this chapter, we began by presenting the theoretical foundations for the complex problem of subsurface microwave imaging of embedded structures. The reviewed theories are not yet complete and the development of theoretical, numerical, and experimental approaches to the problem is ongoing. Experimental and theoretical exploration of the subsurface imaging problems, particularly those that use NSMM, are in the early stages of development and significantly more work has to be done. This is true even more so for fully quantitative, calibrated characterization of subsurface structures. Experimentally, it has been demonstrated that the technique is capable of subsurface, nondestructive detection of embedded objects. Future work must establish calibrated, quantitative characterization of the size and material properties of embedded inclusions, defects, and interfaces.

References

[1]   R. Persico, *Introduction to Ground Penetrating Radar* (Wiley, 2014).
[2]   A. N. Reznik and N. V. Yurasova, "Near-Field Microwave Tomography of Biological Objects," *Technical Physics* 49 (2004) pp. 485–493.
[3]   K. P. Gaikovich, "Subsurface Near-Field Scanning Tomography," *Physical Review Letters* 98 (2007) art. no. 183902.
[4]   R. K. Amineh, A. Khalatpour, H. Xu, Y. Baskharoun, and N. K. Nikolova, "Three-Dimensional Near-Field Microwave Holography for Tissue Imaging," *International Journal of Biomedical Imaging* (2012) pp. 1–11.
[5]   G. S. Shekhawat and V. P. Dravid, "Nanoscale Imaging of Buried Structures via Scanning Near-Field Ultrasound Holography," *Science* 310 (2005) pp. 89–92.
[6]   A. Sommerfeld, *Partial Differential Equations in Physics* (Academic Press, 1964), W. Cho Chew, *Waves and Fields in Inhomogeneous Media* (Van Nostrand Reinhold, 1990).
[7]   H. A. Bethe, "Theory of Diffraction by Small Holes," *Physical Review* 66 (1944) pp. 163–182.
[8]   H. Levine and J. Schwinger, "On the Theory of Electromagnetic Wave Diffraction by a Small Aperture in an Infinite Plane Conducting Screen," *Communication on Pure and Applied Mathematics* 3 (1950) pp. 355–391.

[9] C. J. Bouwkamp, "Diffraction Theory," *Reports on Progress in Physics* 17 (1954) pp. 35–100.

[10] V. V. Klimov and V. S. Letokhov, "A Simple Theory of the Near-Field in Diffraction by a Round Aperture," *Optics Communications* 106 (1994) pp. 151–154.

[11] D. Colton and R. Kress, *Inverse Acoustic and Electromagnetic Scattering Theory* (Springer-Verlag, 1992).

[12] K. P. Gaikovich, "Scanning Near-Field Electromagnetic Tomography," *Journal of Nano and Microsystems Technique* 8 (2007) pp. 50–65 (in Russian).

[13] A. N. Tikhonov, A. V. Goncharsky, V. V. Stepanov, and A. G. Yagola, *Numerical Methods for the Solution of Ill-Posed Problems* (Kluwer Academic Publishers, 1995).

[14] A. N. Tikhonov, A. S. Leonov, and A. G. Yagola, *Nonlinear Ill-Posed Problems Volumes 1 and 2* (Chapman and Hall, 1998).

[15] A. Kirch and R. Kress, "A Numerical Method for an Inverse Scattering Problem." In *Inverse Problems*, (Engl and Groetsch, eds.) (Academic Press, 1987) pp. 279–290.

[16] T. S. Angell, R. E. Kleinman, and G. P. Roach, "An Inverse Transmission Problem for Helmholtz Equation," *Inverse Problems* 3 (1987) pp. 149–180.

[17] D. Colton and P. Monk, "A Novel Method for Solving the Inverse Scattering Problem for Time-Harmonic Acoustic Waves in the Resonance Region," *SIAM Journal on Applied Mathematics* 45 (1985) pp. 1039–1053.

[18] P. M. van den Berg and R. E. Kleinman, "A Contrast Source Inversion Method," *Inverse Problems* 13 (1997) pp. 1607–1620.

[19] K. Belkebir, R. E. Kleinman, and Ch. Pichot, "Microwave Imaging – Location and Shape Reconstruction from Multifrequency Scattering Data," *IEEE Transactions on Microwave Theory and Techniques* 45 (1997) pp. 469–476.

[20] Yu. Gayday, V. Sidorenko, O. Sinkevych, and Yu. Semenets, "Form Preserving Regularization in Near-Field Microwave Microscopy Inverse Problems," *Radiophysics and Electronics* 16 (2011) pp. 17–19.

[21] Yu. A. Gayday, V. S. Sidorenko, and O. V. Sinkevych, "Near-Field Microwave Tomography of Subsurface Dielectric Layers," *Radioelectronics and Communications Systems* 55 (2012) pp. 131–135.

[22] K. P. Gaikovich, P. K. Gaikovich, Ye. S. Maksimovitch, and V. A. Badeev, "Multifrequency Microwave Tomography of Absorbing Inhomogeneities," *Proceedings of Ultrawideband and Ultrashort Impulse Signals* (2010) pp. 156–158.

[23] K. P. Gaikovich and P. K. Gaikovich, "Inverse Problem of Near-Field Scattering in Multilayer Media," *Inverse Problems* 26 (2010) art. no. 125013.

[24] K. P. Gaikovich, P. K. Gaikovich, Ye. S. Maksimovitch, and V. A. Badeev, "Pseudopulse Near-Field Subsurface Tomography," *Physical Review Letters* 108 (2012) art. no. 163902.

[25] N. D. Vdovicheva, A. N. Reznik, and I. A. Shereshevskii, "Numerical Method for the Inverse Problem of Near-Field Microscopy of Layered Media," *International Conference Days on Diffraction 2011*, St. Petersburg, May 30–June 3, 2011.

[26] A. N. Reznik, "Quasistatics and Electrodynamics of Near-Field Microwave Microscope," *Journal of Applied Physics* 115 (2014) art. no. 084501.

[27] A. N. Reznik, I. A. Shereshevskii, and N. D. Vdovicheva, "The Near-Field Microwave Technique for Deep Profiling of Free Carrier Concentration in Semiconductors," *Journal of Applied* 109 (2011) art. no. 094508.

[28] C. P. Vlahacos, R. C. Black, S. M. Anlage, A. Amar, and F. C. Wellstood, "Near-Field Scanning Microwave Microscope with 100 mm Resolution," *Applied Physics Letters* 69 (1996) pp. 3272–3274.

[29] J. J. Kopanski and S. Mayo, "Intermittent-Contact Scanning Capacitance Microscope for Lithographic Overlay Measurement," *Applied Physics Letters* 72 (1998) pp. 2469–2471.

[30] C. Plassard, E. Bourillot, J. Rossignol, Y. Lacroute, E. Lepleux, L. Pacheco, and E. Lesniewska, "Detection of Defects Buried in Metallic Samples by Scanning Microwave Microscopy," *Physical Review B* 83 (2011) art. no. 121409(R).

[31] J. Rossignol, C. Plassard, E. Bourillot, O. Calonne, M. Foucault, and E. Lesniewska, "Imaging of Located Buried Defects in Metal Samples by Scanning Microwave Microscopy," *Procedia Engineering* 25 (2011) pp. 1637–1640.

[32] J. Rossignol, C. Plassard, E. Bourillot, O. Calonne, M. Foucault, and E. Lesniewska, "Non-destructive Technique to Detect Local Buried Defects in Metal Sample by Scanning Microwave Microscopy," *Sensors and Actuators A* 186 (2012) pp. 219–222.

[33] J. Kopanski, L. You, J.-J. Ahn, E. Hitz, and Y. Obeng, "Scanning Probe Microscopes for Subsurface Imaging," *ECS Transactions* 61 (2014) pp. 185–193.

[34] L. You, J.-J. Ahn, Y. S. Obeng, and J. J. Kopanski, "Subsurface Imaging of Metal Lines Embedded in a Dielectric with a Scanning Microwave Microscope," *Journal of Physics D: Applied Physics* 49 (2016) art. no. 045502.

[35] J. Chisum and. Z. Popovic, "Performance Limitations and Measurement Analysis of a Near-Field Microwave Microscope for Nondestructive and Subsurface Detection," *IEEE Transactions on Microwave Theory and Techniques* 60 (2012) pp. 2605–2615.

[36] S. M. Anlage, V. V. Talanov, and A. R. Schwartz, "Principles of Near-Field Microwave Microscopy." In *Scanning Probe Microscopy: Electrical and Electrochemical Phenomena at the Nanoscale, Volume 1*, (Kalinin and Gruverman, eds.) (Springer Verlag, 2007) pp. 215–253.

[37] G. Gramse, E. Brinciotti, A. Lucibello, S. B. Patil, M. Kasper, C. Rankl, R. Giridharagopal, P. Hinterdorfer, R. Marcelli, and F. Kienberger, "Quantitative Sub-surface and Non-contact Imaging Using Scanning Microwave Microscopy," *Nanotechnology* 26 (2015) art. no. 135701.

[38] M. Zhao, X. Gu, S. E. Lowther, Ch. Park, Y. C. Jean, and T. Nguyen, "Subsurface Characterization of Carbon Nanotubes in Polymer Composites via Quantitative Electric Force Microscopy," *Nanotechnology* 21 (2010) art. no. 225702.

13 Dynamics of Nanoscale Magnetic Systems

13.1 Introduction to Magnetization Dynamics

The topic of nanoscale magnetic systems is broad and could easily provide enough material for an entire book on its own. In this chapter, as in previous ones, we will focus on the nanoscale magnetic systems for which NSMM measurement systems offer characterization capabilities that are difficult to achieve by other measurement techniques. Note then that the scope of such systems extends beyond ferromagnetic materials alone. For example, two-dimensional electron gases (2DEGs) in semiconductors, interfaces between complex oxide layers, and low-dimensional materials are also of interest. While building toward a discussion of NSMM measurements of magnetization dynamics, we will also review several complementary, broadband measurement techniques for characterization of micro- and nano-magnetic systems. At present, these complementary techniques are more mature than NSMM and more widely applied to magnetic systems. These include all-electrical measurements and magnetomechanical measurement techniques based on microelectromechanical systems (MEMS).

We begin with a brief summary of the theory of magnetization dynamics. We will limit our discussion to a phenomenological treatment in terms of classical mechanics. Historically, the development of reliable, microwave sources opened the broad research field related to the motion of magnetization. It was discovered that for a given applied, static magnetic field the response of a magnetic material has a characteristic resonance behavior. Magnetic resonances may be measured by a variety of techniques, depending on the origin of the magnetic properties of the material. If the magnetic properties of a material arise from the electron and its spin, then the material may be investigated by electron paramagnetic resonance (EPR) and FMR. If the source of the magnetic response is the nucleus, then the material may be investigated by NMR. Our focus will be on materials where the origin of the magnetic response is the electrons. Therefore, as we review the theory of magnetization dynamics, we will lay the groundwork for NSMM-based EPR and FMR applications.

Consider a single magnetic dipole placed in a static magnetic field. We will initially assume that this magnetic field is uniform. From the Lorentz force law, it follows that this magnetic moment will experience a torque that tends to align this shape and may include the influence of the anisotropy. The relation between external and internal susceptibilities is given by

$$[\chi_e] = ([I] - [\chi][N])^{-1}[\chi],$$ (13.7)

where $[I]$ is the identity tensor. In the small signal limit, the components of the susceptibility tensor can be expressed in terms of the Gilbert damping parameter as

$$\chi_e^{x,y} = \frac{\omega_M\left(\omega_{x,y} + j\alpha\omega\right)}{\omega_x\omega_y - \omega^2 + j\alpha\omega\left(\omega_x + \omega_y\right)} = \chi_e^{\prime\,x,y} + j\chi_e^{\prime\prime\,x,y},$$ (13.8a)

$$\kappa_e = \frac{-\omega_M\omega}{\omega_x\omega_y - \omega^2 + j\alpha\omega\left(\omega_x + \omega_y\right)} = \kappa_e^{\prime} + j\kappa_e^{\prime\prime},$$ (13.8b)

where

$$\omega_{x,y} = |\gamma|[H_0 + M_s\left(N_{x,y} - N_z\right)];\ \omega_M = \gamma M_s;\ \omega_0 = |\gamma|H_0.$$ (13.9)

Further,

$$\omega_{FMR} = \sqrt{\omega_x\omega_y} = |\gamma|\sqrt{[\omega_0 + (N_x - N_z)\omega_M][\omega_0 + (N_y - N_z)\omega_M]}$$ (13.10)

is the well-known Kittel resonance condition [5]. The Gilbert damping parameter is related to the spin-spin relaxation time by

$$\alpha = \lambda/(|\gamma|M_s) = 2/[T_2\left(\omega_x + \omega_y\right)].$$ (13.11)

Note that the damping parameter in ferromagnets is not the same as the relaxation rate. From Equation (13.11) it follows that if the damping parameter is constant then the relaxation rate is shape dependent. Similar expressions can be obtained for the damping terms in other scenarios. In impulse-regime NMR and electron spin resonance (ESR), the Bloch equations have to be solved in the time domain and the usual approach involves transformation into a rotating reference frame. The generalization of the demagnetizing tensor $[N]$ to include the anisotropy contribution can be found in Reference [6].

Uniform precession is a unique, special case among many possible magnetization excitations. If the motion of the magnetization is nonuniform throughout the sample, then it is necessary to consider additional solutions to the equation of motion for the magnetization. If the wavelength of the magnetization excitations is comparable to the sample dimensions, then electromagnetic boundary conditions have to be considered in addition to the equation of motion. As discussed in Chapter 2, Maxwell's equations are complemented by the materials equations. In such a case, the solution to Equation (13.8) enters the relation between the magnetic induction and the magnetic field through Equation (13.6). Most of the time, Maxwell's equations for this particular case are solved in the so-called magnetostatic limit that neglects the electric field contribution, but includes the electromagnetic boundary conditions. The resulting magnetostatic solutions consider the sample shape and allow nonuniform motion of the magnetization throughout the sample volume. Furthermore, inclusion of the exchange interactions between the spins leads to the so-called magnetostatic-exchange solutions. The interested reader can see further details in References [7] and [8].

Another case is presented when the wavelength of the magnetization excitations is much smaller than the sample dimensions. In this case, the electromagnetic boundary conditions at the surface of the sample can be neglected. Due to strong exchange coupling between the neighboring spins, the presence of defects or other material inhomogeneity, or nonuniformity of the external magnetic field will lead to a change of the relative phase in the motion of the neighboring spins. The solution of the equation of motion in this case takes the form of propagating modes. In the phenomenological approach, we express the dynamic variables – the magnetization and the dynamic effective fields – in terms of plane waves

$$\boldsymbol{m}(\boldsymbol{r},t) = \sum_{k \neq 0} \boldsymbol{m}_k(t) e^{-j\boldsymbol{k}\cdot\boldsymbol{r}}, \tag{13.12}$$

where $\boldsymbol{k}$ is a wave vector of the particular excitation. The corresponding propagating waves of magnetization motion are called "spin waves." The introduction of the wave vector $\boldsymbol{k}$ leads to more complicated relations for the frequencies of individual spin excitations that are functions of the wave vector. This gives rise to dispersion relations representing the functional dependence between the frequency (energy) and wave vector (momentum) that characterize nonuniform magnetization motion. Because these relations are inherently nonlinear, such an approach is useful for the investigation of dynamic, nonlinear magnetization effects. A full description of spin waves, their dispersion and interactions is beyond the scope of this work. A comprehensive study of magnetism and magnetization excitations can be found in the series edited by Rado and Suhl [9], and in References [10] and [11]. The equations introduced in the preceding text provide a foundation for basic understanding of the nanoscale magnetic processes and measurements discussed in this chapter. As needed, we will provide additional theoretical descriptions for specific cases.

13.2 Measurements of Linear Dynamics in Microscale and Nanoscale Magnetic Systems

13.2.1 Mechanical Measurement of Magnetization Dynamics

The standard technique for investigation of magnetization dynamics is FMR. In this measurement technique, the material is biased with a static, external magnetic field and simultaneously exposed to a high-frequency electromagnetic field. As the electromagnetic radiation drives the magnetic precession at resonance, energy is absorbed by the system and dissipated via damping. This is manifested as changes in the transmission and reflection of incident radiation when the system is driven through resonance. From FMR measurements, one may subsequently determine the components of the magnetic susceptibility tensor. The FMR technique may be carried out in a variety of experimental environments, including microwave cavities and on-wafer platforms. The microwave cavity approach has been described in numerous journal articles and books, including several referenced earlier. Reviews of on-wafer FMR measurement methods can be found in References [12] and [13], for example.

Here, we will focus on a slightly nonconventional technique for measurement of magnetization dynamics in small magnetic samples. Specifically, we are going to discuss measurement approaches in which a magnetic, thin film is integrated with a microcantilever. While these approaches are not, strictly speaking, scanning probe methods, they illustrate the interaction of RF fields with magnetic probes. Contemporary magnetomechanical experiments are enabled by MEMS technology and are based on the fundamental physical fact that a change of angular momentum over time manifests itself as a torque. In the specific case of magnetic systems, changes in the magnetic moment are transduced to a mechanical torque. These approaches can be applied to both static and dynamic processes in small magnetic samples. Furthermore, these techniques take advantage of the facts that cantilevers are commercially available and that cantilever motion is detectable through interferometry and beam-bounce techniques. We will first introduce the principle of the experimental measurement and then we will show how it can be applied to measurement of magnetization dynamics.

The first measurement of gyromagnetic effects in macroscopic bodies was observed by Einstein and de Haas [14]. They demonstrated that the mechanical rotation of an object could be induced by a change in the object's magnetization. The complementary effect – induced magnetization by mechanical rotation – was demonstrated soon afterward by Barnett [15]. The approach presented here extends the Einstein-de Haas experiment to microscale magnetic systems [16]. In the original experiment [14], the rotation of a macroscopic iron cylinder suspended by a glass wire was induced by applying an alternating magnetic field along the central axis of the cylinder. At the microscale, a thin magnetic film that is deposited on a micro cantilever takes the place of the iron cylinder [16]. In both cases, the alternating magnetic field induces a change in magnetic moment μ and a corresponding change in angular momentum J_{tot}. To satisfy the conservation of angular momentum, the changes in J_{tot} must be offset by changes in the mechanical angular momentum of the magnetic system. Thus, when an alternating magnetic field is applied perpendicular to the length of the cantilever, this change in angular momentum manifests itself as a mechanical torque, leading to bending oscillations of the cantilever.

The absolute value of the magnetomechanical ratio is defined as

$$g' = \frac{2m_e}{e} \frac{|\mu|}{|J_{tot}|}$$ (13.13)

where m_e is the electron rest mass. In Equation (13.13), J_{tot} includes contributions from both the spin and orbital angular momentum, and μ is the magnetic moment. The magnetomechanical ratio is related to the gyromagnetic ratio g, which considers only spin angular momentum, through the following relation introduced by Kittel [17]

$$2 - g' = g - 2. \tag{13.14}$$

The magnetomechanical torque on the cantilever in the presence of the applied alternating field will bend the cantilever. The maximum magnitude of this torque is

$$T_0 = \frac{2m_e \omega}{eg'} \Delta\mu, \tag{13.15}$$

where ω is the frequency of the alternating driving field and $\Delta\mu$ is the change of the magnetic moment. Experimentally, the deflection of the cantilever may be measured by use of optical interferometry, beam-bounce methods, capacitance measurements, or piezoresistive measurements, among other techniques. An example of the measured deflection as a function of the driving alternating field frequency is shown in Fig. 13.1. Note that the sensitivity of the technique is maximized when the alternating field frequency is close to the mechanical resonance frequency of the cantilever (about 13.2 kHz for the data shown in Fig. 13.1).

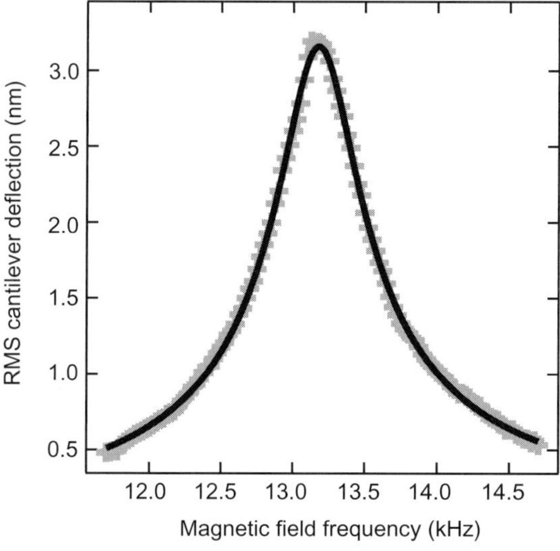

Figure 13.1. Cantilever-based measurement of the Einstein-de Haas effect. Root mean square cantilever deflection measured by optical interferometry as a function of the applied alternating field frequency. The symbols represent the experiment and the solid line is the fit from Equation (13.16). The amplitude of the alternating magnetic field was 367 A/m and the resulting change in the magnetic moment of the film was estimated to be $\Delta\mu = 0.335$ nA m². Reprinted from T. M. Wallis, J. Moreland, and P. Kabos, *Applied Physics Letters* 89 (2006) art. no. 122502, with permission from AIP Publishing.

By modeling the cantilever motion as a forced harmonic oscillator, it is possible theoretically to predict the deflection. In particular, a simple model of a force acting on a point mass at the free end of a rectangular cantilever beam leads to the following expression for the amplitude of the cantilever deflection:

$$z_0 = \frac{F_0 / m_{mod}\,\omega}{\sqrt{(\omega_0^2 - \omega^2)^2 / \omega^2 + \omega_0^2 / Q^2}}, \tag{13.16a}$$

where

$$F_0 = \frac{4m_e}{l_c e g'}\,\Delta\mu\omega, \tag{13.16b}$$

m_{mod} is the modal mass of the cantilever beam, ω_0 is the resonance frequency of the beam, Q is the quality factor, and l_c is the length of the cantilever. Comparison of the predicted deflection with the experiment enables extraction of the magnetomechanical ratio. A more complete theoretical analysis of this problem was presented in Reference [18].

Having introduced the concept of measurement of magnetization dynamics by the Einstein-de Haas effect, we proceed to the measurement of magnetization dynamics by direct transfer of angular momentum from a microwave field to a mechanical system. This approach was introduced in References [19] and [20], in which the authors presented several ways to mechanical measurements of magnetization dynamics. Ferromagnetic resonance was detected through three distinct effects. The first effect is the reduction of static magnetization when FMR conditions are met. The FMR is then detected as a change of magnetostatic torque acting on a sample in a magnetic field, as illustrated in Fig. 13.2(a). The second effect is the damping torque. This torque is the result of the damping process within a given material that produces a counter torque on the precessing spins, as illustrated in Fig. 13.2(b). Note that the first effect causes a bending motion of the cantilever while the second effect causes a torsional, twisting motion of the cantilever, provided that the orientations of the magnetization and external fields are as shown in Fig. 13.2. The final effect is heating of the sample due to the excitation of FMR, which results in the absorption of microwave energy by the spin system. When heated, the bimaterial system formed by the film and the cantilever will undergo a bending motion. The heating of bimaterial cantilever systems was also used in Reference [21] to map high-frequency, magnetic fields.

To understand how the first effect can lead to a mechanical torque, it is instructive to consider the case when the magnetic system is driven close to resonance. In the case of small damping ($\alpha \ll 1$) and near magnetic resonance, it is possible to assume that $\omega^2 \cong \omega_x \omega_y$. Under these conditions, the imaginary parts of the susceptibility tensor dominate. The relation between the magnetization components, m_x and m_y, and the external RF field in the x direction, h_x, can be expressed as

$$m_x = j\chi''h_x, \tag{13.17a}$$

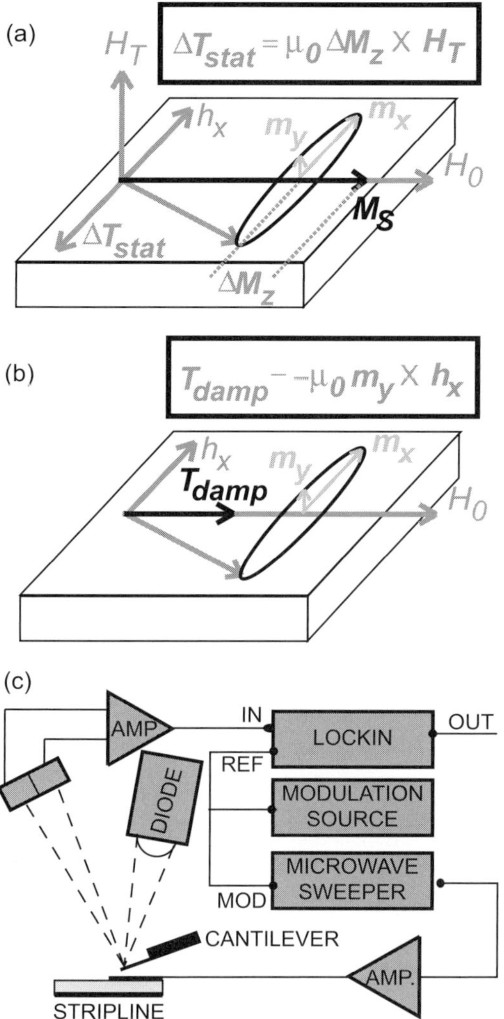

Figure 13.2. Magnetomechanical measurements of magnetization dynamics. Illustrations of two approaches to the mechanical measurement of the magnetization dynamics are shown: (a) the magnetostatic torque in a perpendicular field and (b) the damping torque. (c) Block diagram of instrumentation used for micromechanical, lock-in detection of magnetization dynamics in a magnetic system deposited on a cantilever. In this example, the RF field is provided by a stripline. Adapted from A. Jander, J. Moreland, and P. Kabos, *Journal of Applied Physics* 89 (2001) pp. 7086–7090, with permission from AIP Publishing.

$$m_y = \kappa'' h_x. \tag{13.17b}$$

These two equations describe elliptical motion of the magnetization, schematically shown in Fig. 13.2(a). Note that the z direction is parallel to the long axis of the cantilever. Following Equations (13.8) and (13.9), the average steady state change of the magnetization in the z direction in the presence of an applied static field H_0 can then be expressed as

$$|\Delta M_z| \approx \frac{|m_x|^2 + |m_y|^2}{2M_s} = \frac{M_s h_x^2}{2\alpha(2H_0 + M_s)H_0}.$$ (13.18)

Now, if an additional static magnetic torque field H_T is applied perpendicular to the film plane, then this change in the z component of the magnetization results in a proportional change in the quasi-static magnetic torque

$$\Delta T_{stat} = \mu_0 \frac{H_T}{H_0} \frac{M_s}{2\alpha^2(2H_0 + M_s)})h_x^2 V,$$ (13.19)

where V is the volume of the sample. This torque is directed in the x direction, causing a bending motion about the axis of the RF field. Note that this torque arises from the interaction of the DC component of the magnetic moment with the static torque field H_T.

The second effect that leads to a torque on the film-on-a-cantilever system comes from the direct interaction of the RF field component h_x with the dynamic, precessing component of the magnetization. Mathematically, this interaction takes the form of the cross product $\mathbf{m} \times \mathbf{h}$. If this product is evaluated with the respective components of $\mathbf{m}$ and $\mathbf{h}$, then the average amplitude of the damping torque is obtained as:

$$T_{damp} = \frac{1}{2}\mu_0 \kappa'' h_x^2 V = \frac{\mu_0 M_s}{2\alpha(2H_0 + M_s)}h_x^2 V.$$ (13.20)

This is the torque that the microwave field must exert on the magnetization in order to maintain the precessional motion in the presence of damping. In the equilibrium state, this torque is transferred to the cantilever and can be measured as a twist in the beam, as was shown in Reference [19].

The third and final effect that produces cantilever motion is the heating of the cantilever through FMR. The deposition of a magnetic, metal film on a Si cantilever forms a bimaterial system that will bend when heated due to the unequal thermal expansion of the two materials. The generated heat is proportional to the power absorbed by the magnetic system and can be expressed as

$$P = \frac{1}{2}\mu_0 \chi'' \omega h_x^2 V = \frac{\mu_0 M_s \gamma(H_0 + M_s)}{2\alpha(2H_0 + M_s)}h_x^2 V.$$ (13.21)

A block diagram of the instrumentation used for micromechanical detection of FMR shown in Fig. 13.2(c) and measurements are shown in Fig. 13.3. Figure 13.3(a) shows the torsional motion from the damping torque measured as a function of the static bias magnetic field H_0 at a constant microwave field frequency of 9.15 GHz. The measurement clearly shows the characteristic torque response: the reversal in the sign of the torque signal as the orientation of the static bias field is switched.

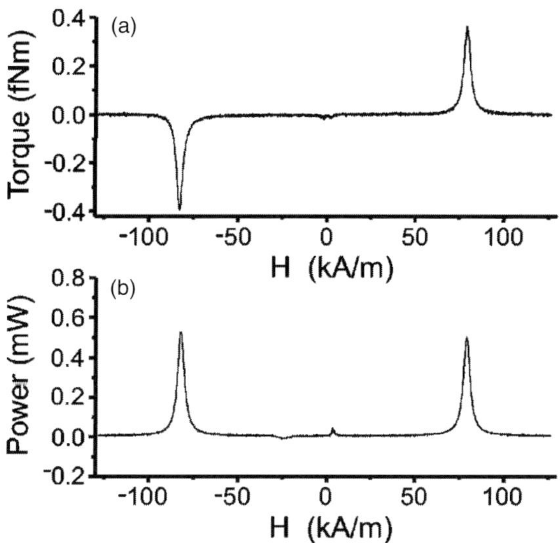

Figure 13.3. Cantilever-based measurements of FMR.
(a) FMR signal detected through measurements of cantilever torsion. (b) FMR signal detected by thermally induced deflection that is proportional to the power absorbed by the magnetic system. Both measurements are shown as a function of the static magnetic field at the fixed microwave excitation frequency of 9.15 GHz.
Reprinted from A. Jander, J. Moreland, and P. Kabos, *Applied Physics Letters* 78 (2001) pp. 2348–2350, with permission from AIP Publishing.

Figure 13.3(b) shows measurements of the bending motion from the power absorption at FMR for the same film. Note that there is no change in the sign of the response with changes in the sign of H_0, as the thermal effect does not depend on the orientation of the static field. The thermal equilibration time of a typical, commercial Si cantilever is about 1 ms. Thus, thermally induced bending is challenging to detect at modulation frequencies above 1 kHz. Therefore, to remove the thermal contribution to torque, measurements may be done at much higher modulation frequencies, usually corresponding to torsional resonant frequency of the cantilever that is on the order of ~ 250 kHz.

Magnetomechanical measurements of dynamics have been extended to investigation of magnetization and spin dynamics in magnetic multilayers in Reference [22]. Specifically, the Einstein-de Haas effect was used to study interfacial spin transport, revealing the influence of the strong spin-orbit coupling in a Pt layer upon transfer of the angular momentum as a function of the applied static magnetic field. The measurements are shown in Fig. 13.4. The figure shows that the addition of a Cu layer to a permalloy (Py) film does little to change the cantilever deflection, but the addition of a Pt layer reduces the cantilever deflection roughly by a factor of two. This suggests a loss of the angular momentum through known spin–orbit interactions within the Pt layer and its subsequent absorption in the Pt layer. In classical FMR experiments, such interfacial interactions translate

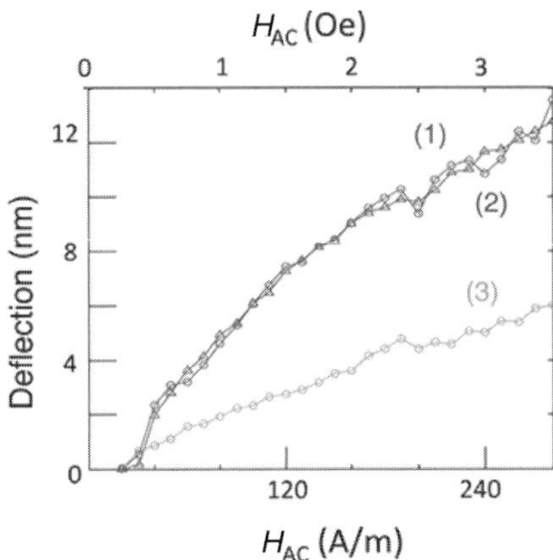

Figure 13.4. Einstein-de Haas effect in multilayered structures.
Root mean square deflection of the cantilever as a function of the root mean square amplitude of a driving AC magnetic field in the presence of a 30 Oe (2.4 kA/m) static bias field for three structures: (1) Single-layer 50 nm Py, (2) Bi-layer 50 nm Py / 8 nm Cu, and (3) Bi-layer 50 nm Py / 8 nm Pt.
Reprinted with permission, from S-H Lim, A. Imtiaz, T. M. Wallis, S. Russek, P. Kabos, Liufei Cai, and E M. Chudnovsky, *Europhysics Letters* 105 (2014) art. no. 37009.

into an increase of the damping parameter. Here, the reduction in deflection amplitude is attributed to a clear transfer of angular momentum into Pt. This result demonstrates the sensitivity of the technique to the pure angular momentum transfer, or "spin flip" due to changes in interfacial transport at the boundary between the ferromagnetic and normal metal layers. The measured data agrees to first order with simple physical models that show that effective spin flip times are inversely proportional to the fourth power of the atomic number of the involved elements.

13.2.2   Time- and Frequency-Domain Measurements of Magnetization Dynamics

Direct, broadband, electrical detection of magnetization dynamics at microscopic and nanoscopic length scales is challenging. The detection sensitivity is strongly dependent on the physical process used for detection. In this section, we will review several broadband and all-electrical methods for the measurement of magnetization dynamics. Complementary methods such as time-resolved Kerr microscopy and x-ray techniques are also extensively used, but will not be addressed here. All-electrical measurement techniques include differential dV/dI resistance [23], [24], spin rectification effect [25], and RF/microwave transmission [26] techniques. These techniques are essentially modifications of broadband time- and frequency-domain measurements of submicron-sized magnetic systems. Often, the test platform takes the form of a coplanar or microstrip waveguide configuration. Time-domain measurements of this type were introduced in Reference [27] and extended to the spectral domain in Reference [28]. In general, all-electrical techniques for characterization of single-layer magnetic systems are based on nonlinear coupling of microwave spin excitations with low-frequency charge currents through the anisotropic magnetoresistance effect.

A widely used approach is measurement of the differential resistance dV/dI of a micromagnetic system in response to a small AC current that alternates at a much smaller frequency than the magnetization precession frequency. The differential resistance is given by: $dV/dI = R + I\, dR/dI$. The first term R represents the DC resistance, while the second term represents reversible processes such as magnetization precession or magnetization switching. For anisotropic magnetoresistance, the resistance depends on the angle between the current direction and the sample magnetization direction. In general, the dependence is quite complicated and has a tensor form for single crystals. For polycrystalline samples, the anisotropic resistance can be expressed as a function of the angle θ between the current and field directions as

$$R_{MA}(\theta) = R_\perp + (R - R_\perp)\cos^2(\theta) \tag{13.22}$$

with R and $R_\perp$ corresponding to θ equal to zero degrees and ninety degrees, respectively. It is important to stress that any motion of the magnetization must be driven by microwave signals, but low-frequency modulation of the microwave signal is a practical strategy to enhance the signal-to-noise ratio by use of lock-in techniques.

Two additional all-electrical strategies for characterization of magnetization dynamics are measurements of the spin rectification effect and measurements of RF transmission. In the presence of a static magnetic field and a microwave field oscillating close at a frequency ω near magnetic resonance, the angle θ is a function of time. Thus, the anisotropic resistance also varies in time. The measured voltage across the sample is then the product of the resistance and the current. This voltage will have the form of a constant term as well as terms with frequencies ω and 2ω. The constant term represents the spin rectification effect. The other terms represent the RF transmission, which can be measured by use of a high-frequency diode and a lock-in amplifier.

A comparison of dV/dI, RF transmission, and spin rectification techniques on a simple microscale magnetic rectangular structure was presented in Reference [29]. All measurements were made with signal modulation and a lock-in technique. Figure 13.5(a) shows the measured response of a rectangular, microscale element acquired with each of the three techniques during up-and-down field sweeps. The measured signals are characteristic of magnetic switching and show significant hysteresis. The abrupt change in dV/dI during the field sweep is attributed to vortex nucleation within the rectangular element. The RF transmission technique yields comparable results, with steps in the RF transmission occurring near the same fields where steps occur in the dV/dI measurements. By contrast, the rectification technique has a similar signal-to-noise ratio to the other measurements, but the response is more complex and the interpretation is more complicated.

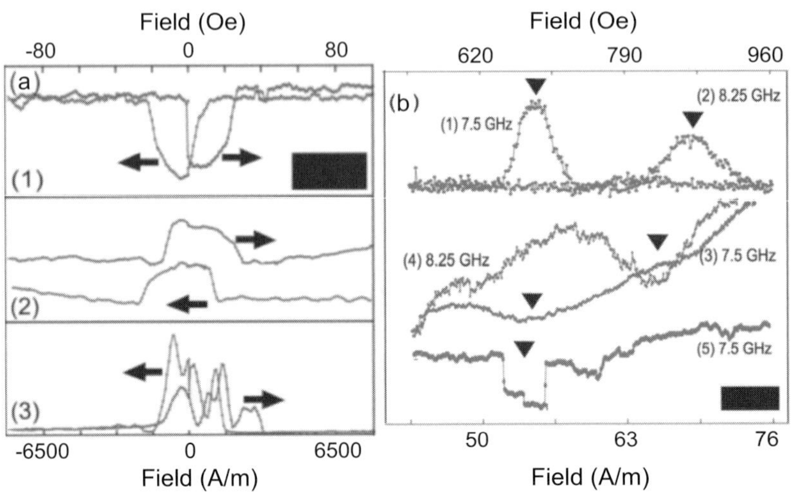

Figure. 13.5. All-electrical measurements of patterned films.
(a) Measurements of a magnetic, 2 μm × 8 μm rectangular element by use of (1) dV/dI, (2) RF transmission, and (3) spin rectification techniques. (b) FMR measurements of the same rectangular element by use of (1), (2) spin rectification measurements, (3), (4) RF transmission measurements, and (5) bridge circuit measurement. Vertical scales have been adjusted to facilitate comparison.
Reprinted from S-H. Lim, T. M. Wallis, A. Imtiaz, D. Gu, P. Krivosik, and P. Kabos, *Journal of Applied Physics* 109 (2011) art. no. 07D317, with permission from AIP Publishing.

Fig. 13.5(b) compares FMR curves measured with each of these techniques. Curves (1), (2) were obtained by use of the rectification technique, curves (3), (4) by use of the transmission technique, and curve (5) by use of an RF transmission technique [29]. The FMR fields are marked by triangles and are found to be in good agreement with the predictions of Kittel's equation. From the comparison of the techniques, it follows that the transmission technique has poorer sensitivity and is the least effective technique for detection of FMR on lossy samples. This weakness is mitigated by implementing a bridge technique. To summarize, the three all-electrical detection techniques on micrometer-size ferromagnetic elements provide complementary information. The different approaches have different sensitivities to specific magnetization dynamics, as shown by the contrasting sensitivities to vortex creation and FMR. Of the three techniques, the spin rectification approach appears to be best suited for the broadband spectroscopy measurements, as it displays the best signal-to-noise ratio in both the low and high field ranges.

13.2.3   Measurements of Magnetization Dynamics in Layered Structures

The number of applications of magnetic multilayer structures exploded following the discovery of the giant magnetoresistance effect by Albert Fert and Peter Gruenberg. In 2007, they were awarded the Nobel Prize in Physics for their discovery. The effect was observed in multilayered structures with alternating ferromagnetic and nonmagnetic layers. When such a multilayer structure is patterned to have nanometer dimensions and an AC or DC current is passed through it, the resulting response is due to both linear and nonlinear effects. These effects, including the giant magnetoresistive effect, provide a useful tool to measure the magnetic properties of these systems and the related magnetization dynamics. In particular, resistance measurements represent an effective, high-bandwidth method to measure magnetization dynamics. The resistance of two adjacent layers separated by nonmagnetic metal or thin dielectric spacer depends on the relative orientation of magnetizations in the neighboring layers. In most applications, the magnetization in one layer, the "fixed layer," is pinned in a fixed direction through exchange coupling while the magnetization in the other layer, the "free layer," is free to move with a trajectory that depends on the coupling between the layers and any applied external fields. The resistance can be then expressed as [30]

$$R_{GMR}(\theta) = R_p + (\Delta R)(1 - cos(\theta))/2 \qquad (13.23)$$

where R_p is the resistance for parallel orientation and ΔR is the difference of the resistance between parallel and antiparallel magnetization orientation. The system of adjacent magnetic layers with one layer free to move with respect to a fixed layer is sometimes referred to as a "spin valve."

Figure 13.6 illustrates several approaches for the measurement of such structures. This versatile experimental configuration covers a broad range of possible measurement regimes, including DC, spectroscopic, frequency-domain, and time-domain magnetization dynamics. The inductor and capacitor in the figure form a bias tee that allows separation of DC / low-frequency signals from high-frequency signals. In this chapter, we specify which equipment and input excitations are utilized for the specific measurements.

We begin with a consideration of the measurement of thermally excited FMR. For this particular measurement, among the components shown in Fig. 13.6, only the source of the DC current I_{DC} and the spectrum analyzer are required. This approach was used in References [31], [32], and [33]. When a DC current is sent through the spin valve, it will heat up slightly. The thermal fluctuations of the magnetization in the layers will lead to fluctuations of the resistivity that appear at the spectrum analyzer as noise. The noise power spectral density can be expressed as [29]

$$S_n(f) = (I_{DC}\Delta R)^2 \frac{2k_B T}{\pi f \mu_0 M_s^2 V} \chi''(f) = \frac{V_n^2}{R}, \qquad (13.24)$$

where $k_B = 1.38064 \times 10^{-23} \, \mathrm{m^2 kg s^{-2} K^{-1}}$ is the Boltzmann constant, V is the volume of the element, ΔR is the resistivity change, and χ'' is the imaginary part of the susceptibility. From this dependence, it follows that the spectral noise density is proportional to the sample susceptibility, which has a resonant character. Therefore, the signal will show a resonance at the corresponding frequency of the natural

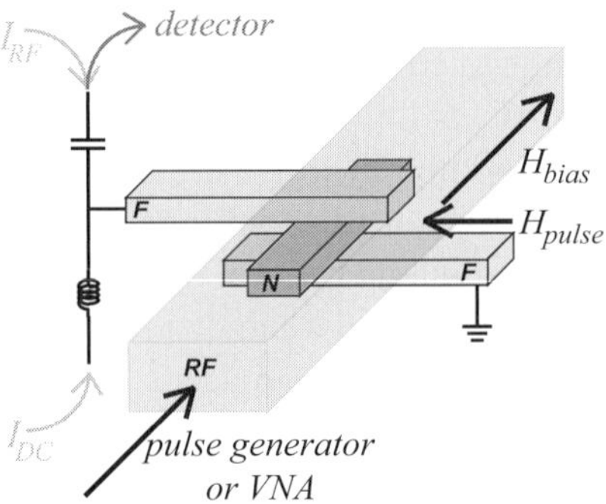

Figure 13.6.   Configuration for magnetoresistance measurements of spin valve structure. The spin valve is formed by a ferromagnet (F) / normal metal (N) / ferromagnet (F) structure. An RF signal line (RF) is placed in close proximity to the spin valve. The RF signal may be supplied by a pulse generator or a VNA. The lower ferromagnetic layer is grounded while the other ferromagnetic layer may be used to supply a DC current (I_{DC}) or AC current (I_{RF}) to the spin valve. This upper layer also provides a detection path for RF signals by use of a VNA, high-speed oscilloscope, or spectrum analyzer. The orientation of static bias field (H_{bias}) and a pulse field (H_{pulse}) are also shown.

FMR of the system. Figure 13.7 shows the results of such measurement on a 20 gigabyte read head from a hard disk drive in the form of a noise spectrum. The resonance frequency near 4 GHz is clearly resolved.

Time-domain or spectral measurements may be performed by introducing an impulse generator or network analyzer as the source in the configuration shown in Fig. 13.6. The "detector" in Fig. 13.6 may be a diode detector, network analyzer, real-time oscilloscope, or sampling oscilloscope. An example of time-domain, inductive measurements is shown in Fig. 13.7(b). The Fourier transform of the time-domain signal has a similar profile to the resonance curve shown in Fig. 13.7(a). From the Fourier transform, one can determine the resonance frequency as well as the losses in the system that are represented by the width at half magnitude of the resonance curve. Further, the resonance linewidth is directly related to damping parameters introduced in Equations (13.3) through (13.5). Figure 13.7(c) shows the results of finite-element modeling of the magnetization response for a 0.8 μm × 4.8 μm spin valve, which is in good agreement with the experimental data.

In the thermally excited FMR measurements described earlier, a relatively small DC current passes through the layers. What happens when this current is increased? If the multilayer structure has lateral dimensions on the order of 100 nm or less, then new phenomena occur when the current exceeds a certain threshold level. As the current passes through the multilayer structure, the fixed layer serves as a spin filter, preferentially allowing electrons with spins parallel to the fixed-layer magnetization to flow through the structure. In the free layer, a spin torque will be exerted on spins that have a magnetization direction different from the fixed layer. This effect can lead to current-controlled hysteretic switching in magnetic nanostructures [34], [35], steady state magnetization precession, or excitation of FMR.

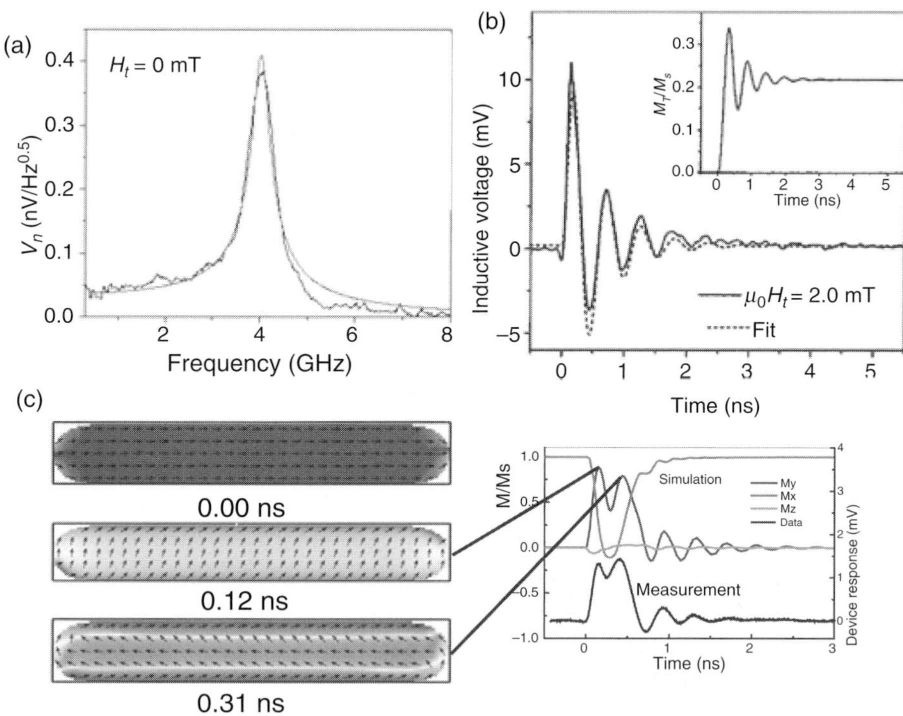

Figure 13.7. Time-domain measurements of magnetization dynamics in a hard disk head. (a) Spectral dependence of the voltage from a 20 Gb hard disk head. (b) Time-domain impulse response. Reprinted from S. Kaka, J. P. Nibarger, S. Russek, N. A. Stutzke, and S. L. Burkett, *Journal of Applied Physics* 93 (2003) pp.7539–7544, with permission from AIP Publishing. (c) The micromagnetic structure of the disk head is simulated by finite-element simulations. The element is saturated when the impulse is applied (0 ns). Simulations are shown at two later times (0.12 ns and 0.31 ns) that correspond to peaks in the y component of the magnetization (M_y). The measurement (lower curve) is proportional to M_y.

Spin-torque switching is currently being explored for use in ferromagnet-based random access memories. Spin-torque switching and precession was initially theoretically predicted by Slonczewski and Berger [36], [37]. The measurement of the dynamic magnetization response of nanoscale multilayer magnetic systems due to spin torque may be carried out in a similar experimental configuration as the configuration described earlier for thermal FMR measurements. The experiment is done by applying a DC current to a structure and measuring the power spectral density with a spectrum analyzer or a diode detection system. The first experimental confirmation of DC current-induced magnetization precession due to this effect was presented in Reference [38]. This opened a broad research area into applications of nanoscale, nonvolatile oscillators. Detailed investigation of FMR induced by DC currents are available in References [39] and [40].

Reference [41] introduced an alternative, particularly efficient method to measure the FMR in magnetic nanostructures. Once again building on the configuration shown in Fig. 13.6, FMR in the nanostructure is driven by an AC current and the detection is done electrically by measurement of the DC voltage across the nanostructure. If both the DC current I_{DC} and RF current from a network analyzer are applied to a magnetic nanostructure at the same time, and the RF frequency is selected to be close to the FMR frequency of one of the magnetic layers, then the magnetization in the layer will precess, resulting in time-dependent resistance $R(t)$. The resulting voltage $I(t)R(t)$ across the nanostructure will contain a DC component that can be expressed as

$$V_{DC} = I_{DC}\left(R_{DC} + \Delta R_0\right) + \frac{1}{2} I_{RF}\Delta R_f \cos(\theta_f). \tag{13.25}$$

θ_f is the phase of the mixing resistance. By modulating the microwave current I_{RF} and using a lock-in technique, the measurement isolates the second term, which depends upon the static magnetic bias field through the change in resistance ΔR_f, which is frequency dependent. In this way, one can measure the FMR in these structures. Assuming a linear response of the system, the mixing voltage will have a simple Lorentzian line shape of the form [40]

$$V_{mix} = \frac{I_{RF}^2 / \Delta_0}{1 + [(f - f_0)/\Delta_0]^2}, \tag{13.26}$$

where f_0 is the resonance frequency without the drive current and $\Delta_0 = \alpha f_0$ represents the damping of the system. A complete, detailed analysis of the line shape of the resonance curve obtained by this method is available in Reference [42].

In addition, electrical detection of magnetization dynamics includes the utilization of the concept of "spin pumping," where the precession of the magnetization of a ferromagnet transfers spins into adjacent, normal-metal layers [43]. This "pumping" of spins produces a voltage in the normal metal that may be used as a measurement of spin wave dynamics in ferromagnetic metal/normal metal structures. Spin pumping has been used to explain observations of FMR in ferromagnet/normal metal structures [44]. Furthermore, electrical characterization of the dynamics in nanomagnetic structures has led to the discovery of new effects such as the spin Hall [45] and inverse spin Hall effects [46]. These phenomena are also extensively used for characterization of ferromagnetic metals and ferrimagnetic dielectrics such as YIG deposited on a normal metal. The experimental techniques that use these effects are essentially the same as described earlier with the difference that the voltage/resistance is measured at two electrodes connected to the normal metal layer. Ideally, the normal metal will have strong spin-orbit coupling as is the case with Pt or Ta. The measured DC voltage is then the function of the magnetization dynamics at the interface between the nonmagnetic and ferromagnetic layers.

13.3   Scanning Probe Measurements of Magnetization Dynamics

In 1991, Sidles [47] proposed the measurement of proton spin through a nontraditional mechanical approach, which was later implemented by Rugar at IBM [48], [49] and culminated in application of this approach to single spin detection [50]. A schematic representation of this experiment is depicted in Fig. 13.8. A small permanent magnet is fixed on the end of a soft cantilever that is scanned over the specimen, an RF excitation field is provided by the coil, and the interaction of the cantilever with the sample is detected by an optical interferometer. The permanent magnet creates a local magnetic field that selects a constant-field slice in the sample within which the driving RF field can resonantly excite the spins. By changing the distance between the magnet and the sample or by changing the RF excitation field frequency, different slices of the sample are sampled and a three-dimensional tomograph may be reconstructed. The technique is known as magnetic resonance force microscopy (MRFM) and the approach triggered wide interest in the scanning probe community, including demonstrated applications to magnetic nanostructures [51], [52], semiconductors, and biological materials. The MRFM technique is one of the few techniques that is capable of local electron paramagnetic resonance detection on nanometer or atomic scale. As of this writing, scanning probe methods that would be able to detect localized ESR signals remain a major challenge. Most of the standard, existing ESR methods are based on cavity or on-wafer methods.

Another early scanning probe approach for measurement of magnetization dynamics is based on the detection of microwave heating of the sample at resonance

Figure 13.8.   "Tip-on-cantilever" configuration for magnetic resonance force microscopy. A small permanent magnet is fixed on the end of a soft cantilever that is scanned over the specimen, an RF excitation field is provided by the coil, and the interaction of the cantilever with the sample is detected by an optical interferometer. The hemisphere represents a slice within which the field is constant. Reprinted by permission from Macmillan Publishers Ltd: Nature. D. Rugar, R. Budakian, H. J. Mamin, and B. W. Chui, *Nature* 430 (2004) pp. 329–332. © 2004.

[53]–[55]. The technique may be considered as a specialized form of scanning thermal microscopy [56], [57]. At present, this approach, known as scanning thermal microscopy ferromagnetic resonance (STM-FMR), is a mature scanning probe technique that enables surface imaging as well as subsurface tomography [56]–[58]. The photoacoustic effect provides the mechanism for detection of the temperature increase in the sample. The photoacoustically detected microwave resonance provides no lateral resolution, but by changing the modulation frequency of the microwave input different subsurface depths can be distinguished. The complementary mirage effect provides a lateral resolution of about 50 μm for the measurement of the surface temperature. Further development of the technique was enabled by the incorporation of a temperature sensitive tip and by detecting local thermal expansion of the sample perpendicular to the surface. Together, these refinements have enabled imaging with 10 nm lateral spatial resolution. Figure 13.9 shows STM-FMR amplitude and phase images of Co stripes on Si taken at the resonance field of the backward volume mode [7], [8], which is defined as a mode that is propagating within the thin film with the direction of propagation along the static bias field. Here, this mode is propagating from the edge of the stripes. The scanned area shows a small cross section of the stripe.

Yet another approach for local measurement of magnetic materials and magnetization dynamics is provided by NSMM and related microwave and RF probing techniques. Though several early implementations of near-field scanning microwave probes were explicitly developed for characterization of magnetic materials [59], [60], the broader application of scanning microwave imaging to magnetic field detection has lagged behind the application to electric field detection. Early near-field probes were formed by opening a small aperture in a microwave waveguide. Following other aperture-based measurement approaches (described earlier in Chapter 8), the material of interest was brought close to the aperture and subsequently scanned back and forth to produce a near-field image. As techniques evolved, the apertures were replaced by specialized probe tips. For example, a magnetic field probe incorporates a specialized micromachined cantilever that incorporates a conducting loop into an AFM system [61], [62], as shown in Fig. 13.10(a).

Figure 13.9.   Scanning thermal microscopy – ferromagnetic resonance images.
STM-FMR Amplitude (left) and phase (right) images of Co stripes on Si obtained by use of STM-FMR at the resonance field of the backward volume mode. The imaging area is near the edge of the 100 μm × 2.5 μm × 0.01 μm Co stripe [58]. © IOP Publishing. Reproduced with permission. All rights reserved.

The 6 μm-diameter loop acts as an inductive antenna that is sensitive to the perpendicular component of the magnetic field.

Figure 13.10(b) shows an experimental configuration for the monitoring of high-frequency fields present above a microwave device by use of a network analyzer. This configuration can also be used for measurement of a dynamic magnetization response due to frequency- or time-domain excitation. This probe configuration enabled mapping of the permeability of the ferromagnetic material with about 100 μm lateral resolution. A similar configuration with a thin film Cu micro-loop deposited at the end of a coaxial cable has been applied in a similar way [63], [64]. At a given applied angular frequency $\omega = 2\pi f$, the probe-sample system can be modeled with a lumped-element model, also shown in Fig. 13.10(b). From this model, it follows that the total impedance at the terminal is

$$Z_T(f,H) \approx j\omega L_0 \left(1-k^2\right) + Z_S(f,H)k^2 \frac{L_0}{L_X}, \tag{13.27}$$

where L_0 is the probe (loop) inductance, L_X is the inductance of the probe's image in the sample, Z_S is the complex surface impedance of the sample, and k is a dimensionless coefficient describing the probe-to-sample coupling expressed through the mutual inductance M: $k = \sqrt{M^2 / L_0 L_X}$. The response of the magnetic sample is represented by the surface impedance that in the thin film limit can be expressed as

$$Z_S(f,H) = j\,\omega t_0\,\mu_0\mu_r(f,H) \tag{13.28}$$

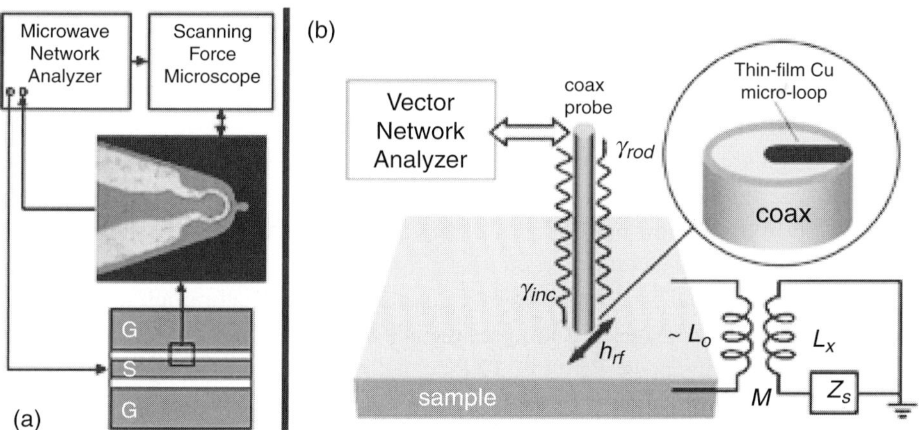

Figure 13.10. Microwave probes for spatially resolved measurements of magnetic systems. (a) Schematic of microwave imaging system, including an SEM image of the microfabricated probe. Reprinted from V. Agrawal, P. Neuzil, and D. W. van der Weide, *Applied Physics Letters* 71 (1997) pp. 2343–2345, with permission from AIP Publishing. (b) Schematic of a coaxial FMR micro-loop probe. The equivalent circuit for the probe–sample interaction is shown in the lower right. Reprinted from D. I. Mircea and T. W. Clinton, *Applied Physics Letters* 90 (2007) art. no. 142504, with permission from AIP Publishing.

where t_0 is the sample thickness and $\mu_r(f,H)$ is the complex permeability of the magnetic sample at a given frequency and static bias magnetic field. The total impedance is obtained from the measurement of the reflection coefficient $S_{11}(f,H)$ and comparing with the standard expression for the reflection coefficient (see Equation (3.3), for example). Note that the measurement can be done either at a constant bias field as a function of frequency or at a constant frequency as a function of the static bias magnetic field.

To reduce the influence of drifts that result from temperature changes in the electrical leads and the network analyzer, and to increase the signal-to-noise ratio, it is useful to obtain the permeability of the sample not from a single measurement, but rather from the difference between the measured total impedance at resonance and the measured total impedance off-resonance. The latter may be measured at a significantly higher or lower field value H_{OFFRES} than the resonance field H_{RES}. This procedure effectively subtracts the background response from the measurement. This difference can be done either with calibrated or non-calibrated scattering parameters obtained by use of the network analyzer. From Equation (13.27), it then follows that

$$\Delta Z_T(f,H) = Z_T(f,H_{RES}) - Z_T(f,H_{OFFRES}). \tag{13.29}$$

The permeability then can be obtained from Equations (13.27) and (13.29) as

$$\mu_r(f,H) \approx k^2 \frac{L_X}{L_0} \frac{1}{\omega t_0 \mu_0} \left[Im\left(\Delta Z_T(f,H) \right) - jRe(\Delta Z_T(f,H)) \right]. \tag{13.30}$$

This technique allows noncontact measurement of local magnetization dynamics and evaluation of a broad range of magnetic material properties. With only minor modifications, the loop-based, near-field probe can be applied to the local characterization of magnetization dynamics in many magnetic systems.

While scanning probe techniques described earlier are based on probes that utilize the loop as the sensing element, it is challenging to fabricate such probes in a way that is compatible with most contemporary NSMM systems, including home-built as well as commercial scanning probe instruments. Figure 13.10(a) shows one method that is compatible with cantilever-based, commercial scanning probe systems. The design and fabrication of loop-based probes that fit common cantilever dimensions is challenging. Often, the loop geometry itself limits the lateral spatial resolution, which is typically tens of nanometers or less.

Therefore, there is strong interest in alternative sensing elements. It is well-known that the tunneling current of a scanned probe can be sensitive to magnetization state of the sample [65], [66]. Broadband measurement capability with scanning tunneling microscopy was introduced in Reference [67]. The techniques described therein enable the study of magnetization dynamics on the atomic scale. In particular, it enables the measurement of electron spin relaxation times of individual atoms adsorbed on the surface using all-electronic, pump-probe techniques [68].

Measurement of single- and few-spin dynamics requires low temperatures and ultra-high-vacuum environments. In this chapter, we will focus on experimental techniques that allow the measurement of the magnetization dynamics of nanoscale materials and devices under ambient temperatures.

A sharp STM or AFM tip can be modeled as an electric dipole. Therefore, in order to measure the magnetic response with a probe tip, the tip either has to be covered by a magnetic material as is the case in a magnetic force microscope (MFM) or the microwave dynamics have to be indirectly measured via the electric field resulting from the motion of magnetization. Unfortunately, readily available, MFM-compatible cantilevers are not suitable for measurements in the GHz frequency range. To circumvent this problem, the authors of Reference [69]

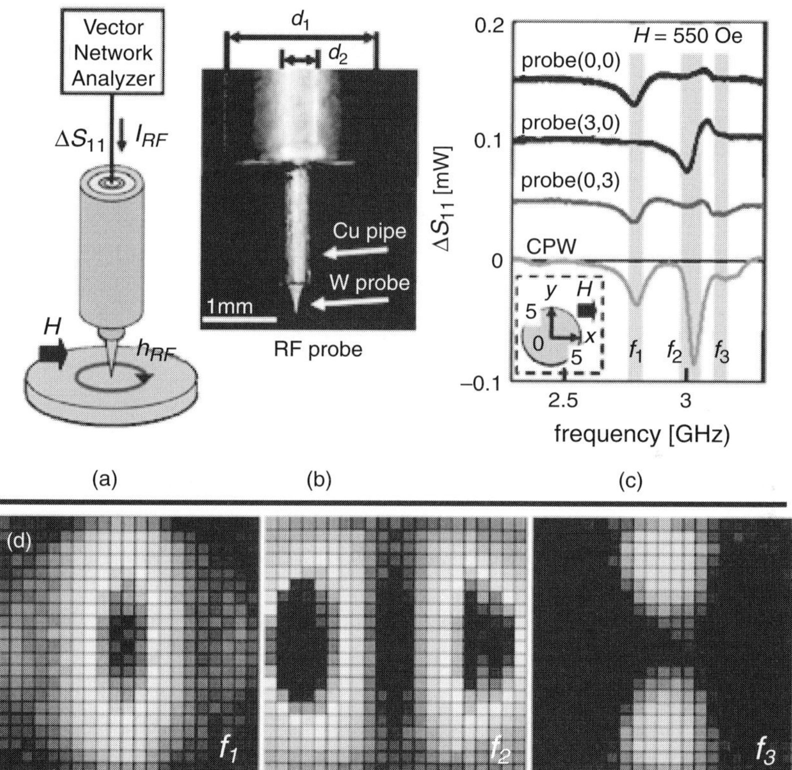

Figure 13.11. Coaxial probe measurements of a YIG disk.
(a) Schematic and (b) photograph of the coaxial RF probe. (c) FMR signals detected from a YIG disk sample in the reflection spectra (S_{11}) with the RF probe positioned above the disk and a CPW positioned under the disk. An in-plane, static magnetic field of 44 kA/m (550 Oe) was applied. The RF probe was located at three different positions above the YIG disk: (x, y) = (0 mm, 0 mm), (3 mm, 0 mm), and (0 mm, 3 mm); (d) reconstructed images of the corresponding modes. © 2010 IEEE. Reprinted, with permission from T. An, N. Ohnishi, T. Eguchi, Y. Hasegawa, and P. Kabos, *IEEE Magnetics Letters* 1 (2010) art. no. 3500104.

positioned a 4 μm- diameter magnetic sphere onto a standard soft cantilever. This detector transduces the electromagnetic interaction of the sphere with the measured dynamic response of the sample into mechanical motion of the cantilever. Specifically, the force acting on the cantilever is proportional to the variation of the longitudinal magnetization component that varies due to excited dynamics in the sample such as FMR.

The other option – measurement of the magnetization dynamics through the electric field component – is experimentally challenging, but it has been demonstrated that the dynamic magnetization response can be registered through the electric field generated by magnetization motion [70]. An open-ended, RF coaxial probe was used to detect the excitations of magnetostatic modes in a YIG disk by measurement of the reflection coefficient. A schematic diagram and an image of the probe are shown in Fig. 13.11(a) and 13.11(b), respectively. Figure 13.11(c) shows frequency dependence of the reflection coefficient at several probe positions above the YIG sample at a constant magnetic field. For comparison, Fig. 13.11(c) also shows frequency dependence of the reflection coefficient for the same sample as obtained by use of a nonlocal, CPW measurement. The agreement of the probe-based measurements with the CPW-based measurements demonstrates the feasibility of measurement of the magnetization dynamics with electric field probe. The critical difference between the CPW measurement and the probe measurement is that the scanning probe signal is obtained locally and the data can be used to reconstruct the spatial profile of the corresponding mode at a given resonance frequency, as shown in Fig. 13.11(d).

References

[1] L. D. Landau and E. M. Lifschitz, "On the Theory of Dispersion of Magnetic Permeability in Ferromagnetic Bodies," *Physikalische Zeitschrift der Sowjetunion* 8 (1935) p. 153.

[2] T. A. Gilbert, "A Phenomenological Theory of Damping in Ferromagnetic Materials," *IEEE Transactions on Magnetics*, 40, No. 6 (2004) pp. 3443–3449 (originally "Equation of Motion of Magnetization," *Armour Research Foundation Technical Report* 11 (1955)).

[3] F. Bloch, "Nuclear Induction," *Physical Review* 70 (1946) pp. 460–473.

[4] N. Bloembergen, "Magnetic Resonance in Ferrites," *Proceedings of the IRE* 44 (1956) p. 1259.

[5] C. Kittel, "On the Theory of Ferromagnetic Resonance Absorption," *Physical Review* 73 (1948) pp. 155–161.

[6] G. Srinivasan and A. Slavin, *High Frequency Processes in Magnetic Materials* (World Scientific Publishing, 1995).

[7] D. D. Stancil, *Theory of Magnetostatic Waves* (Springer, 1993).

[8] P. Kabos and V.S. Stalmachov, *Magnetostatic Waves and Their Applications* (Springer, 1994).

[9] G. T. Rado and H. Suhl, *Magnetism* (Academic Press, 1963).

[10] B. Lax and K.J. Button, *Microwave Ferrites and Ferrimagnetism* (McGraw Hill, 1962).

[11] A. G. Gurevich and G. A. Melkov, *Magnetization Oscillations and Waves* (CRC Press, 1996).

[12] S. S. Kalarickal, P. Krivosik, M. Wu, C. E. Patton, M. L. Schneider, P. Kabos, T.J. Silva, and J. P. Nibarger, "Ferromagnetic Resonance Linewidth in Metallic Thin Films: Comparison of Measurement Methods," *Journal of Applied Physics* 99 (2006) art. no. 093909.

[13] I. Neudeckera, G. Woltersdorf, B. Heinrich, T. Okuno, G. Gubbiotti, and C. H. Back, "Comparison of Frequency, Field, and Time Domain Ferromagnetic Resonance Methods," *Journal of Magnetism and Magnetic Materials* 307 (2006) pp. 148–156.

[14] A. Einstein and W. J. de Haas, "Experimental Proof of the Existence of Ampère's Molecular Currents," *Koninklijke Akademie van Wetenschappen te Amsterdam, Proceedings*, 18 (1915) pp. 696–711 (in English) and *Verh. Dtsch. Phys. Ges.* 17 (1915) p. 152 (in German).

[15] S. J. Barnett, "Magnetization by Rotation," *Physical Review* 6 (1915) pp. 239–270.

[16] T. M. Wallis, J. Moreland, and P. Kabos, "Einstein-de Haas Effect in NiFe Film Deposited on a Microcantilever," *Applied Physics Letters* 89 (2006) art. no. 122502.

[17] C. Kittel, "On the Gyromagnetic Ratio and Spectroscopic Splitting Factor of Ferromagnetic Substances," *Physical Review* 76 (1949) pp. 743–748.

[18] R. Jaafar, E. M. Chudnovsky, and D. A. Garanin, "Dynamics of the Einstein–de Haas Effect: Application to a Magnetic Cantilever," *Physical Review B* 79 (2009) art. no. 104410.

[19] A. Jander, J. Moreland, and P. Kabos, "Angular Momentum and Energy Transferred through Ferromagnetic Resonance," *Applied Physics Letters* 78 (2001) pp. 2348–2350.

[20] A. Jander, J. Moreland, and P. Kabos, "Micromechanical Detectors for Local Field Measurements Based on Ferromagnetic Resonance," *Journal of Applied Physics* 89 (2001) pp. 7086–7090.

[21] S. Lee, Y. C. Lee, T. M. Wallis, J. Moreland, and P. Kabos, "Near-field Imaging of High Frequency Magnetic Fields with Calorimetric Cantilever Probes," *Journal of Applied Physics* 99 (2006) art. no. 08H306.

[22] S.-H. Lim, A. Imtiaz, T. M. Wallis, S. Russek, P. Kabos, Liufei Cai, and E M. Chudnovsky, "Magneto-mechanical Investigation of Spin Dynamics in Magnetic Multilayers," *Europhysics Letters* 105 (2014) art. no. 37009.

[23] S. Kasai, Y. Nakatani, K. Kobayashi, H. Kohno, and T. Ono, "Current-Driven Resonant Excitation of Magnetic Vortices," *Physical Review Letters* 97 (2006) art. no. 107204.

[24] W. Lin, J. Cucchiara, C. Berthelot, T. Hauet, Y. Henry, J. A. Katine, E. E. Fullerton, and S. Mangin, "Magnetic Susceptibility Measurements as a Probe of Spin Transfer Driven Magnetization Dynamics," *Applied Physics Letters* 96 (2010) art. no. 252503.

[25] N. Mecking, Y. S. Gui, and C.-M. Hu, "Microwave Photovoltage and Photoresistance Effects in Ferromagnetic Microstrips," *Physical Review B* 76 (2007) art. no. 224430.

[26] H. Zhang, A. Hoffmann, R. Divan, and P. Wang, "Direct-Current Effects on Magnetization Reversal Properties of Submicron Size Permalloy Patterns for Radio-Frequency Devices," *Applied Physics Letters* 95 (2009) art. no. 232503.

[27] T. J. Silva, C. S. Lee, T. M. Crawford, and C. T. Rogers, "Inductive Measurement of Ultrafast Magnetization Dynamics in Thin Film Permalloy," *Journal of Applied Physics* 85 (1999) pp. 7849–7862.

[28] A. B. Kos, T. J. Silva, and P. Kabos. "Pulsed Inductive Microwave Magnetometer," *Review of Scientific Instruments* 73 (2002) pp. 3563–3569.

[65] R. Wiesendanger, *Scanning Probe Microscopy and Spectroscopy: Methods and Applications* (Cambridge University Press, 1994).

[66] R. Schmidt, C. Lazo, H. Holscher, U. H. Pi, V. Caciuc, A. Schwarz, R. Wiesendanger, and S. Heinze, "Probing the Magnetic Exchange Forces of Iron on the Atomic Scale," *Nano Letters* 9 (2009) pp. 200–204.

[67] G. Nunes, Jr. and M. Freeman, "Picosecond Resolution in Scanning Tunneling Microscopy," *Science* 262 (1993) pp. 1029–1032.

[68] S. Loth, M. Etzkorn, Ch. P. Lutz, D. M. Eigler, and A. J. Heinrich, "Measurement of Fast Electron Spin Relaxation Times with Atomic Resolution," *Science* 329 (2010) pp. 1628–1630.

[69] G. de Loubens, V. V. Naletov, M. Viret, O. Klein, H. Hurdequint, J. Ben Youssef, F. Boust, and N. Vukadinovic, "Magnetic Resonance Spectroscopy of Perpendicularly Magnetized Permalloy Multilayer Disks," *Journal of Applied Physics* 101 (2007) art. no. 09F514.

[70] T. An, N. Ohnishi, T. Eguchi, Y. Hasegawa, and P. Kabos, "Local Excitation of Ferromagnetic Resonance and Its Spatially Resolved Detection with an Open-Ended Radio-Frequency Probe," *IEEE Magnetics Letters* 1 (2010) art. no. 3500104.

14 Nanoscale Electromagnetic Measurements for Life Science Applications

14.1 High-Resolution Optical Microscopy of Nanoscale Biological Systems

14.1.1 Far-Field Techniques

The number of microscopic imaging applications related to biological systems is large and continues to grow. Figure 14.1 shows the dimensions of selected, biological systems. Depending on the length scale of the system, a variety of different microscopy techniques are available. For comparison, Fig. 14.1 also illustrates comparable solid state, electronic systems at various length scales. In this chapter, we will describe methods for imaging and characterization of biological materials at submicrometer length scales. The focus will be on electrical scanning probe imaging methods, but by way of introduction we will discuss selected optical methods first. Our review of optical methods is but a brief survey of a small subset of many available techniques.

Historically, the most prominent techniques for biomaterial imaging have been optical techniques, especially those based on fluorescence imaging. The importance of optical techniques is due in large part to the fact that they allow imaging of living cells with minimal invasiveness. Although electron microscopy has better spatial resolution, optical microscopy still dominates biological research. The resolution of most optical microscopy has historically been diffraction-limited with resolution of approximately $0.6\lambda/NA$ where λ is the wavelength of the light and NA is the numerical aperture of the microscope objective lens. Thus, at the diffraction limit of conventional fluorescent microscopies, resolution is on the order of 200 nm. This is practical in many cases, but if the molecules are separated by distances on the order of 200 nm to 350 nm, then it is not possible to tell them apart. As a result, it has been necessary to develop sub-diffraction-limit techniques in order to image nanoscale biological systems and distinguish the subsystems within them.

One critical breakthrough in the imaging of life science systems was the development of tagging methods for proteins and other biological molecules by use of fluorescent probe molecules. Staining techniques for inducing contrast in optical images date back to the nineteenth century and played a central role in biological imaging of that period, such as the work of 1906 Nobel Prize winner Santiago Ramon y Cajal. More recently, the importance of tagging techniques has been acknowledged through two additional Nobel prizes. The first was in 2008 and was

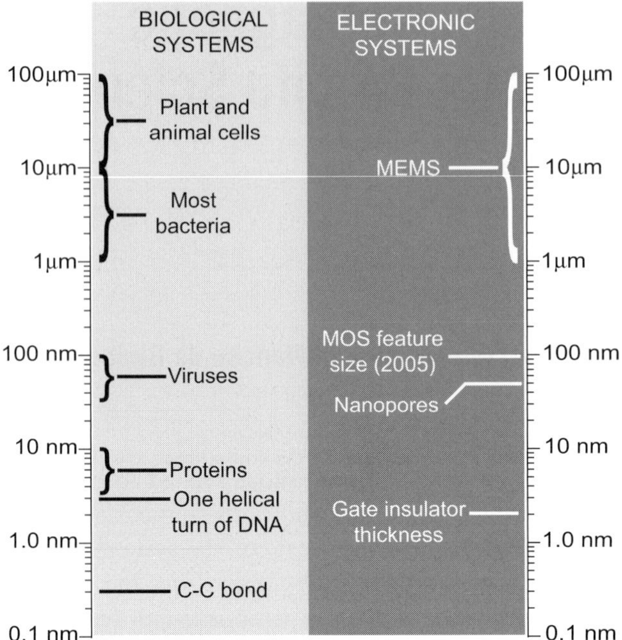

Figure 14.1.   Length scales of biological and electronic systems.
A comparison of length scales of biological (left) and electronic (right) systems that can be
accessed by near-field scanning microwave microscopy and other measurement techniques.

awarded to Shimomura, Tsien, and Chalfie for the discovery and development of
green, fluorescent protein. The second was in 2014 and was awarded to Betzig,
Hell, and Moerner for the development of super-resolution fluorescence microscopy. Indeed, the unique photochemical properties of fluorescent molecules are the
foundation for a number of sub-diffraction limit microscopies.

However, overcoming the diffraction limit is not merely about detecting tagged
molecules, but also about distinguishing one molecule from another when the inter-
molecular separation is on the order of 100 nm or less. This is accomplished in large
part by leveraging stimulated emission, as originally introduced by Hell [1]. Hell's
approach was based on keeping a fraction of the molecules in the observation area
in a dark state while the other, bright-state molecules in the observation area are
detected. One strategy to keep some molecules in a dark, ground state is the use of
stimulated emission. This idea led to stimulated emission depletion (STED) microscopy. The STED technique utilizes two lasers: one to excite the fluorophores and a
second to deplete the excited-state through stimulated emission before the fluorophores emit light through spontaneous fluorescence. In other words, one beam is
used to illuminate the molecules and excite them into the bright state. The second
light beam induces stimulated emission and returns the excited molecules into their
ground (dark) state. By properly shaping the second beam into a ring in the focal
plane, all the molecules but the ones at the center will be turned off. The effect of
this second laser, referred to as a STED beam, is a drastic reduction in the size

of the fluorescence spot. An image can subsequently be acquired by coordinated, overlapped scanning of the excitation and STED laser beams across the sample. At present there are several variants of this approach with 50 nm to 100 nm resolution.

Another approach to nanoscale microscopy, known as superlocalization, was pioneered by Moerner [2]. This approach built upon the work of Betzing, in which photo-switchable fluorescent proteins were employed as an active control mechanism for photoactivated localization microscopy (PALM). Experimentally, the technique relies on simultaneously suppressing background signals while maximizing emission from the molecules of interest. The fundamental analytical idea introduced by Moerner was overlaying the shape of the measured point-spread-function with a Gaussian (Airy) function. This innovation, which is at the heart of superlocalization, reduced the Abbe diffraction limit by a factor of $\sqrt{N}$, where N is the number of measurements. Put simply, two conceptually simple ideas are at the heart of this technique. The first is to use spectrally distinct tags in order to distinguish molecules in "multicolor" images. The second idea is to induce the molecules to emit sequentially such that they are not all emitting simultaneously. When combined with the superlocalization technique, these two ideas enable image reconstruction in which closely spaced molecules can be distinguished. The experimental technique has a number of different acronyms depending on the strategies for controlling emitting concentrations and sequential emission. Application of localization microscopies has improved spatial resolution to 20 nm to 50 nm scale [3].

14.1.2   Near-Field Techniques

So far, we have limited our discussion to sub-diffraction-limit, optical techniques based on far-field observation. In this section, we turn to sub-diffraction-limit optical techniques that are based on near-field interactions. These near-field methods are topically closer to the preceding chapters and near-field scanning microwave microscopy (NSMM), in particular. One example of a near-field optical technique is total internal reflection fluorescent microscopy (TIRF) [4], which utilizes evanescent waves to selectively illuminate and excite fluorophones in a specimen immediately adjacent to a glass-liquid interface. When the light is obliquely incident from the glass (substrate) side at an angle equal to or greater than the critical angle of refraction given by Snell's law, the light is totally reflected. During this process, a surface evanescent wave is generated at the interface. This effect is well known and is called total internal reflection. Due to exponential decay of the amplitude of this wave from the surface the probed penetrated area of the specimen is given by

$$d = \frac{\lambda}{4\pi n_2}\left(\frac{\sin^2\theta}{\sin^2\theta_c} - 1\right)^{-1/2}, \tag{14.1}$$

where θ is the angle of incidence, $\theta_c = \sin^{-1}\left(\frac{n_2}{n_1}\right)$ is the critical angle of incidence, λ is the free space wavelength of the incident light, n_2 is the refractive index of liquid medium, and n_1 is the refractive index of the substrate. For a typical system such as a glass-water interface, this represents a penetration depth of about 100 nm into the sample. The technique has been used to observe the fluorescence of a single molecule with the relative intensity of the molecular fluorescence depending on the distance of the molecule from the interface.

Near-field scanning optical microscopy (NSOM) is another, near-field optical technique that was introduced after the discovery of scanning probe microscopes. It was originally developed by Betzig and Trautman [5]. As originally conceived, NSOM relies on optical radiation incident upon a sub-wavelength aperture. Thus, it shares its fundamental ideas with early NSMM. In one early realization of this technique, Ash and Nicholls used illumination of a small aperture that was a part of an open resonator to go beyond the diffraction limit [6]. Building upon this initial demonstration, Betzing and Trautman designed a successful NSOM by use of a glass capillary coated with aluminum. This first, near-field optical microscope surpassed the diffraction limit by a factor of two. Significant improvement was achieved when the capillary was replaced by a tapered optical fiber, leading to more efficient transmission of the light to the NSOM probe and improved coupling to the evanescent optical modes at the probe tip. At present, NSOM is used for a variety of applications, including optical data storage, super-resolution photolithography and nano-spectroscopy.

14.2   Electrical Characterization of Biological Systems

14.2.1   The Measurement Problem

In addition to NSOM, there are a variety of other techniques to image biological objects by use of standard or modified STMs and AFMs. The properties of proteins, nucleic acids, membranes, and cells have been studied by use of a broad range of AFM- and STM-based approaches [7]–[9]. The ultimate objective of this chapter is to describe the application of electrical, scanning-probe-based techniques, including NSMM, to life science metrology. However, before addressing the specific challenges related to NSMM of biological systems, we must answer several broader questions about the electrical characterization of biological systems. How do we model the interaction of electromagnetic radiation with biological systems? What are the relevant material parameters and measurands of interest? What modifications must be made in order to extend the measurements to heterogeneous systems? Are any special considerations required to account for measuring in a liquid environment?

14.2.2   Microwave Antenna Probes

Interest in the application of microwave techniques to medical diagnostics is driven in part by the fact that microwave techniques are noninvasive. In fact, some tissues are transparent to microwaves. Potential applications include microwave heating therapy and cancer detection. The latter application is based on the fact that the conductivity of cancerous tissue is about a factor of five higher than that of healthy tissue [10]. Thus, the identification of malignant tissue is based on a temperature difference between malignant and healthy tissue. The temperature is measured indirectly through the spectral radiance of the electromagnetic radiation, which is given by

$$I(f,T) = \mu\varepsilon \frac{2hf^3}{e^{\frac{hf}{k_BT}} - 1}, \tag{14.2}$$

where μ is the permeability of the medium, ε is the permittivity of the medium, h is Plank's constant, f is the radiating frequency, k_B is the Boltzmann constant, and T is the temperature. Clearly, in order to realize such medical applications, one must understand the interaction of electromagnetic radiation with biological systems.

As we have previously noted in this book, accurate probe-sample models are required in order to extract material parameters from microwave probe measurements. The first such model that we will consider here is an *in vivo* microwave antenna probe. The operating principle of such a probe is based on the antenna modeling theorem, which can be expressed as

$$\frac{Z(\omega, \varepsilon^*)}{\sqrt{\mu_0/\varepsilon^*}} = \frac{Z(n\omega, \varepsilon_0)}{\sqrt{\mu_0/\varepsilon_0}} \tag{14.3}$$

where Z is the terminal impedance of the antenna probe, ε^* is the complex permittivity of the medium, $n = \sqrt{\varepsilon^*/\varepsilon_0}$ is the complex index of refraction of the medium, and $Z_0 = \sqrt{\mu_0/\varepsilon_0}$ is the impedance of free space. The left-hand side of Equation (14.3) corresponds to the probe in the dielectric medium while the right-hand side corresponds to the probe in free space. Equation (14.3) is derived under the assumption that the antenna is entirely contained within the medium and that the surrounding medium is infinite. This relationship can be applied to any type of microwave probe, provided that the terminal impedance of the probe is known. For a short monopole probe with length $\lambda/10$ or less, as is the case for the NSMM, the antenna impedance in free space is given by [11]

$$Z(\omega, \varepsilon_0) = A\omega^2 + \frac{1}{j\omega C}, \tag{14.4}$$

where the constant A and the capacitance constant C are determined from the antenna's physical dimensions. When the antenna is in a lossy medium, it follows from Equations (14.3) and (14.4) that the impedance is

$$Z(\omega, \varepsilon^*) = R + jX, \tag{14.5}$$

where

$$R = \frac{sin2\delta}{2\varepsilon'\omega C} + A\sqrt{\varepsilon'}\omega^2\sqrt{\frac{sec\delta+1}{2}}, \tag{14.6a}$$

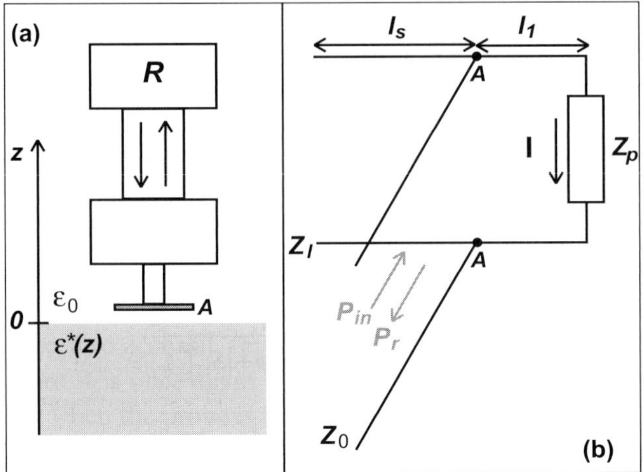

Figure 14.2.   Model of an external probe.
(a) Schematic of the external, RF probe antenna interacting with a sample. The height-dependent complex permittivity of the sample is $\varepsilon^*(z)$. (b) The corresponding transmission line equivalent probe circuit. The position that is labeled "*A*" in the schematic corresponds to the circuit nodes labeled "*A*" in the circuit diagram. Adapted from A. N. Reznik and N. V. Yurasova, *Journal of Applied Physics* 98 (2005) art. no. 114701, with permission from AIP Publishing.

combination of an open-ended line with length l_s that serves as the matching circuit and a line of length l_1 that is loaded with the input impedance Z_p of the probe. The characteristic impedance of these lines is Z_l and propagation constant $\gamma_1 = jk_l$, where k_l is corresponding wave number of the propagating wave. The lines are assumed to be lossless. The parallel combination of the probe and the matching circuit is fed by a line with characteristic source impedance Z_0. The feed line also serves as the detection path. The total input impedance of the circuit shown in Fig. 14.2(b) follows from standard transmission line theory and is

$$Z_{in} = \frac{Z_1 Z_2}{Z_1 + Z_2},\qquad(14.13)$$

where

$$Z_1 = -jZ_l \cot\left(k_l l_s\right),\qquad(14.14a)$$

$$Z_2 = Z_l \frac{Z_l - jZ_p \cot(k_l l)}{Z_p - jZ_l \cot(k_l l)}.\qquad(14.14b)$$

If the matching line is lossy, appropriate modifications have to be applied to the model. Note that Equations (14.14a) and (14.14b) depend upon the input impedance of the probe, including the effects of the sample properties and the interaction between the probe and the material. Simple forms of this impedance have been discussed for a variety of cases in Chapter 9. More complex expressions can be found in Reference [14] that take into account depth-dependent permeability and field distributions of the excited modes. As in the case of the inserted probe, the external near-field probe requires reference samples with known permittivity in order to obtain calibrated, quantitative measurements.

14.2.3　Multilayer Systems

Modeling the probe-sample impedance becomes more challenging for biological samples, most of which are heterogeneous. For example, a more complex model for a two- or three-layer system representing the response of mammalian skin was presented in Reference [15]. In the two-layer model, the upper layer is a good conductor and represents the combined effects of the epidermis. The underlying second layer represents subcutaneous fat, which has a low dielectric constant. In the three-layer model, the epidermis is split into the stratum corneum and the dermis. The stratum corneum is about 30 micrometers thick and has a low dielectric constant. The dermis is about 1 mm to 2 mm thick and has a high dielectric constant. Following this modification, the third layer now represents the subcutaneous fat. The coupling of an external probe to this system is considered to be capacitive and may be analytically determined through a combination of the propagating TEM mode and evanescent TM modes, as discussed in Chapter 9. As we show in this chapter, though the two- and three-layer models are conceptually simple, the analysis required to extract quantitative information from such a system is complex and challenging. Though the following analysis is framed in terms of the specific example of mammalian skin, this multilayer model form can be generalized to represent a variety of layered biological structures.

The problem at hand is to calculate the capacitance of a coaxial probe in contact with the layered biological medium. This capacitance is related to the effective permittivity (dielectric constant) through a simple relation $C_p = \varepsilon C_0$, where C_0 is once again the capacitance of the probe in free space. If the inner and outer conductors of the coaxial probe are a and b, respectively, then this model applies only under the condition that $2a/\lambda \ll 1$. Clearly, this condition limits the frequency range within which this approach is valid. If it is assumed that the potential of the inner conductor of the coaxial probe is V_0, then the potential inside of the cable can be expressed as [15]

$$V_p(z,\rho) = V_0 \frac{\ln(\rho/b)}{ln(a/b)} + V_0 \sum_{i=1}^{N} g_i T_i(\rho) \exp(p_i z); \qquad a \le \rho \le b \qquad (14.15)$$

with

$$T_i(\rho) = A_i J_0(p_i \rho) + B_i Y_0(p_i \rho). \qquad (14.16)$$

$$P_{03} + g_1 P_{31} + g_2 P_{32} + g_3 P_{33} = -V_0^2 \frac{\varepsilon_t}{\varepsilon_1} g_3 p_3 t_3. \tag{14.32c}$$

Similarly, the capacitance for $N = 2$ can be written as

$$\frac{C_p}{\varepsilon_0} = \frac{1}{V_0^2} 2\pi\varepsilon_1 \left(P_{00} + 2g_1 P_{01} + 2g_2 P_{02} + 2g_1 g_2 P_{21} + g_1^2 P_{11} + g_2^2 P_{22} \right) + 2\pi\varepsilon_t \sum_{i=1}^{2} g_i^2 p_i t_i$$

$$\tag{14.33}$$

The solution of these equations requires the numerical evaluation of the integrals in Equation (14.29), which is nontrivial. For some simpler cases, approximate formulae can be used. For a bilayer structure the permittivity seen by the probe can be approximated as

$$\varepsilon_p = \left(\varepsilon_1 - \varepsilon_2 \right)\left(1 - e^{-qd} \right) + \varepsilon_2 \qquad , \; \varepsilon_1 > \varepsilon_2 \tag{14.34}$$

or

$$\frac{1}{\varepsilon_p} = \left(\frac{1}{\varepsilon_1} - \frac{1}{\varepsilon_2} \right)\left(1 - e^{-qd} \right) + \frac{1}{\varepsilon_2} \qquad , \; \varepsilon_1 < \varepsilon_2 \; . \tag{14.35}$$

In Equations (14.34) and (14.35), q is a constant that depends on the probe size. These equations can be extended for the three-layer skin model with $\varepsilon_1 < \varepsilon_2$ and $\varepsilon_2 > \varepsilon_3$. It is useful for the three-layer case to distinguish the adjustable constants for stratum corneum $q = q_{sc}$ and epidermis/dermis $q = q_{ed}$. One can then write

$$\varepsilon_{d+sf} = \varepsilon_d \left[1 - \exp(-q_d d_d) \right] + \varepsilon_{sf} \exp(-q_d d_d) \tag{14.36}$$

and

$$\frac{1}{\varepsilon_p} = \frac{1 - \exp(-q_{sc} d_{sc})}{\varepsilon_{sc}} + \frac{\exp(-q_{sc} d_{sc})}{\varepsilon_d \left[1 - \exp(-q_d d_d) \right] + \varepsilon_{sf} \exp(-q_d d_d)}. \tag{14.37}$$

The indices sc, d, and sf correspond to stratum corneum, dermis, and fat layers.

Though the mathematical details of this example are somewhat complex, it is important to stress that the objective is conceptually simple. In general, the interaction of the probe with the sample is modeled by calculating an impedance. In this particular case, as in most cases where a dielectric sample is being characterized at microwave frequencies, the task is further simplified to a calculation of a capacitance, as in the foregoing Equations (14.28) and (14.33).

14.2.4   Heterogeneous, Liquid Systems

Having established a framework for modeling the electromagnetic probe–sample interaction and having applied that framework to the particular case of multilayer systems, we turn to characterization of heterogeneous, biological systems in a liquid environment. Quantitative measurement of local electromagnetic properties within a liquid environment by use of a coaxial cavity was addressed in Reference [16]. As with several other resonant-cavity-based systems discussed throughout this book, the measured parameters are the shift of the resonance frequency and the change in the quality factor. Once again, these changes are quantitatively related to the presence of the sample material within the near field of the probe. In a physiological, liquid environment, further consideration must be given to the dielectric description of the biological tissue, including various relaxation phenomena and water binding. Reference [16] was among the first to address the fact that the application of higher frequencies enables separate extraction of the free water relaxation. In addition, in cases where an inclusion is present in a host sample medium, the mixture may be described by an effective permittivity given by the Maxwell-Garnett mixing formula [17]

$$\varepsilon_{eff} = \varepsilon_h + f_V \varepsilon_h \frac{\varepsilon_i - \varepsilon_h}{\varepsilon_h + (1-f)(\varepsilon_i - \varepsilon_h)}, \tag{14.38}$$

where ε_h is the permittivity of the host, ε_i is the permittivity of the inclusion, and f_V is the fraction of the total volume of the inclusion. Many biological media consist of two components: a dry component with permittivity ε_i and water with permittivity ε_w. In this case, the permittivity of such a mixture satisfies the relationship [18]

$$V_w \frac{\varepsilon_w\left(f_V,T,S\right) - \varepsilon_{eff}}{\varepsilon_w\left(f_V,T,S\right) + 2\varepsilon_{eff}} + \left(1 - V_w\right)\frac{\varepsilon_i - \varepsilon_{eff}}{\varepsilon_i + \varepsilon_{eff}} = 0. \tag{14.39}$$

Here, V_w is the volume of water. The relative permittivity of water is a function of the frequency, temperature, and salinity. It can be calculated from [19]

$$\varepsilon_w = \varepsilon_\infty + \frac{\varepsilon_s\left(T,S\right) - \varepsilon_\infty}{1 + j\omega\tau(T,S)} - j\frac{\sigma(T,S)}{\omega\varepsilon_0}, \tag{14.40}$$

where ε_0 is the permittivity of vacuum and $\varepsilon_\infty = 4.9$ is the relative permittivity of the water at optical frequencies. Formulas for the static permittivity ε_s, relaxation time τ, and conductivity σ as a function of temperature T and salinity S can be found in the literature [19].

Though effective dielectric constants can be important in determining the structure and the thermodynamics of electrostatically interacting systems under different electrostatic conditions, it is difficult to isolate and estimate the dielectric constant of specific constituent parts of biological macromolecules such as DNA when they are embedded in a host environment. This estimation is further complicated due to the presence of ions in the system. At low ion concentration, the problem can be reduced to an analysis of Coulomb forces. At high ion concentration, the interactions can only be described numerically by calculations that include correlations. The ions' influence on the dielectric constant in turn influences the resulting forces and the Debye screening length. In this process, one has to distinguish between the static dielectric constant of the solution and the static dielectric constant of the solvent.

Thus, when describing the dielectric constant of biological macromolecules in an ionic solution, it may be necessary to include the statistical behavior of the system. One reasonable approach is to separate the overall dielectric constant into self-terms for individual components (component dielectric constants) and cross-terms between components. This approach is well defined mathematically. Reference [20] showed that the cross-terms are negligible compared to the self-terms when the components are grouped appropriately. Although the dielectric constant of an inhomogeneous system is an aggregate property combining the effects of all the constituents of the system, it is possible to separately calculate the contributions of a macromolecule and the solvent to the total dielectric constant by use of molecular dynamics simulations. Furthermore, such simulations enable the evaluation of any cross-terms, should they be non-negligible. The analysis underlying such simulations is as follows. The dielectric constant of a system ε surrounded by an infinite dielectric medium with a dielectric constant ε_e is related to dipole fluctuations by [20]

$$\frac{(\varepsilon-1)(2\varepsilon_e+1)}{(2\varepsilon_e+\varepsilon)} = \frac{\langle M \rangle^2 - \langle M \rangle^2}{3\varepsilon_0 V k_B T} = \alpha. \tag{14.41}$$

M, V, and T are the total dipole moment, the volume, and the temperature of the system, respectively. Two special cases are of interest. First, as $\varepsilon_e \to \infty$ then

$$\varepsilon_1 = 1 + \alpha, \tag{14.42}$$

where ε_1 is the dielectric constant of the system. Second, when $\varepsilon_e = \varepsilon$ then

$$\varepsilon_2 = \frac{1}{4}(3\alpha + 1 + \sqrt{9\alpha^2 + 6\alpha + 9}) \tag{14.43}$$

where ε_2 is the dielectric constant of the system.

The total dipole moment of a system of N particles is defined as

$$M = \sum_{i=1}^{N} q_i r_i \tag{14.44}$$

where q_i is the partial charge and r_i is the position vector of the *ith* particle. For a system composed of particles of m different groups it is possible to decompose M into m components

$$M = \sum_{k=1}^{m} M_k \quad \text{with} \quad M_k = \sum_{i=N_{k-1}+1}^{N_{k-1}+n_k} q_i r_i \tag{14.45}$$

Here n_k is the number of particles in group k and $N_k = \sum_{l=1}^{k} n_l$. Following this decomposition, the mean square deviation of dipole moment can be expressed as

$$\Delta(M) = \langle M \rangle^2 - \langle M \rangle^2 = \sum_{k=1}^{m} \left(\langle M_k^2 \rangle - \langle M_k \rangle^2 \right)$$
$$+ 2 \sum_{l=k+1}^{m} \sum_{k=1}^{m} \left(\langle M_k \cdot M_l \rangle - \langle M_k \rangle \cdot \langle M_l \rangle \right). \quad (14.46)$$

This formulation can be used to calculate the dielectric constant of many complex systems. For example, it was used in [21] to calculate the internal dielectric constant of DNA through molecular dynamic calculations.

14.3 Electrical Scanning Probe Microscopy of Biological Systems

14.3.1 General Considerations

Ideally, electrical scanning-probe measurements are carried out under experimental conditions that match physiological conditions [22]. In order to characterize fragile biological systems such as soft membranes and molecular complexes in their native environments, AFMs are often operated in "gentler" modes, such as dynamic force microscopy. Dynamic microscopy modes are usually less destructive than direct contact mode, but require driven mechanical oscillation of the cantilever or sample. For example, an external magnetic field may be used to drive the oscillation of a cantilever in a solution. Dynamic microscopy modes provide a rich set of information. Measurements of the phase of the cantilever oscillations and measurements of higher order vibrational modes of the cantilever reveal features of the investigated material that can't be obtained from topographic images alone. In many cases, the immobilization of objects on a surface is critical to AFM imaging. This requires proper selection of the substrate as well as careful attention to electrical charging of the surface. Mica is a commonly used substrate for this purpose, as it is smooth and easy to cut. Furthermore, its surface can be modified to be either positively or negatively charged and thus can be used to immobilize both negatively and positively charged samples.

In the following subsection, we focus primarily on electrical scanning probe microscopy, but scanning-probe-based, mechanical measurement techniques are also important for life sciences metrology. For example, monitoring the inter- and intra-molecular forces is an area where the application of AFM has proven to be quite useful. Molecular-scale force measurements provide insight about structure dynamics within investigated species. The range of these forces is from several femtonewtons to several nanonewtons, roughly corresponding to the forces necessary for the rupture of covalent bonds.

14.3.2 Electrostatic Force Microscopy

Numerous electrical, AFM-based metrology tools, such as scanning electrochemical microscopy and spreading resistance microscopy, are based on the measurement of electric currents. Others, such as electrostatic force microscopy and Kelvin probe microscopy, are based on the measurement of induced electrostatic forces. Each of these techniques provides valuable, complementary insight into electronic and transport properties of material systems. In the specific application of these techniques to biological objects, measurement in a relatively low-frequency regime requires capacitance measurements with sub-attofarad sensitivity. In addition, such measurements may be complicated by operation in a liquid environment. There are very few experimental techniques that can fulfill these experimental requirements.

One such technique is amplitude-modulated, electrostatic force microscopy (AM-EFM), which is a slight modification to conventional EFM that operates at MHz frequencies and allows dielectric imaging in liquid media with nanoscale spatial resolution [23]. The critical aspect of this approach is that the application of a tip voltage at frequencies above 1 MHz introduces a strong electrostatic force at the apex of the probe, which is sensitive to local dielectric properties of the sample under study. At the lower operating frequencies typically used in conventional EFM, nonlocal contributions from the cantilever and other supporting structures are sensed in addition to the localized contribution from the probe tip. The measured force is exerted by a sinusoidal voltage $v(t) = V_{ac}\cos(\omega t)$ between the tip and the bottom of the sample and is given by

$$F_{els}(z,t) = \frac{1}{2}\frac{dC}{dz}V_{ac}^2(\cos(\omega t))^2 = -\frac{1}{4}\frac{dC}{dz}V_{ac}^2 + \frac{1}{4}\frac{dC}{dz}V_{ac}^2\cos(2\omega t). \quad (14.47)$$

For excitation frequencies greater than 1 MHz that are well beyond the resonance frequency of the cantilever only the first term on the right-hand side of Equation (14.47) can be detected as a static bending of the cantilever, which depends on the excitation amplitude V_{ac}. To improve detection accuracy, the signal amplitude is modulated by $V_{ac}(t) = V_0\cos(\omega_{mod}t)$. By use of a lock-in amplifier, one can measure the effective capacitance gradient

$$\frac{dC}{dz} = 4\left|F_{els}(z,t)\right|_{f_{mod}} / V_0^2, \quad (14.48)$$

where V_0 is the amplitude of the modulation signal, f_{mod} is the frequency of the modulated signal typically set to a few kHz, z is the tip-sample distance, and t is time. Importantly, the effective capacitance gradient depends on the frequency, solution conductivity and the dielectric properties of the sample.

A qualitative physical understanding of AM-EFM be obtained from an equivalent circuit model in which a simple parallel plate model represents the tip-sample system in one dimension [23]. A schematic of this model is shown in Fig. 14.3(a) and a corresponding equivalent circuit model is shown in Fig. 14.3(b). The model accounts for the capacitance and resistance of the solution, the sample capacitance, C_{smpl}, and the double layer capacitance(s), C_{dl}, that may form close to the electrodes in the solution. This model can be further simplified if the double layer and sample capacitances are combined into an equivalent capacitance C_{eq}, as it is shown in Fig. 14.3(c) (note that typically C_{dl} can also be neglected since it is much larger than

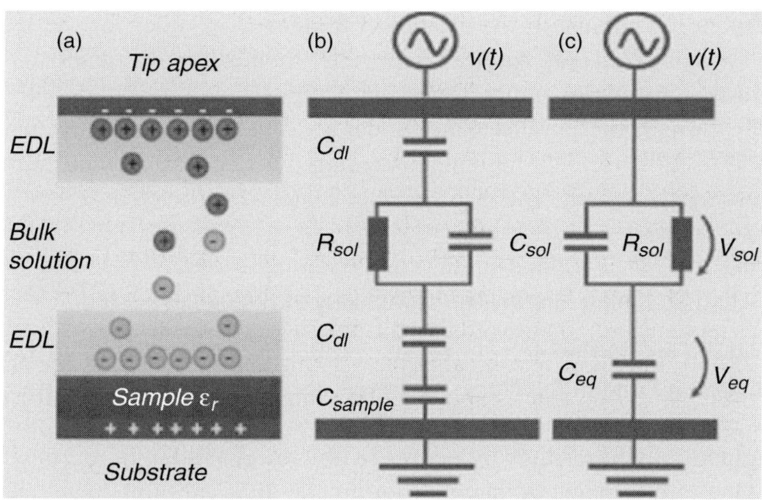

Figure 14.3. Models of an electrostatic force microscope operating in liquid. (a) Schematic of the tip-solution-sample system, including the tip apex, electrochemical double layers (EDLs), bulk solution, and sample. (b) Equivalent electric circuit model for the system. (c) Simplified electric circuit model [23]. © IOP Publishing. Reproduced with permission. All rights reserved.

and in series with C_{smpl}). The capacitance and resistance values can be calculated from the corresponding material properties by $C_{smpl} = \varepsilon_0 \varepsilon_{r,smpl} / h$, $C_{sol} = \varepsilon_0 \varepsilon_{r,sol} / z$ and $R_{sol} = \dfrac{z}{\Lambda c}$ where Λ is the molar conductivity, c is the salt concentration, $\varepsilon_{r,i}$ are the relative permittivities of the solution and the sample, z is the tip-sample distance, h is the sample thickness, and ε_0 is the permittivity of free space. Following the simplified model in Fig. 14.3(c), the electrostatic force can be expressed as [23]:

$$\left[F_{els}(z,f)\right]_{f_{mod}} \approx \frac{1}{4} \frac{\varepsilon_0 \varepsilon_{r,sol} f^2 \left(\dfrac{2\pi\varepsilon_0 \varepsilon_{r,smpl}}{\Lambda ch}\right)^2}{1 + f^2 \left(\dfrac{2\pi\varepsilon_0 \varepsilon_{r,smpl} z}{\Lambda ch} + \dfrac{2\pi\varepsilon_0 \varepsilon_{r,sol}}{\Lambda c}\right)^2} V_0^2. \tag{14.49}$$

The force depends strongly on frequency. The practical implication of this is that the force can only be at frequencies above a certain actuation frequency, which is given by [24]:

$$f_{act} = \frac{c\Lambda}{2\pi\varepsilon_0 \varepsilon_{r,sol}} \left(\frac{\varepsilon_{r,smpl}}{\varepsilon_{r,sol}} \frac{z}{h} + 1\right)^{-1}. \tag{14.50}$$

For an ideal case of a pure dielectric solution ($c = 0$), Equation (14.49) becomes independent of the frequency and $f_{act} = 0$. Note that f_{act} is dependent on the tip-sample distance z. Thus, changes in the probe-sample separation will require a corresponding change in the excitation frequency in order for the local tip apex force to be detected [23]. Another important implication of Equation (14.49) is that electrostatic force rises much slower with decreasing z in liquid compared to air. Finally, though the parallel plate model gives valuable qualitative insights, comparison to finite-element modeling is required for accurate, quantitative interpretation of measurements and estimation of material parameters [25].

14.3.3   Near-Field Scanning Microwave Microscopy of Biological Systems

Measurements of the electromagnetic properties of single cells at frequencies in the microwave and millimeter-wave ranges are important for microwave-frequency-based diagnostics and therapies. Furthermore, such measurements provide insight into potentially hazardous influences of microwave radiation on biological material. One electromagnetic material parameter that captures many characteristics of biological specimens is complex permittivity. From an experimental point of view, there is particular interest in the measurement of transmission, reflection, and absorption of microwave radiation by biological materials, such as tissue, cells, proteins, and bacteria. Up to now, most measurements of the complex permittivity of biological materials have been enabled by integrating the sample with microelectrodes or microfluidic devices. The resolution of such approaches is inherently limited at small length scales and falls short of the subcellular level. An alternative, nondestructive experimental approach for accessing the electromagnetic properties of biological samples with high spatial resolution is offered by NSMM. One further advantage of NSMM is that the penetration depth of microwaves into these materials is significant. Therefore, the technique offers a unique opportunity to simultaneously produce high resolution maps of both surface and subsurface electromagnetic properties.

One area where NSMM is especially useful is the characterization of samples that combine cells or other soft matter with artificial nanostructures. One of the first applications of evanescent microwaves as a nondestructive, high-resolution imaging technique for biological systems was pioneered by Wei [26]. This was followed by the work of Tabib-Azar [27], which pushed the spatial resolution from about 5 μm to $\lambda/750,000$, which corresponds to 0.4 μm at 1 GHz and 0.04 μm at 10 GHz. This early work clearly demonstrated the capability to measure biological and botanical specimens with NSMM. Since this early work, the development of NSMM techniques for biological applications has grown significantly.

14.3.4   Topographic Artifacts in Microwave Microscopy

In addition to addressing the challenges of operation in a heterogeneous, liquid environment, it is necessary to address the convolution of topographic effects with material measurements during NSMM measurements. This is a general challenge to NSMM image interpretation and is not limited to biological materials. As the probe is scanned over nonuniform topography, changes in the probe-sample distance and the sample curvature modify the probe-sample capacitance. Thus, overall changes to the probe-sample capacitance represent the combination of changes due to geometry and changes of the electromagnetic properties of the investigated material. As a result, the geometrical contribution can mask the local dielectric response. It is an important challenge that has to be met in order for NSMM to be an effective technique for the quantitative characterization of materials and devices. Research into this problem is ongoing, but there are several approaches that can be used to mitigate this problem, though none of them is universal.

Reference [22] addresses the topography "cross talk" problem for nonplanar, biological samples, namely the single *E. coli* bacterial cells. The objective is to remove the effects of topography in order to measure the nanoscale permittivity of a single cell. The first step is to calculate theoretical NSMM capacitance images using numerical, three-dimensional methods that account for the tip-sample geometry as well as the electromagnetic material properties. The model assumes that the tip possesses a spherical apex and that the cell is an ellipsoid with uniform relative permittivity ε_r. As a first approximation, this approach gives a reasonable estimate of the geometrical contribution to the capacitance. Within such a model, it is possible to generate simulated NSMM images through numerically calculated, virtual scans at a constant tip-sample distance. Figure 14.4(a) shows such a virtual scan over a single cell with $\varepsilon_r = 4.0$. Figure 14.4(b) shows a calculated line scan for the same bacterial cell. The calculation illustrates that when the tip is moving over the bacterial cell there is a significant decrease in the capacitance due to increased distance between the tip and the underlying substrate. This example also demonstrates that the contribution of the topography-dependent capacitance dominates the contribution of the dielectric response of the sample, especially for small tip apex sizes below 100 nm. Such a complication is common in NSMM capacitance images of soft, dielectric matter. Special design of the probes and their shielding will mitigate but not completely remove this contribution. To obtain the intrinsic dielectric response of the sample, the topographic cross-talk contribution has to be subtracted from the measured capacitance image as shown in Fig. 14.4(c). At present, this approach represents a practical route to local, topography-free, quantitative permittivity characterization.

An alternative approach to removal of the topographic cross-talk is based on the use of measured experimental data in conjunction with an empirical model. This approach does not require a priori numerical modeling based on system geometry. The topography-dependent contribution to the measured reflection signal S_{11} is approximated by selecting a subregion of an image within which the material properties are known not to vary. The difference between the measured and predicted values represents an estimate of the contribution due to material property variations alone.

The empirical model approach is demonstrated with representative NSMM images of a GaN nanowire shown in Fig. 14.5. A topographic image and an image of the magnitude of the raw reflection coefficient S_{11} are shown in Figs. 14.5(a) and 14.5(b), respectively. Fig. 14.5(c) shows an S_{11} image that has been

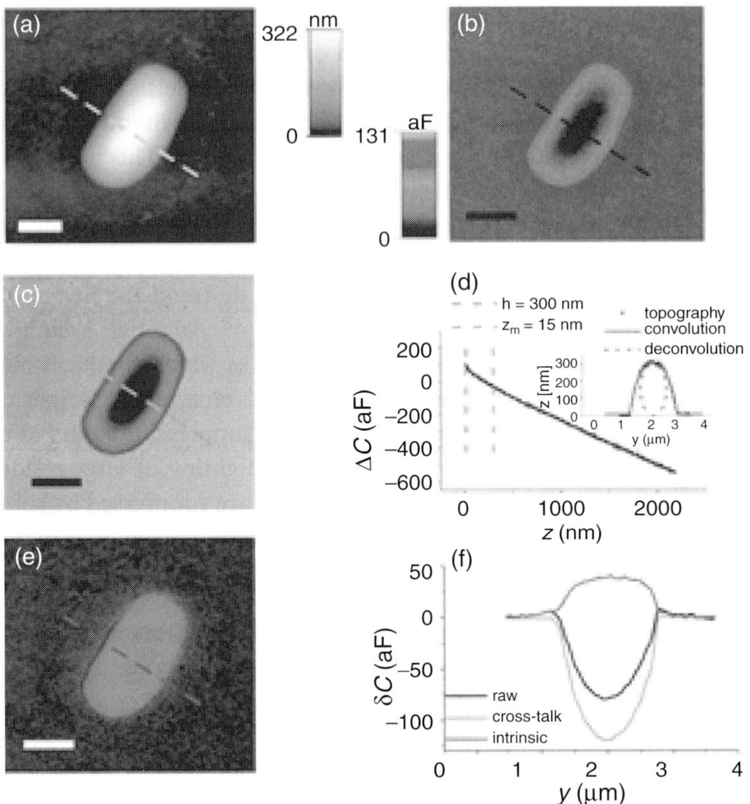

Figure 14.6.   Near-field scanning microwave microscopy of a bacterial cell.
(a) Topography, (b) calibrated capacitance, and (c) estimated topographic cross-talk images for the same bacterial cell in Fig. 14.4. (d) Change in probe-sample as a function of tip height as the tip approaches a metallic sample. The curve has been offset in order to level it with the background substrate capacitance in the image. (d) Inset: topographic cross-section profile along the line in (a). (e) Intrinsic capacitance image, as determined by the process illustrated in Fig. 14.4. (f) Transverse line cuts referenced to the substrate and taken along the dashed lines in (b), (c), and (e). Reprinted with permission from M. Ch. Biagi, R. Fabregas, G. Gramse, M. Van Der Hofstadt, A. Juárez, F. Kienberger, L. Fumagalli, and G. Gomila, *ACS Nano* 10 (2016) pp.280–288. © 2016, American Chemical Society.

an STM feedback mechanism based on RF signals was demonstrated in Reference [34], offering the possibility to extend this approach to nonconducting materials. The STM tip is capacitively coupled with the broadband and does not rely on a resonant cavity. As in conventional STM, a feedback loop keeps the probe at a constant distance from the sample by adjusting the height to keep the tunneling current constant. The combined STM-microwave microscope simultaneously images the topography and the reflection coefficient.

To calibrate STM-based microwave microscopes, three arbitrary, but known loads are measured [32], [33]. The error matrix model is shown in Fig. 14.7(a). It differs from the one-port calibration procedure discussed in Chapter 7 in that the unknown error admittance matrix $[Y_e]$ is a two-port circuit positioned between the tip and the microwave source. The tip is considered to be a sphere with the calibration reference plane crossing the center of the sphere, as shown in Fig. 14.7(b). Note that single mode propagation is assumed. In the case of strong sample–tip interactions there could be far-field contributions and one would have to apply a multimode approach. Furthermore, due to the chosen position of the reference plane, the lower half of the tip sphere also contributes to the measured response of the sample. In principle, one could choose the reference plane at the very bottom of the sphere. The choice of the reference plane position is arbitrary, but not inconsequential in that it does not influence the calibration procedure, but does determine which parts of the test platform are embedded with the sample properties in the calibrated, measured images. If it is assumed that the error admittance matrix is reciprocal, then only three known loads are required to fully characterize the matrix elements. One way to produce three different loads is to perform measurements at three different heights of the tip above the sample. An image of an individual sarcomere obtained by use of this calibrated, STM-based microscope is shown in Fig. 14.7(c). The actin in I-band and myosin in the A-band possess different densities and thicknesses, leading to the banded appearance of features in Fig. 14.7(c) [35].

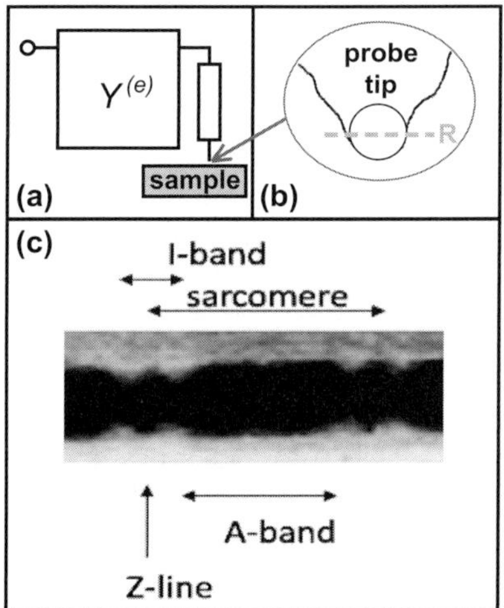

Figure 14.7. Radio frequency scanning tunneling microscopy of a sarcomere. (a) Schematic of the error matrix $Y^{(e)}$ positioned between the source and the STM-based, RF probe. (b) Reference plane position (R). The tip is assumed to terminate in a sphere and the reference plane crosses the center of the sphere. © 2011 IEEE. Adapted, with permission from M. Farina, D. Mencarelli, A. Di Donato, G. Venanzoni, and A. Morini, *IEEE Transactions on Microwave Theory and Techniques* 59 (2011) pp. 2769–2776. (c) RF STM image of a sarcomere. © 2015 IEEE. Reprinted, with permission from M. Farina, A. Di Donato, D. Mencarelli, G. Venanzoni, A. Morini and T. Pietrangelo, *Proceedings of the 45th European Microwave Conference* (2015) pp. 666–669.

Finally, a recently introduced approach allows NSMM measurements of biological samples and other liquids by enclosing the samples within molecularly impermeable, chemically inert enclosures. Such enclosures, known as environmental cells, have been used with NSMM [36]. The enclosure is made from an 8 nm to 50 nm thick SiN or SiO_2 membrane. The thickness of this wall has to be chosen such that it can sustain the pressure difference between ambient air pressure and the interior. It must also be able to withstand the force due to probe tip contact with the membrane. The imaging depth is on the order of few multiples of probe-membrane radii. Though environmental cells have been used with other techniques such as scanning electron microscopy, the advantage of the microwave technique is that the microwaves represent a low-energy, less-invasive form of penetrating radiation. With the energies in the range of tens of micro-electronvolts, microwave radiation does not affect either electronic states or chemical bonds, thus eliminating destructive effects that may occur in electron, X-ray or optical microscopies.

References

[1] S. W. Hell, "Nobel Lecture: Nanoscopy with Freely Propagating Light," *Reviews of Modern Physics* 87 (2015) pp. 1169–1181.

[2] W. E. Moerner, "Nobel Lecture: Single-Molecule Spectroscopy, Imaging, and Photocontrol: Foundations for Super-resolution Microscopy," *Reviews of Modern Physics* 87 (2015) pp. 1183–1212.

[3] S. Habuchi, "Super-resolution Molecular and Functional Imaging of Nanoscale Architectures in Life and Materials Science," *Frontiers in Bioengineering and Biotechnology* 2 (2014) art. no. 20.

[4] D. Axelrod, "Cell Substrate Contacts Illuminated by Total Internal Reflection Fluorescence," *Journal of Cell Biology*, 89 (1981) pp. 141–145.

[5] E. Betzig, "Nobel Lecture: Single Molecules, Cells, and Super-resolution Optics," *Reviews of Modern Physics* 87 (2015) pp. 1153–1167.

[6] E. A. Ash and G. Nicholls, "Super Resolution Aperture Scanning Microscope," *Nature* 237 (1972) pp. 510–512.

[7] L. Tetard, A. Passian, K. T. Venmar, R. M. Lynch, B. H. Voy, G. Shekhawat, V. P. Dravid, and T. Thundat, "Imaging Nanoparticles in Cells by Nanomechanical Holography," *Nature Nanotechnology* 3 (2008) pp. 501–505.

[8] R. Guckenberger, M. Heim, G. Cevc, H. F. Knapp, W. Wiegräbe, and A. Hillebrand, "Scanning Tunneling Microscopy of Insulators and Biological Specimens Based on Lateral Conductivity of Ultrathin Water Films," *Science* 266 (1994) pp. 1538–1540.

[9] J. J. Hoh and P. K. Hansma, "Atomic Force Microscopy for High Resolution Imaging in Cell Biology," *Trends in Cell Biology* 2 (1992) pp. 208–213.

[10] Aman Setia and S. Kishore Reddy, "Advancements in Microwave Breast Imaging Techniques," *IJREAS* 2 (2012) pp. 1679–1690.

[11] E. C. Burdette, F. L. Cain, and J. Seals, "In vivo Probe Measurement Technique for Determining Dielectric Properties at VHF Through Microwave Frequencies," *IEEE Transactions on Microwave Theory and Techniques* MTT-28 (1980) pp. 414–427.

[12] T. W. Athe, M. A. Stuchly, and S. S. Stuchly, "Measurement of Radio Frequency Permittivity of Biological Tissues with an Open-Ended Coaxial Line: Part I," *IEEE Transactions on Microwave Theory and Techniques* MTT-30 (1982) pp. 82–86.

[13] M. A. Stuchly, T. W. Athey, G. M. Samaras, and G. E. Taylor, "Measurement of Radio Frequency Permittivity of Biological Tissues with an Open-Ended Coaxial Line: Part II – Experimental Results," *IEEE Transactions on Microwave Theory and Techniques* MTT-30 (1982) pp. 87–92.

[14] A. N. Reznik and N. V. Yurasova, "Electrodynamics of Near Field Probing: Application to Medical Diagnostics," *Journal of Applied Physics* 98 (2005) art. no. 114701.

[15] E. Alanenyz, T. Lahtinenyx, and J. Nuutiueny, "Measurement of Dielectric Properties of Subcutaneous Fat with Open-Ended Coaxial Sensors," *Physics in Medicine Biology* 43 (1998) pp. 475–485.

[16] X. Wu and O. M. Romahi, "Near-Field Scanning Microwave Microscopy for Detection of Subsurface Biological Anomalies," *Antennas and Propagation Society International Symposium* 3 (2004) pp. 2444–2447.

[17] A. Sihvola, *Electromagnetic Mixing Formulas and Applications* (Institution of Engineering and Technology, 1999).

[18] A. N. Reznik and N. V. Yurasova, "Detection of Contrast Objects Inside Biological Media by Near-Field Microwave Diagnostics," *Technical Physics* 51 (2006) pp. 86–99.

[19] L. A. Klein and C. T. Swift, "An Improved Model for the Dielectric Constant of Sea Water at Microwave Frequencies," *IEEE Transactions on Antennas and Propagation* AP-25 (1977) pp. 104–111.

[20] L. Yang, S. Weerasinghe, P. E. Smith, and B. M. Pettitt, "Dielectric Response of Triplex DNA in Ionic Solution from Simulations," *Biophysical Journal* 69 (1995) pp. 1519–1527.

[21] A. Cuervo, P. D. Dans, J. L. Carrascosa, M. Orozco, G. Gomila, and L. Fumagalli, "Direct Measurement of the Dielectric Polarization Properties of DNA," *PNAS* (2014) pp. E3624–E3630.

[22] M. Ch. Biagi, R. Fabregas, G. Gramse, M. Van Der Hofstadt, A. Juárez, F. Kienberger, L. Fumagalli, and G. Gomila, "Nanoscale Electric Permittivity of Single Bacterial Cells at Gigahertz Frequencies by Scanning Microwave Microscopy," *ACS Nano* 10 (2016) pp. 280–288.

[23] G. Gramse, M. A. Edwards, L. Fumagalli, and G. Gomila, "Theory of Amplitude Modulated Electrostatic Force Microscopy for Dielectric Measurements in Liquids at MHz Frequencies," *Nanotechnology* 24 (2013) art. no. 415709.

[24] G. Gramse, M. A. Edwards, L. Fumagalli, and G. Gomila Lluch, "Dynamic Electrostatic Force Microscopy in Liquid Media," *Applied Physics Letters* 101 (2012) art. no. 213108.

[25] G. Gramse, A. Dols-Perez, M. A. Edwards, L. Fumagalli, and G. Gomila, "Nanoscale Measurement of the Dielectric Constant of Supported Lipid Bilayers in Aqueous Solutions with Electrostatic Force Microscopy," *Biophysical Journal* 104 (2013) pp. 1257–1262.

[26] T. X. Wei, X. -D. Xiang, W. G. Wallace-Freedman, and P. G. Schultz, "Scanning Tip Microwave Near-Field Microscope," *Applied Physics Letters* 68 (1996) pp. 3506–3508.

[27] M. Tabib-Azar, J. L. Katz, and S. R. LeClair, "Evanescent Microwaves: A Novel Super-Resolution Noncontact Nondestructive Imaging Technique for Biological Applications," *IEEE Transactions on Instrumentation and Measurement* 48 (1999) pp. 1111–1116.

[28] M. Farina, A. Lucesoli, A. di Donato, D. Mencarelli, L. Maccari, G. Venanzoni, A. Morini, and T. Rozzi, "Algorithm for Reduction of Noise in Ultramicroscopy and Application to Near-Field Microwave Microscopy," *Electronics Letters* 46 (2010) pp. 50–52.

[29] K. J. Coakley, A. Imtiaz, T. M. Wallis, J. C. Weber, S. Berweger, and P. Kabos, "Adaptive and Robust Statistical Methods for Processing Near-Field Scanning Microwave Microscopy Images," *Ultramicroscopy* 150 (2015) pp. 1–9.

[30] LOCFIT package can be obtained via www.locfit.info/

[31] G. Gramse, M. Kasper, L. Fumagalli, G. Gomila, P. Hinterdorfer, and F. Kienberger, "Calibrated Complex Impedance and Permittivity Measurements with Scanning Microwave Microscopy," *Nanotechnology* 25 (2014) art. no. 145703.

[32] S. Fabiani, A. Lucesoli, A. di Donato, D. Mencarelli, G. Venanzoni, A. Morini, T. Rozzi, and M. Farina, "Dual-Channel Microwave Scanning Probe Microscopy for Nanotechnology and Molecular Biology," *Proceedings of the 40th European Microwave Conference* (2010) pp. 767–770.

[33] M. Farina, D. Mencarelli, A. Di Donato, G. Venanzoni, and A. Morini, "Calibration Protocol for Broadband Near-Field Microwave Microscopy," *IEEE Transactions on Microwave Theory and Techniques* 59 (2011) pp. 2769–2776.

[34] G. P. Kochanski, "Nonlinear Alternating-Current Tunneling Microscopy," *Physical Review Letters* 62 (1989) pp. 2285–2288.

[35] M. Farina, A. Di Donato, D. Mencarelli, G. Venanzoni, A. Morini, and T. Pietrangelo, "Imaging of Biological Structures by Near-Field Microwave Microscopy," *Proceedings of the 45th European Microwave Conference* (2015) pp. 666–669.

[36] A. Tselev, J. Velmurugan, A. V. Ievlev, S. V. Kalinin, and A. Kolmakov, "Seeing through Walls at the Nanoscale: Microwave Microscopy of Enclosed Objects and Processes in Liquids," *ACS Nano* 10 (2016) pp. 3562–3570.

Index